P9-CCQ-532

A MAN ON THE MOON

ANDREW CHAIKIN

ONE GIANT LEAP

A MAN ON THE MOON

ANDREW CHAIKIN

I

ONE GIANT LEAP

Commemorating
the 30th Anniversary
of the first landing on the moon,
July 20, 1969

BY ANDREW CHAIKIN AND THE EDITORS
OF TIME-LIFE BOOKS, ALEXANDRIA, VIRGINIA

THE AUTHOR

Born in 1956, Andrew Chaikin grew up in Great Neck, New York, with a fascination for the heavens and space exploration. At age 12, he made his first visit to Cape Canaveral, where he was lucky enough to meet several astronauts, including Jim Irwin *(below).* While studying geology at Brown University, he participated in the Viking missions to Mars at the NASA/ Caltech Jet Propulsion Laboratory. After graduating in 1978, he became a researcher at the Smithsonian's Center for Earth and Planetary Studies at the National Air and Space Museum in Washington. In 1980 he joined the staff of *Sky & Telescope* magazine, where he was an editor until 1986. Chaikin is now a contributing editor for *Popular Science* and has authored numerous articles for *Air & Space/Smithsonian, Discover, Popular Science, World Book Encyclopedia,* and other publications. He is a commentator for National Public Radio's Morning Edition and served as a consultant on the HBO miniseries *From the Earth to the Moon.* When he is able to take time out from writing, Chaikin pursues songwriting and performing. He lives in Arlington, Massachusetts.

CONTENTS

FOREWORD

by Tom Hanks

I was captured by the spirit of the Apollo program when I was twelve years old. I had been following the space program throughout the Mercury and Gemini flights, building model kits and watching the launches from my mom's house in northern California. We had an above-ground pool in the backyard, and I would put a brick in the back of my swim trunks to hold me down on the bottom, suck in air through a garden hose, and lie there with my arms and legs adrift, pretending I was walking in space. I was, of course, eagerly anticipating the Apollo missions to the moon, because that would give me more models to build. But it wasn't until the lunar orbit flight of Apollo 8 at Christmastime, 1968, that I really recognized what was happening. Mankind was leaving Earth, for the first time in history. I thought about the three astronauts traveling all that distance, and wondered what they felt and what they must have seen. I'll never forget the telecast they made as they circled the moon Christmas Eve, and their reading from the book of Genesis. It was a perfect moment that captured the romantic, epic, historic adventure of it all.

I returned to that adventure in 1994 for the filming of *Apollo 13.* In preparing for the role of astronaut Jim Lovell (who, appropriately, had also been aboard Apollo 8), I realized there was a great deal about the Apollo program that had never been brought to light, things that I did not know. I wanted to understand the events that enabled Neil Armstrong and Buzz Aldrin to make the first landing on the moon. I also wanted to know what went on when other men, like Pete Conrad, Al Bean, Dave Scott, Jack Schmitt, and Gene Cernan, made their footprints in the lunar soil in the five landings that came afterward. I wanted the whole story of mankind's exploration of the moon.

I found it in Andy Chaikin's impressive and illuminating book, *A Man on the Moon.* It had just come out; I picked it up and read it straight through. It was fascinating, just great stuff. What I enjoyed most was the way Andy captured the personalities of the men themselves—not just what they did, but who they were and what made them take on the job of flying to the moon. At the same time, he did a great job of describing the nuts and bolts of how we got there. I am, quite frankly, not a rocket scientist. (I only played one in a movie.) I had been confused by the technology of Apollo. But here it was, explained in a way that I could understand and that was, above all, entertaining. Even the moon itself—full of mystery and spectacular scenery—was an ongoing character in a grand scientific detective story.

A Man on the Moon showed me that each of the missions was fraught with peril, and each accomplished things that were of stellar importance to humanity. Each was an example of the best part of us all. The only thing that is disappointing about the Apollo story is that it comes to an end; our work on the moon was halted after the Apollo 17 mission in December 1972. (I have the secret hope that the story is only Volume One, with Volume Two coming soon.)

Late in the winter of 1995, after *Apollo 13* had finished filming, I read Andy's book again, this time for sheer pleasure. By then, I had met many of the people I was reading about, and that gave the stories new meaning and immediacy. And I began playing with the idea of bringing the entire Apollo saga to television, in a dramatic, episodic event—something like a combination of *Brideshead Revisited* and *The Civil War* by way of all those official NASA mission re-cap films. Not long afterward I called Andy, whom I had met on the set of *Apollo 13.* By summer, work on the series for HBO *From the Earth to the Moon* had begun.

As my colleagues became immersed in the task of putting Apollo on film, excerpts from *A Man on the Moon* were the first things in our reference binders. The big challenge we faced was to make each new mission as interesting as the one before, to give each episode of the series its own unique dramatic narrative and theme. Often these themes came right out of what Andy writes about, or hints at, in the book's pages. When there was a question of accuracy or accomplishment, someone on the staff always asked the most obvious question: "What does Chaikin say about this?"

What I have hoped to convey with *From the Earth to the Moon* is what Andy's captured so well in this book—just how magnificent an undertaking Apollo really was. That going to the moon was not just a technological endeavor, but an artistic one, like Michelangelo's frescoes on the Sistine chapel ceiling. The same kind of imagination that allowed Michelangelo to produce the crowning achievement of his era helped NASA's engineers build their moonships. Just as Michelangelo needed faith in his own abilities to sustain him during the long years of his effort, so faith was at the heart of what it took to put men, and their shoes and socks, and pictures of their children, on the surface of the moon.

Above all, Apollo was a voyage of inspiration. The thing that still fuels me in my day-to-day life, and what I want to convey to my children, and to the audience, is that if mankind can figure out a way to put twelve men on the moon, then, honestly, we can solve anything. That's why I believe the six Apollo landings are six of the seven greatest stories ever told.

PREFACE

The sixties were a time of cultural earthquakes: the horror of the Kennedy and King assassinations, the arrival of four mop-topped singers from Liverpool, the flower-fragrant Summer of Love, the din of protest, and—most of all—the violence of the war in Vietnam. And something else extraordinary happened: On the night of July 20, 1969, two Americans walked on the moon. In what seemed like a miracle of technology, we witnessed it live on television. Across the country, 400,000 people who had worked to make it happen celebrated their triumph. TV commentators and editorial writers proclaimed that five hundred years from now our century would be remembered for those footsteps, when human beings left their home planet to explore the universe.

But in our own time their impact was fleeting. Even as astronauts returned to the moon for bolder and more ambitious explorations, our attention was diverted. While the nation was absorbed in the war, the environment, and unrest at home, NASA was expanding the reach of human beings on another world. On Apollo 11, the first lunar landing, Neil Armstrong and Buzz Aldrin had spent little more than a day on a bland acre of moonscape. In their single moonwalk—which lasted a bit longer than a feature-length film—they never ventured more than a couple of hundred feet from their lander. By 1972, the final pair of moonwalkers, Gene Cernan and Jack Schmitt, were living for three full days on the moon, exploring a lunar valley ringed by spectacular mountains whose rocks would provide a key to the origin of the solar system. During three moonwalks, each lasting more than seven hours, they drove more than ten miles in a battery-powered car. Their activities were broadcast live, in color, and with remarkable clarity, by a TV camera that was remotely controlled from earth. But we had stopped watching.

We have never really come to terms with what we saw on that summer night twenty-five years ago, or with the events that followed. In large measure, it is because we never really understood what had happened. TV showed us what the astronauts did on the moon, but could not transmit the immensity of the venture. The astronauts knew this, and when they returned to earth, they struggled to describe their experiences. But astronauts are not communicators, and with rare exceptions, their words could not bridge the gap between the high-tech realm of spaceflight and everyday experience. The real impact of Apollo—the experiences of the first men to visit another world—remained, like the moon itself, beyond our grasp.

I was thirteen when Armstrong and Aldrin landed on the moon; like countless other space-struck teenagers, I kept a daily vigil in front of the TV, surrounded by press kits, lunar maps, and scale models of rockets and spacecraft. That passion stayed with me when I embarked on a science journalism career, writing articles on astronomy and space exploration as an editor of *Sky & Telescope* magazine. The Space Shuttle era had arrived, but for me, it was Apollo that held special fascination. By 1984 I had interviewed Gene Kranz, the flight director who had been in the trenches of mission control during the first lunar landing, and Harvard geologist Clifford Frondell, who had been present when the first box of moon rocks was opened inside the windowless expanse of NASA's Lunar Receiving Laboratory. For an article on space motion sickness I had experienced weightlessness aboard a special NASA cargo plane used for astronaut training. I had tried on a lunar space suit and flown the Apollo command module simulator. Without realizing it, I had been preparing myself to become a vicarious participant in the Apollo adventure.

By the summer of 1984, the idea of a book about the moon experience began to take shape. To write it, I would need to talk at length to each of the twenty-three surviving moon voyagers (Apollo 13 astronaut Jack Swigert had died in 1982). The following year I conducted the first interviews for the project, with the crew of Apollo 12, Pete Conrad, Dick Gordon, and Alan Bean. In 1986, I began to work on the book full-time, and my house became a museum of lunar exploration: panoramas of the moon on the walls, mission transcripts and debriefings on the bookshelves, Apollo videotapes in the VCR.

At first a number of the astronauts were reluctant to get involved; some of the interviews came about only after years of perseverance. Not until 1992, seven years after I began, did the last of the lunar veterans, Apollo 16's Ken Mattingly, sit down with me to tell his stories. The encounter was typical of the interviews on this project: We talked for six hours, in which time Mattingly vividly recounted his own unique lunar experiences.

During the past eight years, when people asked me what my book was about, I would say that it was the story of the lunar voyages that the astronauts never wrote. These men are such loners that the thought of all of them sitting down to write a book together seems more far-fetched than the idea that they have been to the moon. I hope that my efforts will serve in their stead, to tell the story of a unique handful of men who have been to the edge of human experience. If we can know what it was like for them—if we can sense the men inside the space suits—then *we* can look back and see what really began on that July night twenty-five years ago: We touched the face of another world, and became a people without limits.

Andrew Chaikin, December 1993

ACKNOWLEDGMENTS

This book would not have been possible without the generous assistance of the NASA history offices at the Johnson Space Center and at NASA Headquarters. My debt to Janet Kovacevich and her staff in Houston is immeasurable. In Washington, Lee Saegesser and his colleagues were tirelessly helpful. Nor can I give enough thanks to Diana Ryan and Peter Nubile in JSC's sound department, who provided me with the audiotapes I used in my research, and who were as valuable to me as the astronauts themselves. For Apollo photographs, Mike Gentry and his staff were always helpful, always professional.

Outside NASA, many friends and colleagues made essential contributions. Mark Washburn was a voice of experience and encouragement. Tony Reichhardt sustained me during the darkest days of the project with thoughtful readings of the manuscript and many helpful discussions. Don Wilhelms and Paul Spudis guided me through the intricacies of Apollo lunar science and lent their great expertise to the reading of several chapters. Spudis was also invaluable in exploring Apollo's significance. Eric Jones, Apollo historian and kindred spirit, was an invaluable sounding board in the final year of the project. My former colleagues at *Sky & Telescope,* and especially Kelly Beatty, were most generous with the use of their offices, research materials, and laser printer, throughout the project. I thank Richard Maurer for his inspiration and for his perspectives on Apollo as history, a key contribution to this book.

Gregg Linebaugh provided videotapes of Apollo television transmissions. At Brown University's planetary data center, Deborah Glavin was generous in her support. Video Vision Associates supplied a laser disc of Apollo mission photography. Judy Mintz of XyQuest introduced me to the word-processing joys of XyWrite. Marcia Bartusiak, Marsha Cohen, David Cooper, Douglas Dinsmoor, Tom Finn, Steve O'Meara, and Donna Donovan O'Meara were sources of inspiration and insight. Kelly Beatty, Hank Bonney, Chip Cohen, Rick Friedman, Holly Hanson, Richard Maurer, Timothy McCall, Mark Washburn, and Frank White provided helpful comments on draft chapters. Lisa Clark was unfailing in her advice and encouragement.

At Viking, Dan Frank saw the potential of this project and gave me the chance to make it a reality. Al Silverman was a steady and supportive helmsman in bringing the book to completion. I owe my greatest debt to Connie Roosevelt: quite simply the best editor anyone could ask for. This book is what it is because of her wisdom, her enthusiasm, and her patience.

Of course, I am grateful to the astronauts, and everyone whom I interviewed, for their time and recollections. Many of them read draft chapters and gave helpful feedback. Most of all, they allowed me to tell their stories.

Any author, especially a first-time one, finds the task of writing a book to be bigger than he or she expected. By the time this one was finished my family and friends—who had tolerated my ups and downs and, hardest of all, cajoled me away from the keyboard from time to time—were as glad to see the ordeal finished as I was. To them, I give my heartfelt thanks: for their patience, their support, and their love.

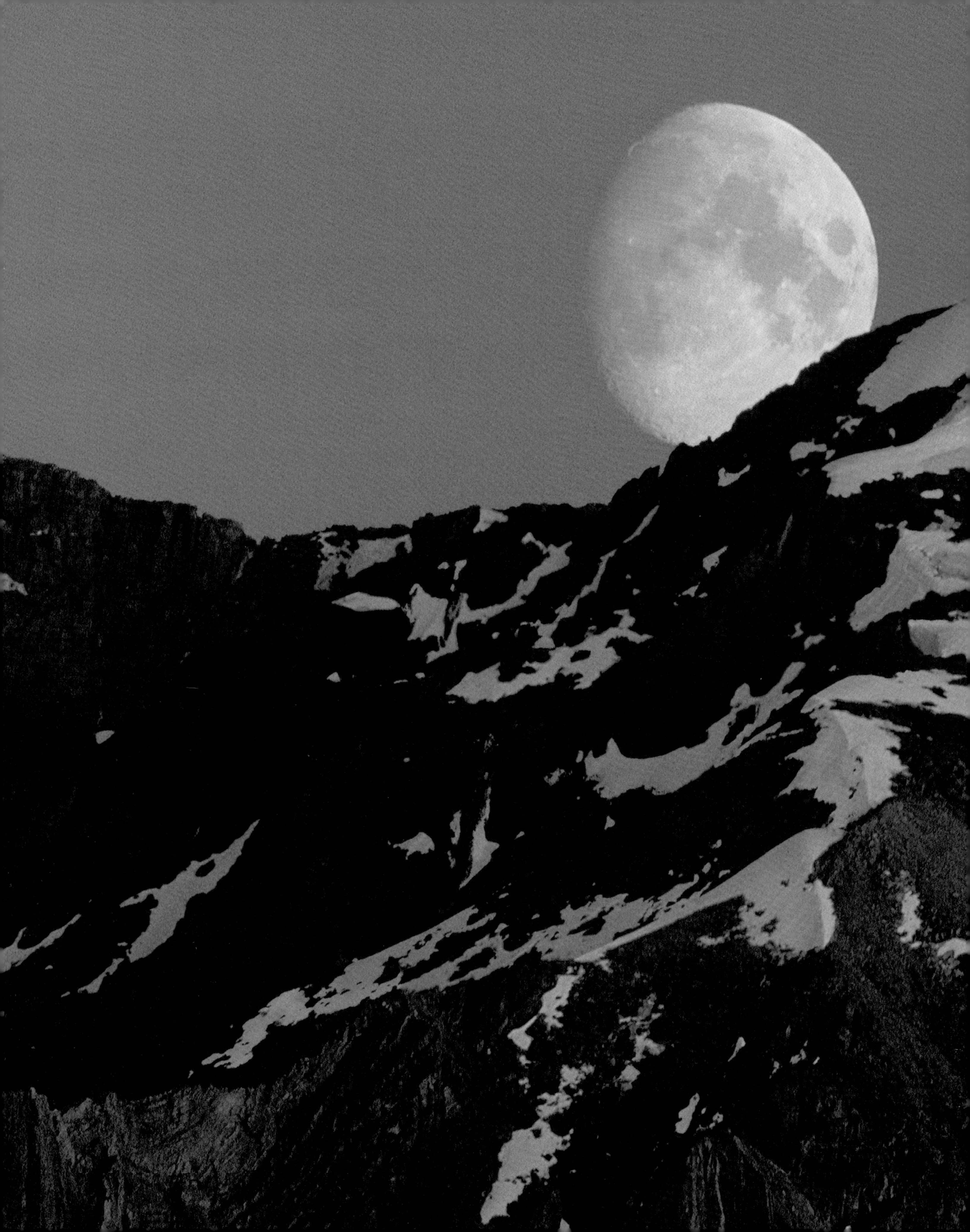

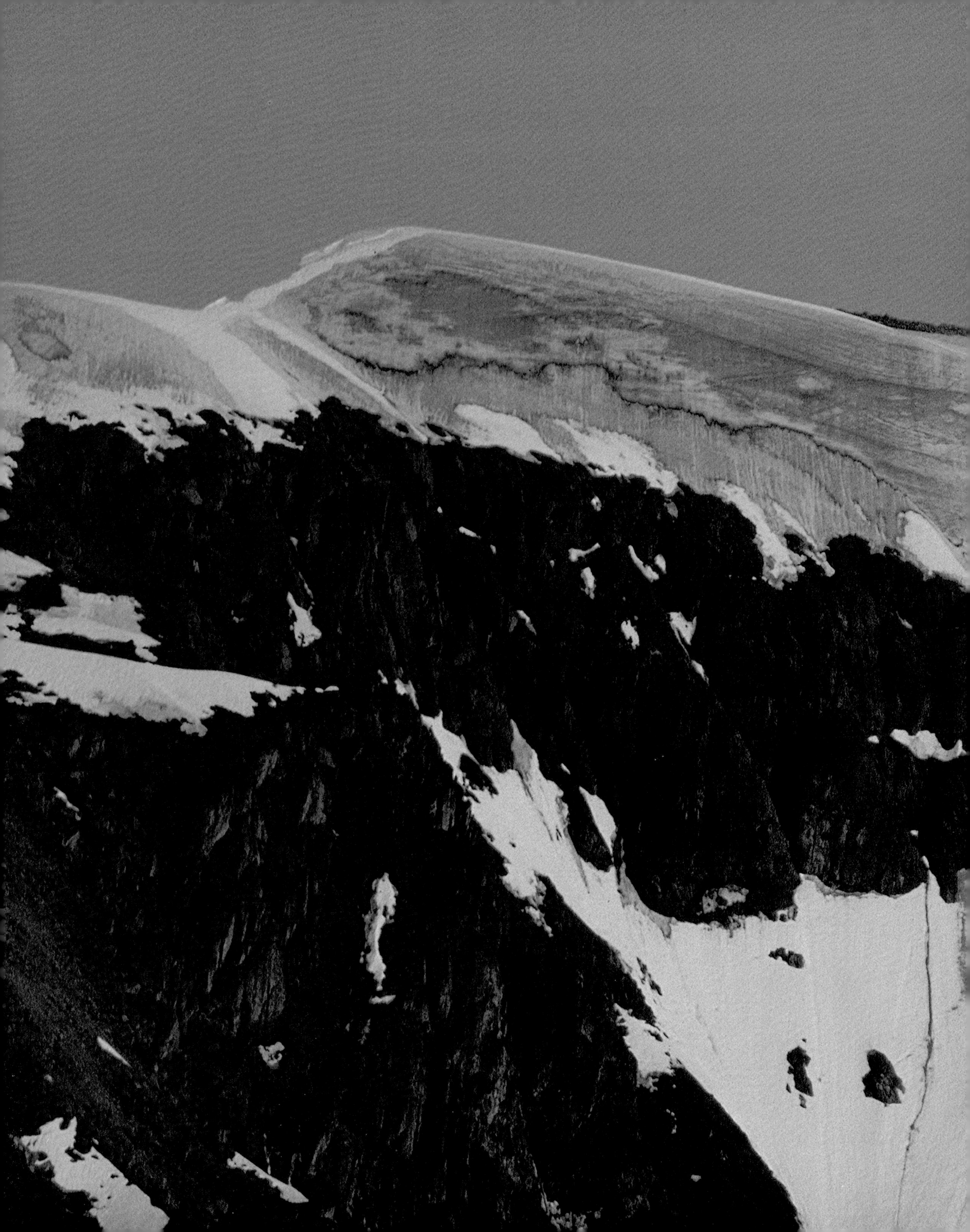

SEAL OF THE PRESIDENT OF THE UNITED STATES

PROLOGUE

MORNING
RICE UNIVERSITY, HOUSTON, TEXAS

John Kennedy stood at the podium at Rice University stadium in the heat of a Texas sun. He had come to Rice, on the outskirts of Houston, to help dedicate the National Aeronautics and Space Administration's new Manned Spacecraft Center, 22 miles away. NASA's burgeoning facilities were proof that America's space program was being transformed from an experimental venture into a bold reach a quarter of a million miles across the void. And it was happening because Kennedy had said it should.

Eighteen months after embarking the United States on a mission to the moon, John Kennedy spoke at Rice University: "Why choose this as our goal?" he asked. "Why climb the highest mountain? . . . Why does Rice play Texas?" Then came the clincher. "We choose to go to the moon in this decade and do the other things—*not* because they are easy, but because they are *hard*."

Nearly sixteen months earlier, in May 1961, the young president had stood before a joint session of Congress, and in the calmest of words, had stated the unthinkable: "I believe this nation should commit itself to achieving the goal before this decade is out, of landing a man on the moon and returning him safely to the earth." Whatever Kennedy's words lacked in emotion, their impact was immediate. Human beings had barely taken their first toddling steps off the planet. NASA had just lofted Alan Shepard on a fifteen minute suborbital flight. Now Kennedy was giving them less than nine years to get to the moon. But the stakes were high: Space was the new battleground of the Cold War, and the Soviet Union was winning.

In April the Soviets had sent the first man into space; for Kennedy that embarrassment was compounded a week later by the Bay of Pigs fiasco. Suddenly nothing mattered more than beating the Soviets in space, and Kennedy's advisers told him only a race to land on the moon offered the U.S. a chance of victory. Within days after Shepard's flight, he had made his decision. ☾

In the past sixteen months, as Kennedy's vision had materialized, so had the clarity of his purpose. To those who questioned this audacious venture, he would now give his answer.

"Why choose this as our goal?" Kennedy asked his audience. "And they may well ask, Why climb the highest mountain? Why, thirty-five years ago, fly the Atlantic?"

When President Kennedy committed the United States to a lunar landing before the end of the 1960s, a dilapidated cowshed and windmill occupied this forlorn patch of coastal prairie near Houston, Texas, the future site of the Manned Spacecraft Center.

☾ Throughout this volume, a crescent at the end of a paragraph signals an author's note at the end of the book.

"Why," he added without missing a beat, "does Rice play Texas?"

The crowd sat quietly in the heat, fanning themselves, mopping their brows, as Kennedy spelled out the technological hurdles that would have to be cleared to build the Apollo spacecraft and its Saturn V booster. With no hint of unease he laid out the staggering costs of the venture: already the space budget had increased to forty cents per person per week for every man, woman, and child in the United States—more than the allocations of the previous eight years combined—and soon it would be more than fifty cents. The effort would spawn new jobs, new knowledge, new technology, Kennedy said, but ultimately, the first voyages to another world would be "in some measure an act of faith and vision, for we do not know what benefits await us." And if the road to the moon seemed long in 1962, there were men and women in this audience who would dedicate themselves to traveling it, and Kennedy's words were fuel for their young, ambitious hearts. ☾

LIFE
19 PAGES
ON SPACE
INSIDE YURI'S CAPSULE
CELEBRATION IN MOSCOW
IMPACT ON WASHINGTON
U.S. SPACE FUTURE
Gagarin Greets Khrushchev
APRIL 21, 1961 20 CENT

Soviet Communist Party chief Nikita Khrushchev greets the world's first space traveler, Yuri Gagarin, who made a single orbit of the earth on April 12, 1961. Moved by his accomplishment, Gagarin said upon return, "The feelings which filled me I can express with one word: joy."

"We choose to go to the moon! We choose to go to the moon in this decade and do the other things—not because they are easy, but because they are *hard.*" His voice rang with energy and confidence; his words soared above the sound of applause. "Because that goal will serve to organize and measure the best of our abilities and skills, because that challenge is one that we are willing to accept, one we are unwilling to postpone, and one which we intend to win . . ."

AFTERNOON
MIRAMAR
NAVAL AIR STATION,
SAN DIEGO, CALIFORNIA

The howl of jet engines enveloped Pete Conrad as he climbed out of his Phantom supersonic fighter, the energy of the flight still inside him. It was past four in the afternoon, and Conrad planned to change out of his flight gear and head straight home. But when he reached the ready room there was a message waiting for him from NASA's Deke Slayton in Houston. Conrad wouldn't let himself believe the thought

Aglow with success, Alan Shepard crosses the deck of the carrier USS *Lake Champlain* after splashing down near Bermuda. Shepard's 15-minute suborbital flight, made in the Mercury spacecraft behind him, helped convince John Kennedy to aim for the moon.

that flashed through his mind, that Slayton's call would change his life.

At age thirty-two, navy lieutenant Charles "Pete" Conrad had already made a name for himself as one of the most gifted test pilots in the country. He'd come to Miramar from the navy's Flight Test School at Patuxent River, Maryland, where he was known as a superb stick-and-rudder man and a fine instructor. Recently, he'd requalified for night carrier landings, the hairiest, most demanding flying in the navy or anywhere else, for his money. It was carrier flying that prompted navy fliers to call themselves "aviators" instead of just pilots. Now Conrad was cleared for combat operations; soon he and the rest of Fighter Squadron 96 would deploy with the aircraft carrier Ranger on a training mission in the Pacific.

Since joining the squadron here at Miramar, Conrad had begun to think about leaving the navy. Mostly it came down to money. He had four boys to put through college someday, and the pay for test pilots was better in industry. Back in June, he'd sent inquiring letters to several aerospace companies, but in 1962 there were few industry jobs for test pilots. His only offer came from North American Aviation in Columbus, Ohio, where they were building the new A-3J Vigilante attack plane. Conrad thought hard about the job and decided it was too risky. If he went to Columbus and the Vigilante program didn't come through, he'd end up in pilot purgatory, flying a twin Beechcraft for some corporate president for the rest of his days. In the end,

Conrad began to think NASA was looking for superb medical specimens even more than piloting skills.

he'd figured, the navy wasn't such a bad deal after all. For one thing, there was job security, and the flying—that was wonderful. Nothing in life mattered more to Conrad than flying.

If anyone had lived the made-to-order childhood of a fighter pilot, Pete Conrad had. At the age of five his father took him to a country fair in Ambler, Pennsylvania, where someone was selling rides in a small Waco cabin plane. The boy badgered his father into paying for a ride, then scrambled into the front seat ahead of everyone else. When the old Waco climbed into the blue, Conrad felt as if he belonged there. From then on, flying became his obsession. Growing up in the Main Line suburbs of Philadelphia, he would sit for hours in cockpits he made from grocery crates and upended chairs, pretending to be Lindbergh flying the Atlantic, or Eddie Rickenbacker in a World War I dogfight. As a teenager he hung around at the local airport, doing odd jobs in exchange for flying lessons—a little too often, in fact, for his parents, who shipped him off to boarding school for a year to keep him out of the sky. But he returned in time to earn his private pilot's license on his seventeenth birthday. As an undergraduate at Princeton, Conrad entered the navy with an eye toward becoming a naval aviator. After graduation he had started down the road that brought him here, to the ranks of the top military pilots in the country. But that wasn't as high as a pilot could go, not any more, and therein lay the one disappointment in Conrad's past.

In 1959, Conrad had been one of sixty-nine young fliers who received secret orders to report to Washington, where they were told about Project

PETE
WARNING
EACH EJECTION SEAT IS LOADED
WITH EXPLOSIVE CARTRIDGES.

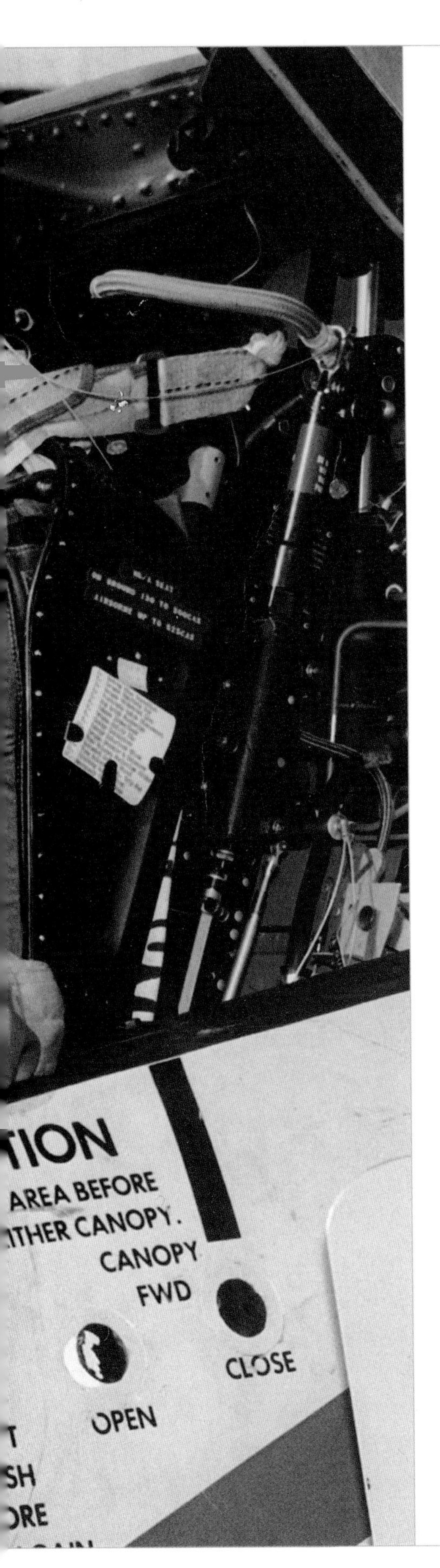

In September 1962, Pete Conrad was a 32-year-old navy fighter pilot stationed at Miramar in California. Turned down for the astronaut program in 1959, Conrad began to wonder where his career was headed—until he got a phone call offering him a chance to go to the moon.

Mercury, the country's high-priority effort to put a man in space. Thirty-seven of the pilots decided not to volunteer. The remaining thirty-two, including Conrad, went on to the Lovelace Clinic in Albuquerque, New Mexico, for a week of exhaustive medical tests. Conrad ran the treadmill. He breathed into a bag while pedaling a stationary bike. His body was subjected to every conceivable invasion and indignity. They measured everything from his IQ to the length of his lower intestine. Conrad began to think NASA was looking for superb medical specimens even more than piloting skills. And his body did not fail him.

Having run the doctors' gauntlet, Conrad moved on to the psychologists, at Wright-Patterson Air Force Base in Dayton, Ohio. For them he stared at inkblots and pictures and made up stories about them. Looking back, he came to realize it was the psychologists who did him in; he didn't take all their hocus-pocus seriously enough. When they showed him a blank card, he studied it for a moment and deadpanned, "It's upside down."

Then came endurance tests of all sorts. He was locked in a dark room, baked in a heat chamber, frozen, shaken, and whirled. When it was all over, Conrad had made it to the final cut, along with an old classmate from Pax River, Wally Schirra. Finally NASA announced that seven men had been chosen as the nation's first astronauts. Wally Schirra was one of them; Pete Conrad was not. Maybe it was just as well. During the selection process Schirra and some of the other navy pilots had gotten to talking, and they had doubts about the whole astronaut program. The stated objective of Project Mercury was to demonstrate that human beings could survive in space. The astronauts would be America's—and perhaps the world's—first space travelers. But would they really be pilots, or just passengers? And what about after the project was over? They'd go back to the navy, and who could say what effect the whole episode would have on their flying careers? Conrad, for one, wasn't sure that having flown around the earth a couple of times would do him much good. He returned to Pax and put the whole astronaut business behind him.

In the years that followed he watched from the sidelines as the Mercury astronauts soared into the limelight and then into space. If he had any lingering doubts about the program, they vanished on February 20, 1962, the day

John Glenn became the first American in orbit. While Glenn circled the earth once every ninety minutes a hundred miles up, higher and faster than any jet pilot, Conrad was flying his Phantom over the Pacific. When Glenn splashed down there was an announcement over the navy radio, and Conrad, in his supersonic jet, felt a sharp pang of envy.

By that time the astronaut program was headed places Conrad never dreamed of. NASA was going to the moon. By the spring of 1962 NASA put out a call for more astronauts; this time applications were to be made on a

Then, in the quiet, understated tone that fighter pilots often use, Slayton asked Conrad the question he would remember for the rest of his life: "How would you like to come fly for us?"

volunteer basis. The new selection amounted to a filling of the roster for the most challenging and historic test flights of all time. However slim his chances might be, Conrad couldn't pass up the opportunity to apply. At Brooks Air Force Base in San Antonio, Conrad was pleasantly surprised to find the psychologists much less in evidence; they seemed to have been replaced by the FBI, if the number of background checks was any indication. Still, when it was all over Conrad headed back to Miramar fully expecting to be rejected a second time. The moon would be for John Glenn and Wally Schirra, but not for him.

●●◐○○○◑●

The ready room—a large space with enough well-worn chairs for the entire squadron of thirty during a briefing—was nearly deserted. One wall was all windows, looking out on the flight line. At the opposite wall a couple of pilots lingered by a row of gray equipment lockers. The duty officer was at his desk. Everyone knew about the impending astronaut selection, and when Conrad came in they excitedly told him about the phone call from Slayton. He thought it was a joke—until he held the message in his hand. Conrad sat on the edge of the duty officer's desk and borrowed the phone to call Houston. Slayton came on the line, and the two men exchanged some brief pleasantries. Then, in the quiet, understated tone that fighter pilots often use, Slayton asked Conrad the question he would remember for the rest of his life:

"How would you like to come fly for us?"

Conrad almost fell over. He was not to tell anyone, Slayton instructed him; NASA wanted everything kept quiet until the official announcement the following week. He was to report to Houston on October 1. Conrad thanked Slayton and hung up, doing his best to hide his elation. He said nothing to the men in the ready room; he did not phone his wife. He changed into his khakis, got in his car, and started the 18-mile drive home. In his excitement he almost wrecked the car.

On that drive, it hit him for the first time that he was going to have a chance to go to the moon. He thought about the others who had been selected. He didn't know who they were, but he found himself wondering: If they really did this thing, how would it affect them? The Mercury pilots walked in the white-hot glare of celebrity even before they flew. When they splashed down they rode into a shower of ticker tape; they mingled with the Kennedys. What would it be like for the ones who came back from the moon? Conrad drove on, thinking about the incredible path his life had just taken. Back when he was at Pax River, a woman who lived across the street started calling him "Moon Man." And that was before there even was an astronaut program. Back then he was just a test pilot who loved to fly. Now he was aiming for the moon, after all. And he promised himself, *It's not going to change me.*

John Glenn and his wife, Annie, share the spotlight with Vice President Lyndon Johnson in a New York ticker-tape parade. On February 20, 1962, Glenn became the first American to orbit the Earth and returned to adulation rivaling that given Charles Lindbergh after his solo transatlantic flight 35 years earlier.

1

APOLLO
NASA

"FIRE IN THE COCKPIT!"

When the moon rises beyond the Atlantic shore of Florida, full and luminous, it seems so close that you could just row out to the end of the water and touch it. In January 1967, the moon seemed to draw nearer by the day to the hard, flat beaches of Cape Kennedy. Seen from there it was no longer the governess of the tides, the lovers' beacon, the celebrated mistress of song; it was a target, a Cold War beachhead in the sky. It was NASA's moon.

Almost six years had passed since John Kennedy's challenge for a lunar landing by decade's end. The moon program had grown into an effort whose size and complexity dwarfed ever the Manhattan Project. At aerospace contractors around the country, 400,000 people were hard at work on the moonships for Project Apollo. Meanwhile, Project Gemini had just come to a spectacular finale. These two-manned missions had bridged the gap between the pioneering Mercury flights and the challenge of the lunar landing. For the first time in the race to the moon, the United States appeared to have pulled ahead of the Soviet Union. With the first manned, earth-orbit Apollo flight scheduled for mid-February, all seemed on target to make Kennedy's vision a reality. But one evening late in January, that soaring optimism sud-

The ill-fated crew of Apollo 1 meets the press a few weeks before their scheduled mission. Apollo 1 commander Gus Grissom *(left)* was a veteran of Mercury and Gemini; he was to be accompanied by Gemini spacewalker Ed White and rookie Roger Chaffee. Behind them is Pad 34, where their mission was to begin.

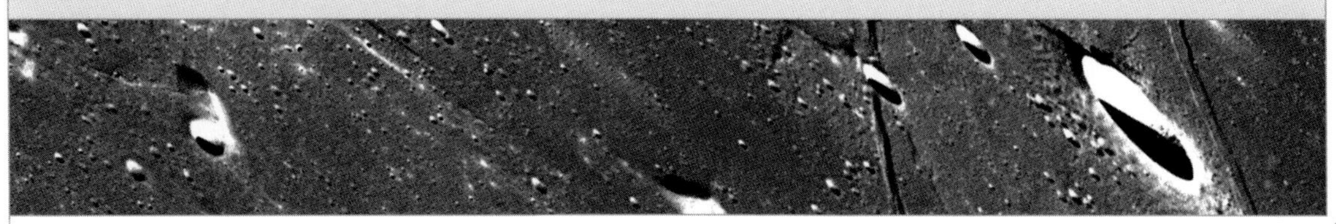

JANUARY 27, 1967
MERRITT ISLAND
LAUNCH AREA,
KENNEDY SPACE CENTER
CAPE KENNEDY, FLORIDA

denly, terribly fell to earth. "How are we going to get to the moon if we can't talk between three buildings?"

Gus Grissom's voice was low and calm, but with an unmistakable edge of irritation. A senior astronaut, veteran of two space missions, Grissom was the commander of the first manned Apollo flight set for February 1967. On this warm January afternoon Grissom and his crew, veteran astronaut Ed White and a rookie named Roger Chaffee, were participating in a simulated countdown, the kind of routine test that preceded every mission. They were sealed inside the cone-shaped Apollo 1 command module, high atop a huge Saturn 1B booster rocket at Pad 34, one of dozens of launch complexes that lined the beach at Cape Kennedy. A few hundred yards away, inside the concrete bunker called the Saturn blockhouse, some two hundred members of the launch team heard Grissom's words. At the Capsule Communicator, or "Stony," console, a young rookie astronaut named Stuart Roosa tried in vain to answer.

"Apollo 1, this is Stony; how do you read?"

In 1967 there were so many astronauts that Grissom and Roosa hardly knew each other, but the younger man could hear the barely contained exasperation in Grissom's voice:

"I can't hear a thing you're saying. Jesus Christ . . . I said, how are we going to get to the moon if we can't talk between two or three buildings?"

No one who knew Gus Grissom took him lightly. Small and powerful, he was known as a fierce competitor. When Grissom was a young air force fighter pilot in the Korean War, the fliers would ride an old school bus from the hangar to the flight line. Only those who had been in air-to-air combat could sit down; the uninitiated pilots had to stand. Grissom stood only once. He brought the same hard-driving determination to his spaceflight career, as one of the Mercury astronauts, known as the Original 7. Even among these most elite fliers, the Seven had a status approaching royalty. Despite their rivalries they had a unique bond that came from being the first Americans to venture into the heavens. Everything the Original 7 did was energized with competition, from flying airplanes to their impromptu drag races on straight Florida roads to their adventures in the nightspots of Cocoa Beach, and Grissom was always a zealous participant. But even to other astronauts, Grissom was not an easy man to know; he was a loner among loners.

In 1967 Gus Grissom stood at the top of the active roster of astronauts,

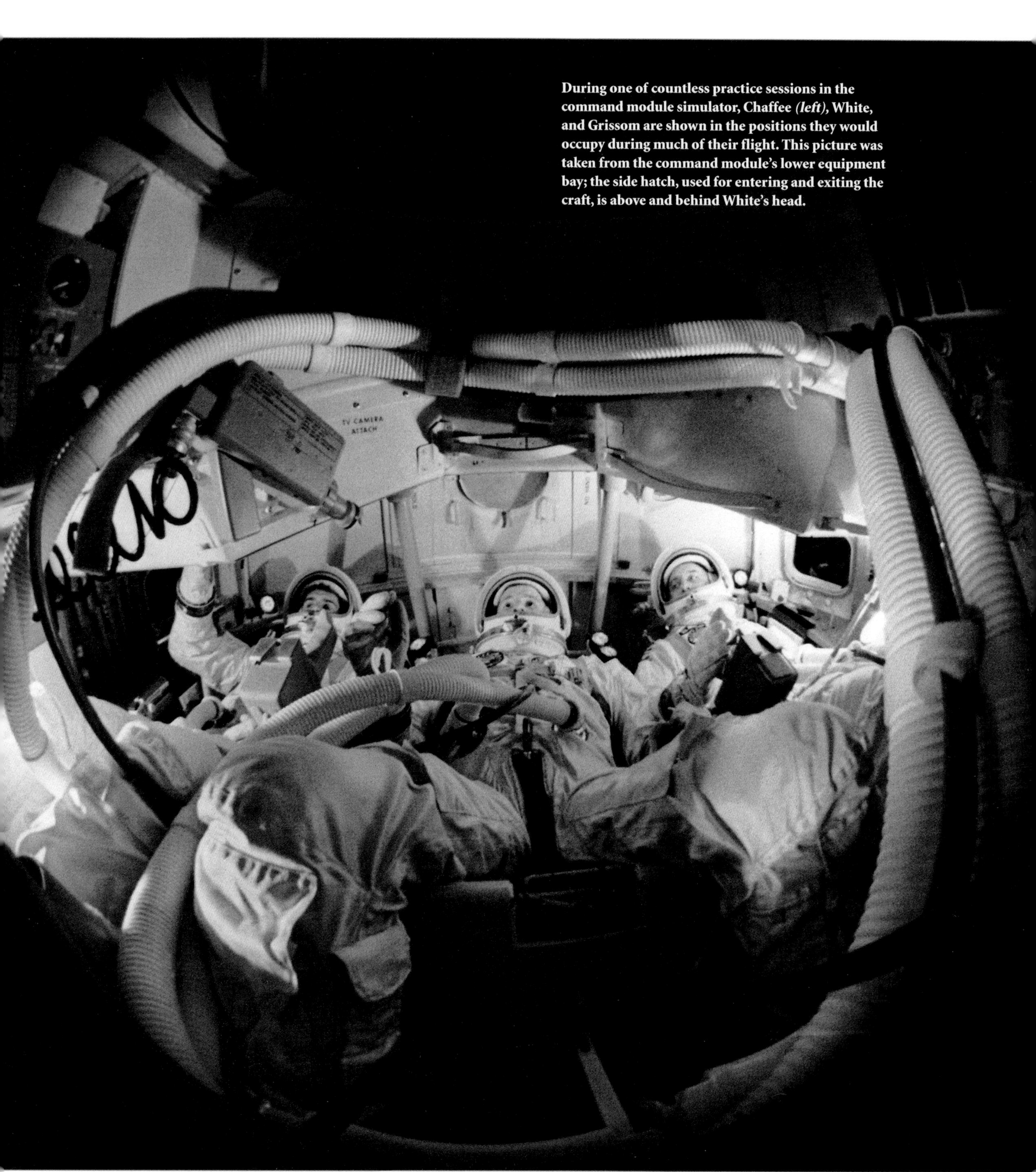

During one of countless practice sessions in the command module simulator, Chaffee *(left)*, White, and Grissom are shown in the positions they would occupy during much of their flight. This picture was taken from the command module's lower equipment bay; the side hatch, used for entering and exiting the craft, is above and behind White's head.

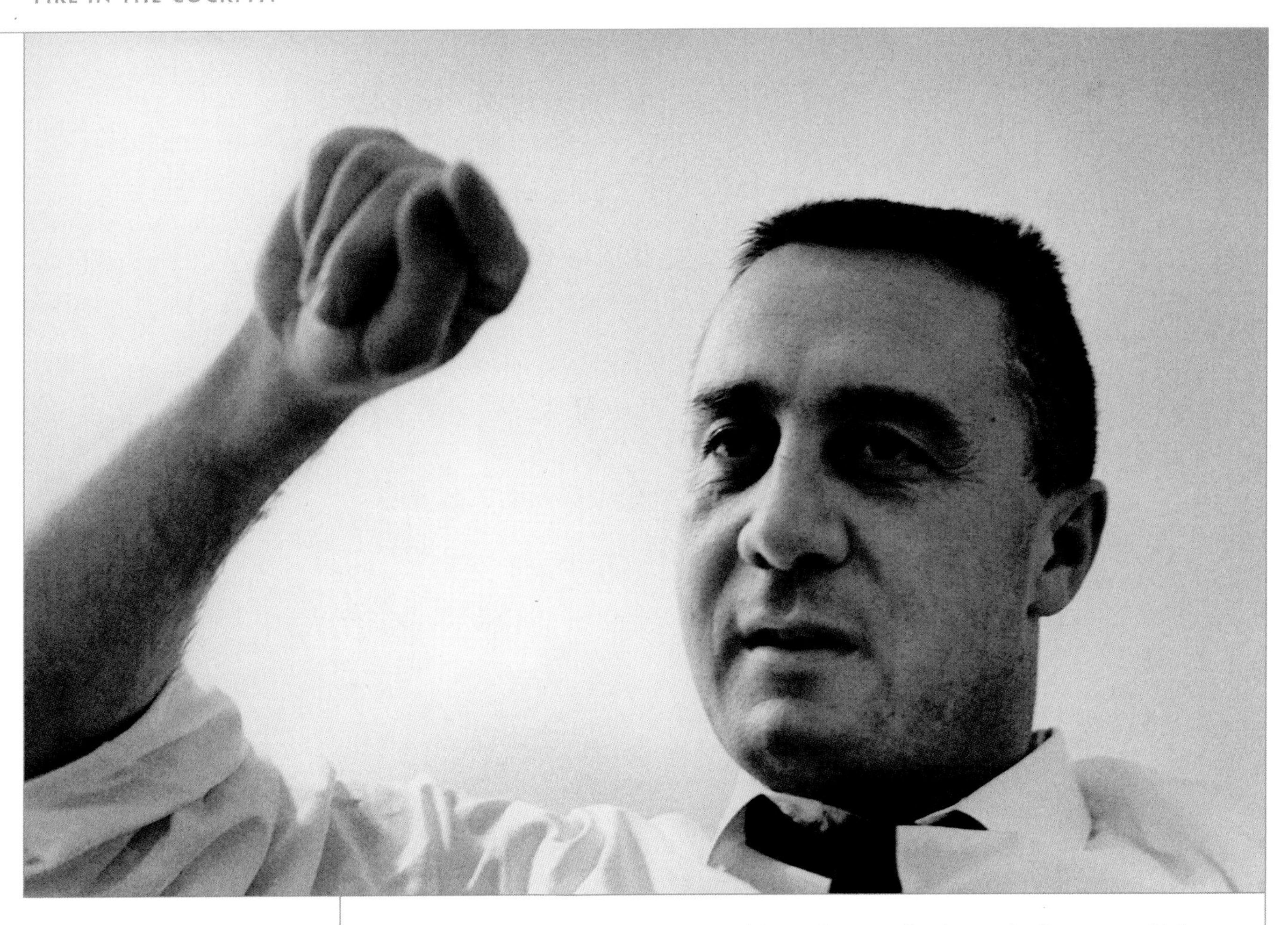

Gus Grissom, seen here after his Gemini flight in March 1965, had a reputation for being tough and competitive. His gruff manner caused some of his fellow astronauts to keep their distance, but they respected him as a skilled test pilot.

but he had started out as something of an underdog. His first spaceflight, a suborbital Mercury mission in 1961, ended in near-disaster when the hatch of his tiny spacecraft *Liberty Bell 7* blew off prematurely after splashdown in the Atlantic. *Liberty Bell* sank; Grissom narrowly escaped drowning. He maintained it had been a malfunction, that the hatch had blown off by itself, but somehow there had been a lingering skepticism—in the press, at NASA, even among other astronauts. The doubts infuriated him. He fought to make up for that image. In 1965, after helping to design the two-man Gemini spacecraft, he commanded its successful first flight. Now, at age forty, after more than a year immersed in the development and testing of the Apollo command module, he would fly its maiden voyage in earth orbit. Commanding the first flight of any new craft was always a prize assignment, but Grissom's ambitions didn't end there. He had his sights on the most coveted mission of all, the lunar landing. More than anything, Grissom wanted to be the first man on the moon. ☾

But on January 27, the moon seemed a long way off. For several months now, Grissom had worked to help ready his spacecraft for its mission, and he had become more and more displeased with the way things were going.

When engineering problems came up, he pushed for better solutions using his experience from Mercury and Gemini, but no one seemed to be listening. It made him so mad, he confided to an interviewer, he couldn't see straight. By the time the Apollo 1 command module left the factory in Downey, California, last August, it still had dozens of separate discrepancies, some of them serious. To make matters worse, the command module simulator here at the Cape was a constant source of difficulty. Just a few days ago, to show his frustration, he'd hung a big Texas lemon on it. Today, his patience was being strained once again by a trouble-plagued test. Despite all the frustrations, Grissom was pushing to get Apollo 1 into space on schedule, not because he was reckless—Grissom hadn't lived this long by being reckless—but because problems were to be expected in any new flight program, whether a new airplane or, especially, a moonship. ☾

Today's simulated countdown was nothing new; it wasn't considered dangerous—the Saturn booster was not fueled—or even difficult. But there was trouble almost from the time Grissom and his crew climbed into the command module cabin, around 1 P.M. First there was an unidentified odor in the breathing oxygen that reminded Grissom of sour milk; that alone had held up the test for an hour. Finally the problem was solved, and at 2:45 P.M. the pad crews installed the command module's heavy, two-piece hatch and sealed it shut. The spacecraft was pressurized with pure oxygen, just as it would be on launch day. Then came the communications trouble. By late afternoon Roosa was able to converse with the men inside the sealed spacecraft, but there were problems with the voice link to the Manned Spacecraft Operations Building 5½ miles away. In the blockhouse, controllers weighed the decision of whether to abort the test. They decided to continue. ☾

In the blockhouse, seated next to Roosa at the Stony console, Deke Slayton listened as technicians tried to fix the faulty communications. Slayton was forty-three years old. He had been a civilian for several years now, but he still carried himself with the quiet, serious demeanor he'd had as a young air force fighter pilot. Another member of the Original 7, Slayton was one of Gus Grissom's best friends, but their fortunes could not have been more different. Grissom was now a veteran of two space missions, while Slayton had been grounded since 1962 for a minor heart irregularity and was still waiting for his first chance to fly in space. Now, as chief of the Manned Spacecraft Center's Flight Crew Operations Directorate, Slayton's role included following astronauts through tests like this one. ☾

Slayton knew only too well that Grissom wasn't happy with the way things were going. He'd had breakfast with Grissom, White, and Chaffee this morning, along with Joe Shea, the hard-driving NASA manager in charge of the command module effort, and they were all running through the litany of troubles with Grissom's spacecraft: Malfunctions in the environmental control system. Coolant leaks. Faulty wiring. Grissom bemoaned the communications trouble that plagued almost every test. "If you don't believe it," Grissom told Shea, "you ought to get in there with us."

Shea declined, largely because there was some question about whether technicians would be able to rig up an extra communications headset for him in time. But Slayton gave the matter some thought. There would be space for him to sit, in shirtsleeves, in the command module's lower

A recovery helicopter plucks Grissom from the sea after his Mercury flight on July 21, 1961. His spacecraft sank when its side hatch blew off prematurely; Grissom narrowly escaped drowning. The incident led to the design of a hatch for the Apollo command module that was secured by bolts that had to be undone by hand.

equipment bay beneath the footrests. He was still weighing the idea around midday as Grissom, White, and Chaffee suited up. Slayton rode with them in the transfer van to Pad 34, and by the time they arrived he had made his decision. He would be better off in the blockhouse, he told Grissom, where he could keep a close eye on the test. Years later, he would still wonder whether or not he made the right choice. ☾

By evening, as dusk settled onto the marshlands, technicians continued to troubleshoot the faulty communications. Searchlights came on, bathing the giant Saturn rocket in white light. Meanwhile other events in the countdown continued. Apollo 1 was on its own electrical power now, just as it would be in the final minutes before liftoff. But the communications troubles were still unsolved, and at T minus 10 minutes, the test director called a hold in the count.

At 6:31 P.M., eleven minutes into the hold, Slayton was looking over the test schedule when he heard a brief, clipped transmission from Apollo 1. It sounded like "Fire."

●◐◐○○○◑◑●

On the other side of the country, at the North American Aviation plant in Downey, California, veteran astronauts Tom Stafford, John Young, and Gene Cernan were sealed inside the second Apollo command module in their own spacecraft test. Besides being the most experienced space crew yet assembled—Cernan had flown once before on Gemini, while Stafford and Young each had two missions under their belts—they had a camaraderie seldom matched on a space crew. Now the three men were assigned as the backup team for the second manned Apollo mission.

This command module, like Grissom's, was a prototype called Block I. It was never built to go to the moon, but was instead designed only to fly in earth orbit, on the first Apollo flights. An improved version, called Block II, was already being developed for the lunar missions, and last month, NASA had decided that there would be only one Block I mission, Apollo 1. Stafford and his crew were here at North American, the prime contractor for the command module, to help provide engineering support for the activities at the Cape. ☾

The work on Block I had been frustrating, in part because of the atmosphere here at Downey. North American had some very competent people, but they had never built a manned spacecraft before. Bundles of wire on the command module floor were unprotected, making them susceptible to damage. There were so many changes made within the cabin

On a veranda near Los Angeles, the primary and backup crews for Apollo 1 review the flight plan. From left to right: Gus Grissom, Ed White, Rusty Schweickart, Roger Chaffee, Jim McDivitt, and Dave Scott.

that the workmen could barely keep track of them all. Wires were constantly being rerouted, black boxes replaced. And lately, astronauts had felt a strain in their relationship with the North American engineers, who had begun to resist their suggestions. ☾

It had been different during Gemini. That spacecraft, like Mercury, was the product of the McDonnell Aircraft Corporation in St. Louis, which had forged a harmonious relationship with NASA. The astronauts who visited the factory had no trouble making inputs into the design process. And if they had a problem, they could always take it to "Mr. Mac" himself, and he would get results. But Apollo wasn't like Gemini. The intimacy was gone now. There was no single boss who could respond to the astronauts' concerns; now there were bosses scattered throughout this massive operation. ☾

Even within NASA, there was a disturbing lack of coordination. In Houston, some of the engineers in the Apollo program office acted as if Gemini had never existed. They had an arrogance that seemed to say, *We know this business better than you do.* Experience from Gemini—Stafford's or anyone else's—seemed to make no impression on them. They rolled their eyes at "those Apollo astronauts and their Gemini war stories." Trying to get through to them was like talking to the wall.

All of them knew of the trouble with Block I, but in the back of everyone's mind was the end-of-the-decade deadline for the lunar landing. However dissatisfied they were, Stafford and the other astronauts had been willing to put up with Block I for the first couple of flights in order to stay on schedule. Their feeling was, "Just get us airborne; we'll fly it."

But on this day, Stafford wasn't so sure. Their test, too, was in trouble: Leaking coolant lines. Short-circuits. At one point the hatch fell on Cernan's foot.

"Go to the moon?" Stafford growled, "This son of a bitch won't even make it into earth orbit."

Finally Stafford stopped the test. When he climbed out of the spacecraft, an emergency call from the Cape was waiting for him.

●◐◐○○○◑●

When Deke Slayton heard the report of fire from Apollo 1, his gaze shot to a nearby closed-circuit television monitor. It showed the picture from a camera pointed at the command module's hatch window. The window was filled with bright flame.

Suddenly there was another message from the spacecraft, this time quite clear, in a voice of contained urgency: "We've got a fire in the cockpit!"

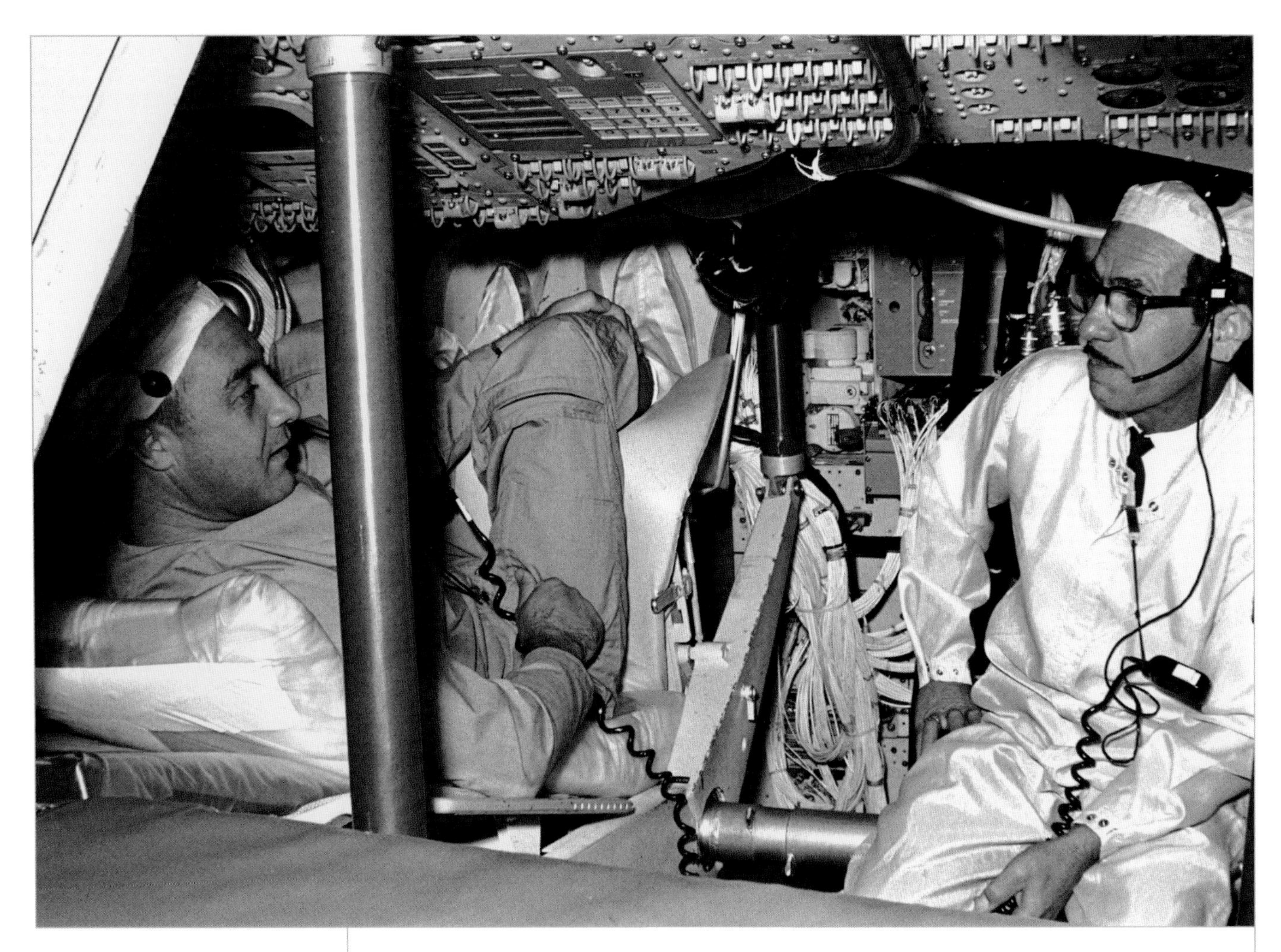

Grissom inspects the cabin of his command module with an engineer at North American Aviation in Downey, California. A key player in the design of the Gemini spacecraft, Grissom felt he was often ignored in the effort to solve command-module problems, a source of intense frustration to the astronaut.

Slayton recognized the voice as Roger Chaffee's. Chaffee was on the right side of the spacecraft, where the radio controls were. It was his job, in an emergency, to maintain contact with the blockhouse.

On the television monitor Slayton could see Ed White's arms reaching back over his head, trying to undo the bolts that held the side hatch shut. Neither Slayton nor anyone else in the blockhouse fully understood what was happening; Slayton would say later that his main concern was not fire, but smoke. But now Slayton heard another voice, clearly frantic: ☾

"We've got a bad fire. . . . We're burning up!"

At first Slayton thought it was Pad Leader Don Babbitt, stationed next to the spacecraft, calling for help. Later, on the tapes, listeners identified the voice as Chaffee's. Seconds later—less than half a minute after the first report of fire—Slayton and the horrified controllers heard the last transmission from Apollo 1. It was a brief cry of pain.

Long seconds passed. Now the communications loop surged with activity as technicians struggled to get the hatch open. "It's too hot," you could hear them say. On the television monitor, through dense smoke, Slayton could see

In the soot-blackened cabin of Apollo 1, seen through the open hatchway a day after the fire, a checklist lies on Ed White's couch. All three crew members died, not from heat, but from asphyxiation.

the pad crews approach the hatch only to be driven back by the intense heat. Roosa tried several times to reestablish contact with the crew, with no response. Several long minutes elapsed before the hatch was finally opened, and a short time after that the pad leader came on the communications loop with a terse and ominous transmission: "I'd better not describe what I see."

Physicians Fred Kelly and Alan Harter were in the blockhouse, and Slayton instructed them to go to the pad. Then he put in a call to the Manned Spacecraft Center in Houston, to set up a command center and get word to the families in case things were as bad as he feared. After several minutes the call came from the doctors, confirming what everyone had dreaded. Slayton made another call to Houston; then he and Roosa left the blockhouse and headed to Pad 34.

●◐◐○○○◑◐

It was a slow Friday afternoon in the Astronaut Office at the Manned Spacecraft Center. Here, twenty-five miles from downtown Houston, built on a flat, coastal prairie that had once been owned by a Texas oilman, was a collection of modern buildings, all dark glass and white stone, set among neatly manicured lawns and artificial duck ponds. Here and there a windowless training facility or a laboratory with massive cryogenic storage tanks hinted at the true nature of this place, but for the most part it looked more like the campus of a community college than a place where engineers, scientists, and astronauts were waging an all-out assault on the moon.

The Astronaut Office was located on the top floor of the three-story structure called Building 4. On this Friday afternoon it was nearly deserted. Most of the astronauts were out of town, chasing down some piece of the Apollo effort at contractors' plants around the country. But a young rookie named Alan Bean was here. Although he had been an astronaut since 1963—like Roger Chaffee, he was a member of the third astronaut group—Bean was still waiting for his first chance to fly in space. A few months ago, Deke Slayton had named him as the Astronaut Office representative on the manned space station project that was planned for the early 1970s. Bean didn't always hear about the details of what was going on in Apollo, but he knew Grissom, White, and Chaffee were at the Cape, getting ready for their mission.

Sometime before 6 P.M. the phone rang. It was one of the support people at the Cape. What he said sounded so strange that Bean did not at first understand:

"We've lost the crew."

The man's voice was quiet. Bean heard no anguish in it. He had to stop and think about the words. "The crew" was surely Grissom's; were the people at the Cape having trouble finding them? *Lost the crew?*

Bean answered, "Where do you think they've gone?"

The voice stumbled over more words that didn't make sense; he just didn't seem to want to tell Bean what had really happened. It took a long time for him to say it: Grissom, White, and Chaffee were dead.

Bean had barely hung up the phone when it rang again; this time it was

On January 28th, a shaken Martha Chaffee ventured out before news cameras, supported by Gene Cernan. A close friend of the Chaffees, Cernan took on a traditional role of military aviators—to console the family after a death in the line of duty.

Mike Collins, in Deke Slayton's office up in the administration building. Collins, acting as the Astronaut Office representative for the Friday afternoon staff meeting, had just heard the same news. Now there was no miscommunication, for it was one pilot talking to another and both knew what needed to be done. They agreed that Bean would coordinate astronauts and wives to go to the homes of the dead pilots. Bean called his wife, Sue, and sent her to the home of Martha Chaffee until Collins could get there. Wally Schirra's wife, Jo, and Chuck Berry, the space center's chief physician, would go to Betty Grissom's. Neil Armstrong's wife, Jan, would go to her next-door neighbor Pat White's, and Bill Anders, another member of the third group, and like Bean, a rookie, would follow.

El Lago was one of a handful of planned communities that had sprung up around the space center, scattered with ranch houses and crisscrossed with tidy streets winding through the greenery. Here was the space community's own suburb: astronauts and NASA engineers and managers all lived next door to each other. And aside from the tourists who occasionally came by looking for some sign of an astronaut as if they were on a tour bus in Beverly Hills, and the mobs of reporters who, during missions, stood watch on an astronaut's lawn as if the Fischer quintuplets had just been born inside, there was nothing remarkable about the way it felt to live here. That was precisely the point: the residents of El Lago, and of nearby Nassau Bay, and Timber Cove, and Clear Lake, clung to normalcy in the midst of the most extraordinary enterprise of the twentieth century.

It was already dark when Bill Anders arrived at the ranch house belonging to Ed and Pat White. Normally there wouldn't have been any reason for Anders to know White very well; the two had never been on a crew together, and they belonged to different astronaut groups, White a Gemini veteran from the second group, Anders a rookie from the third. Anders and his wife, Valerie, had gotten to know Ed and Pat White mostly because they lived one street apart.

Even among the astronauts, Ed White had always stood out; a strapping six-footer who had barely missed becoming an Olympic hurdler, he was known as one of the finest physical specimens in the Astronaut Office. And perhaps more than any astronaut except John Glenn, White subscribed to their all-American image. In 1965, after he became the first American to walk in space, White easily wore the mantle of a national hero. There appeared to be no limit to how far he might go. Ed and Pat seemed perfectly matched. Few women were so devoted to their husbands.

Now, Bill Anders would have to tell Pat White that her husband was dead.

Bill Anders had lived with death for most of his adult life, first as an air force fighter pilot, then as an astronaut. He had been to his share of funerals. And it was at funerals that Anders had noticed something about himself that made him feel different from other people. He did not shed tears. While he felt sadness for the family, he did not grieve for the man who was killed. When he'd decided to become a fighter pilot, he'd accepted deadly risk as part of the bargain. And he knew the dead pilot would have felt the same way: "Sure, it's a shame to lose a good man—but he knew what he was getting into. He flipped a coin, and he lost."

By 1967 death had become a part of the astronaut life. Just last year Charlie Bassett and Elliot See had been killed in a plane crash, and two years before that, it was Ted Freeman who perished the same way. And none of this had made Anders hesitate to climb into the cockpit of a supersonic jet fighter, or trust his life to a pressure suit in the vacuum of a test chamber. It didn't blunt his desire to ride a moon rocket. Like all the astronauts, Anders accepted the risks of the job, and it wasn't difficult for him to do that. What was difficult—the hardest thing Anders would ever do—was to go to the house of this attractive woman in her thirties who was raising two children, bearing bad news. A short while ago Pat had picked up her daughter from a ballet lesson. When she arrived home her neighbor Jan Armstrong was waiting silently for her. She must have been surprised, then confused—after all, Ed was at the Cape, he wasn't flying tonight—and then she must have filled with dread. But it wasn't up to one astronaut wife to tell another that her darkest nightmare had come true. That task most often fell to another astronaut. Anders rang the doorbell.

●◐◐○○○○◑◑

When Slayton and Roosa arrived at Pad 34 ambulances waited in vain at the base of the huge launch tower, their rotating beacons flashing in the night. Entering the steel gridwork of the gantry, the two men boarded a small elevator and rode it to Adjustable Level A-8, 218 feet up, and headed across the swing arm to the small enclosure at the other end, called the White Room. Even before they arrived they were assaulted by the stench of burned electrical insulation and incinerated plastics. At last, reaching the end of the swing arm, they could see it: the command module's square hatchway, flush against the side of the White Room. Something was hanging from the open hatch; it was an arm, clad in a white space suit. Kelly and Harter, the two doctors, were here; they told Slayton the dead men were still inside, snared in a web of melted nylon netting that had once hung within the cabin. The doctors had been unable to remove them.

Slayton gripped the rail just above the hatchway and leaned into the cabin of Apollo 1. He could see the familiar configuration of the command module cabin: three couches, side by side; the broad center instrument panel, amid a forest of switches, knobs, and controls. Warning lights still glowed amber on the blackened panel. Much of the once spotless cabin was covered with soot. On the right side of the cabin Slayton could see Chaffee, his space-suited form motionless in his couch, still strapped in. The other two couches were empty. Slayton looked down, below the edge of the hatchway, and spotted two helmeted heads, both with clear faceplates still closed. Right below the hatch were a pair of legs, doubled up, from which the layers of space suit material had been burned off. It was impossible to tell who was Grissom and who was White. ☾

Slayton told Kelly and Harter not to do anything more until photographs could be taken. Then he and Roosa turned away from the blackened cabin of Apollo 1, carrying the smell of fire and death with them. For the rest of their lives, even after the terrible images had faded in memory, they would remember that smell.

●◐◐○○○◑◑●

The astronauts had always known it was only a matter of time. Gemini had had its share of close calls, none worse than when Gemini 8 began tumbling out of control: Neil Armstrong and Dave Scott had narrowly escaped with their lives. And the Gemini pilots had taken some calculated risks. For example, everyone knew that the Gemini ejection seat and parachute that served as the only means of escape in a launch emergency was effective only under very limited conditions. And there were other risks, not only on Gemini but on the comparatively primitive Mercury flights. Looking back, it was not only superb hardware and outstanding people, in space and on the ground, that had averted tragedy, it was also luck. But the lunar missions were even more complex, and among the astronauts there was an unspoken feeling that it was only a matter of time before their luck ran out.

Gus Grissom had known that. Sometime during Gemini, he had told his wife, "If there's ever a serious accident in the program, it's probably going to be me." And a few weeks ago, at a press conference, he had said, "If we die, we want people to accept it. We're in a risky business...." But he was talking about dying in space. If Grissom, White, and Chaffee had burned up in reentry, if they had perished in the fireball of an exploding booster, if their parachutes

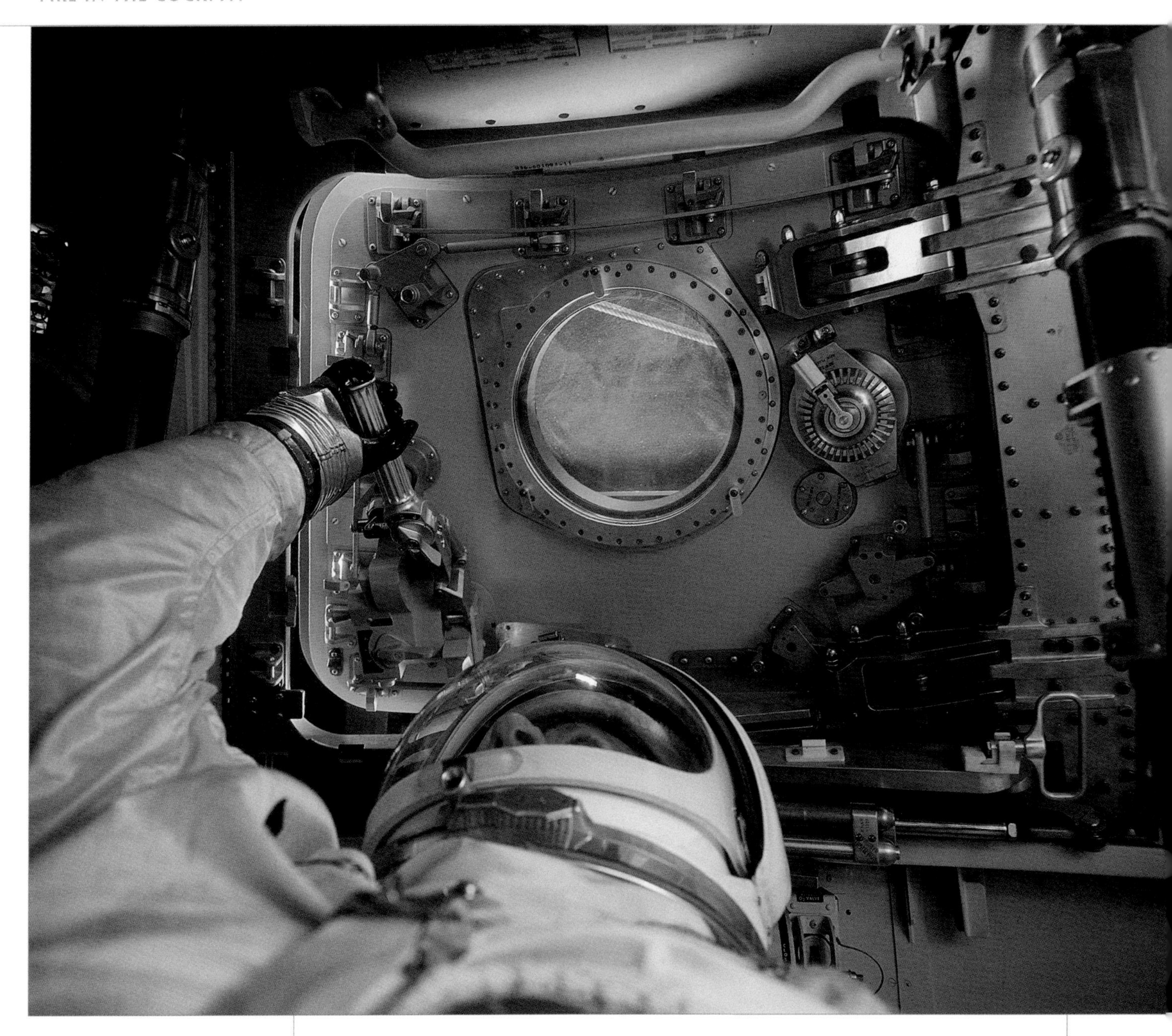

By January 1968, a year after the fire, a hinged hatch, which swung outward and could be opened in three seconds, supplanted the unwieldy hatch that had prevented escape from the fire. The new hatch was only one feature of a redesigned, fireproof command module that was taking shape.

hadn't opened and they had plummeted into the ocean—any of those fates would have been easier to accept. The terrible shock of this January night, and the irony, was that they died while their spacecraft was sitting on the pad, with technicians all around them, safety just on the other side of the hatch. And yet, no one had been able to save them. ☾

What went wrong? Even years after investigators began to sift through the wreckage of Apollo 1 piece by piece, no one could say exactly. But within weeks, the general picture became clear; the fire was a disaster waiting to happen. During the test, the command module was pressurized with pure oxygen at 16.7 pounds per square inch, slightly above sea-level atmospheric pressure. Pure oxygen can be a fire hazard even at low pressure, but at 16.7 psi the danger grows to frightening proportions. And yet, the practice of pressurizing with oxygen on the pad had been used dozens of times during Mercury and Gemini, without mishap.

NASA had chosen pure oxygen for Apollo for the same reasons it had been used in Mercury and Gemini: it eliminated the weight and complexity that would have been required for an oxygen/nitrogen mixture. Pure oxygen was essential while in orbit, when the cabin pressure was only 5 psi. Somehow, no one had absorbed the realization that on the pad, the pressure was kept slightly above sea-level conditions. And if oxygen carried a fire risk, the command module's designers thought they had removed all possible sources of ignition from the command module cabin. ☾

But they were wrong. Shortly after 6:31 P.M. on January 27, the review board concluded, there was a spark inside Apollo 1, probably in the vicinity of some damaged wires in the lower equipment bay at the foot of Grissom's couch. Perhaps aided by flammable fumes leaking from a nearby coolant pipe, the spark ignited some nylon netting that had been installed underneath the couches to catch dropped equipment, and the fire spread quickly. Other flammable items, including foam pads that were there to protect the interior finish during the test, fueled the blaze. Even materials normally considered flame-resistant burned as if they had been doused in kerosene. On the walls Velcro fasteners, a favorite means of securing loose gear in weightlessness, exploded in a shower of fireballs. In seconds, as the temperature soared to 2,500 degrees Fahrenheit, the command module became an incinerator, and Grissom, White, and Chaffee never had a hope of escape—because the hatch had become impossible to open. ☾

The command module's side hatch was one of the inevitable design compromises. It was a two-piece affair, with an outer hatch and an inner hatch that opened inward, into the cabin. Some NASA engineers and astronauts, and even engineers at North American, had questioned the design, calling for a one-piece hatch that could be swung open, like the one used in Gemini. But the Apollo managers, and Joe Shea in particular, always had sound reasons for vetoing the change. Mostly it was a matter of weight: Each pound of payload cost many times its own weight in propellant to haul it off the surface of the earth and send it to the moon. The two-piece hatch was not only the most lightweight design, it was also the simplest. And to anyone concerned about air leaks during a two-week trip to the moon and back, an inward-opening hatch solved the problem: cabin pressure would keep it tightly sealed. But even under the best conditions it was very difficult to open. The inner hatch was a heavy, cumbersome metal plate secured by a set of bolts. The man in the center couch had to reach back over his head, undo the bolts using a special tool, and then lower it out of the way. Ed White had been in Apollo I's center couch, and no astronaut surpassed him for sheer physical strength. For exercise, White and his backup, Dave Scott, used to practice opening the hatch; it was like pressing a couple of hundred pounds at the gym. But only seconds after the fire started neither Ed White nor any other human being would have been strong enough to open it. As the fire progressed the buildup of hot gases sealed the hatch shut with thousands of pounds of force. As it was, White never had a chance even to undo the bolts. ☾

The blaze would undoubtedly have consumed the three men had it continued, but within fifteen seconds after the first report of fire, the pressure in the cabin soared to nearly twice sea-level atmospheric pressure—high enough to rupture the command module's hull. Hot gases rushed through the breach with a loud whoosh that startled the pad crews, leaving the cabin enveloped in thick smoke. But by then, the horror of being trapped in an inferno was nearly over for Grissom's crew. Seconds later their oxygen hoses burned through and carbon monoxide forced its way into their space suits. Fifteen to 30 seconds after that, the medical examiners estimated, the men lost consciousness. Within four minutes there was no hope of reviving them. Grissom, White, and Chaffee did not burn to death; they were asphyxiated.

North American Aviation president Lee Atwood *(right)* and Vice President Dale Myers testify at a congressional inquiry into the Apollo 1 fire. Although they had harbored misgivings about the use of pure, high-pressure oxygen in Apollo—an extreme fire hazard—the executives accepted a share of the blame for the disaster.

The greatest irony was that Gus Grissom, who had almost drowned after his Mercury mission because of a hatch that opened prematurely, was claimed by a hatch that could not be opened at all. ☾

Even the astronauts who did not see the charred spacecraft or smell its acrid odor in the Florida night were stunned by the news from the Cape that Friday evening. They were appalled to realize how much they had overlooked. They had talked about what they would do if a fire broke out while they were in space, how the flames might propagate in zero gravity. But none of them had even considered a fire on the pad. Like so many things about this disaster, it was almost beyond comprehension.

In hindsight, there was enough blame to go around. Some astronauts, in anger, singled out North American, saying its engineers, yielding to schedule pressure, had taken shortcuts. And as the investigation of the fire progressed, there were charges of mismanagement and shoddy workmanship. Nor was NASA without fault: All of this had happened under the agency's supervision.

But there was another, more forgiving view. True, North American wasn't like McDonnell. But no one had ever tried to assemble a moonship before. The Apollo command module was the most complex flying machine ever devised, an intricate package crammed full of state-of-the-art equipment. It would have been naive not to expect all kinds of things to go wrong the first time they put one together. And no one, at NASA or North American, had knowingly compromised the astronauts' safety.

And in the next weeks the astronauts came to admit what Stafford, Young, and Cernan did over a few drinks one night, that there was a hidden blessing in this disaster: the wreckage of Apollo 1 was there for the accident board to examine, not a silent tomb circling the earth or drifting in the translunar void. Although three men had died, three or perhaps six more lives had probably been saved.

On the cold, wet January day when they buried Grissom, White, and Chaffee, most of the astronauts put the tragedy of the fire behind them with an acceptance that was difficult for outsiders to understand. Ultimately, when Grissom had spoken of accepting death in a dangerous, critical undertaking, he had been speaking for all of them. Their biggest concern now was making Apollo fly. The immediate causes of the fire had to be fixed—and now there was a much needed chance to correct the long list of other inadequacies. It would take time, but they had to recover and move on. Before them was the most extraordinary goal of the twentieth century, and it came with a deadline.

Three riderless horses symbolizing the fallen Apollo 1 crew pull the caisson bearing Roger Chaffee's casket in Arlington National Cemetery on January 31, 1967. Gus Grissom was also laid to rest at Arlington that day; Ed White was buried at West Point.

Betty Grissom watches as the flag from her husband's casket is folded. She later settled a lawsuit against NASA and North American Aviation for negligence in the fire.

BORMAN
LOVELL
NASA

THE OFFICE

APRIL 1967
MANNED SPACECRAFT CENTER, HOUSTON

Deke Slayton stood in the small conference room on the third floor of Building 4, facing the collection of astronauts he had called together. In the spring of 1967, NASA was moving on from the tragedy that would always be known as "the Fire." Amid lingering criticism of the agency, the accident review board had made its report to Congress. At the North American plant in California, a redesigned, fireproof Block II command module was under development, and now there was an astronaut—Gemini veteran Frank Borman—working hand in hand with new management. And here in Houston, it was time to get back to the business of making the end-of-the-decade deadline for the lunar landing. Eighteen astronauts waited for the meeting to begin.

Slayton was not one for fanfare. "The guys who are gonna fly the first lunar missions," he said, "are the guys in this room."

At first glance, there was nothing remarkable about these men. Sometimes, when they walked the halls of the space center, clad in NASA-blue flying coveralls, they were easy to spot; here, dressed mostly in sports shirts and slacks, they wore a casual anonymity. But their trim bodies hinted at

Getting a taste of weightlessness, three members of the second group of astronauts—the New 9—tumble inside a KC-135 training aircraft in 1963. From left, Frank Borman, Tom Stafford, and Jim Lovell sample squeeze tubes of food designed to be eaten in zero gravity.

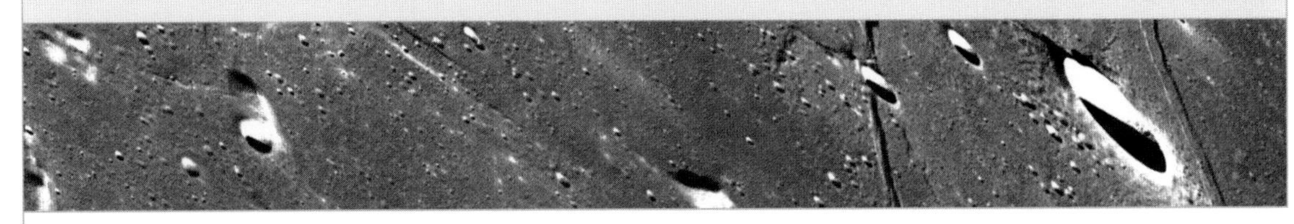

Deke Slayton *(left)* and his astronaut corps gather around a conference table in 1963. Grounded because of an irregular heartbeat, Slayton assumed the task of assigning crews to Gemini and Apollo missions. Seated to Slayton's right is Frank Borman; then, clockwise around the table are: Ed White, Neil Armstrong, Wally Schirra, Scott Carpenter, John Glenn, Gordo Cooper, Pete Conrad, Jim Lovell, Tom Stafford, John Young, Elliot See, Gus Grissom, Al Shepard, and Jim McDivitt.

lives of self-discipline, and their eyes, quick and bright, revealed an unwavering concentration born of countless adventures in the air, in a profession in which the mind simply does not wander. To the people who worked with them—engineers, simulation instructors, flight controllers—they were "just regular guys," but even they could not deny that there was something remarkable about who these men were, where they had been and, most of all, where they were going. Most of these eighteen were veterans of the two-man Gemini flights of 1965 and 1966. Those missions, as bold as they were successful, had been their training ground for the moon.

If the astronauts had been a squadron, Slayton would have been their wing commander; sometimes, in conversation with each other, they called him Father Slayton. In 1959—a lifetime ago, it seemed—there were just seven men who called themselves astronauts, and Slayton had been one of them. He was considered one of the most gifted pilots of the Original 7, but even now Slayton was still waiting for his first spaceflight. In 1962, only weeks before he was scheduled to become the second American in orbit, Slayton was abruptly grounded by the doctors. He had a condition called idiopathic paroxysmal atrial fibrillation: every so often, for unknown reasons, his heartbeat became irregular. Slayton's protests that he felt absolutely fine and that his performance was unaffected were to no avail. NASA's administrator, James Webb, ruled that the agency couldn't take the risk of sending him into space. And that was not the end of Slayton's frustration; he was banned from flying an airplane by himself.

Since then, while waging a lonely battle to get back on spaceflight status, Slayton had served as the space center's director of Flight Crew Operations, managing the affairs of every pilot at the center, including the astronauts. From his office on the ninth floor of the administration building, he oversaw just about everything the astronauts did, from training to business trips to public appearances. In the most difficult of ironies, it was Slayton who picked the crew for every mission. He would have given anything to join them, to trade his executive suite for a government-issue desk in the Astronaut Office, and they knew it.

There is no record of how the astronauts reacted to Slayton's announcement; probably they said nothing. To some, Slayton's words came as welcome reassurance; for others they were not news. For a handful—the rookies—just being in this room at all was reward enough; it meant an end to years of struggling to make it onto a space crew. But none of them, rookie or veteran, missed the significance of what Slayton said. They were finalists in an unofficial and largely unspoken competition in which the prize was the ultimate test flight, the first lunar landing. Most of them saw it the way Pete Conrad did. He knew he belonged in this room; he had known he would be here all along.

No one in the Astronaut Office was more of a natural than Pete Conrad. At the age of thirty-six, he was not only one of the most seasoned pilots in the Office, but one of the most flamboyant. He appeared to enjoy everything he did. He liked to race Formula-V cars and was given to wisecracks and practical jokes. He used foul language in the simulator. He enjoyed giving the other pilots nicknames they hated. It was Conrad, for example, who had bestowed the monicker "Shaky"—anathema for a test pilot—on his good friend Jim Lovell. For every Conrad scheme that the world knew about, there were probably three others he never pulled off. Returning to earth after Gemini 5, he planned to do a somersault across the carrier deck, just to show the world that eight days in space hadn't harmed him—but at the last moment, spotting the red carpet, the admirals, and the bands, he decided against it. Conrad was the kind of astronaut the press loved to write about.

Now Conrad had his own Apollo crew, and with characteristic enthusiasm he was ready to lead them into training as backups for an early Apollo flight in earth orbit. Eventually, he knew, his team would rotate into a prime crew assignment, and although nobody could predict what his mission might be, he had his sights on commanding a lunar landing flight—and if possible, the first one. He felt sure the other men in this room, especially the mission commanders—Wally Schirra, Jim McDivitt, Frank Borman, Tom Stafford, and Neil Armstrong—shared his ambition. But no one could predict who might make that first landing, not even Slayton, and that was part of his message on this April day. The Block II command module wouldn't be ready for its first manned flight until next year at the earliest. After that, NASA would have to clear a string of daunting technical hurdles before anyone could attempt a moon landing. Astronauts would have to check out the lunar lander and practice space rendezvous, perhaps several times, in a succession of earth-orbit missions. Then there would be test flights in lunar orbit. And then, only if everything before had gone well, the next crew would try for the landing.

But that plan could change at any time. If something went wrong on one mission, the next flight would have to make up those missed objectives. And fate could intervene without notice, just as it had with Grissom, White, and Chaffee, and a backup crew would have to step in. Slayton had put together these six crews so that no matter what happened there would be a team of

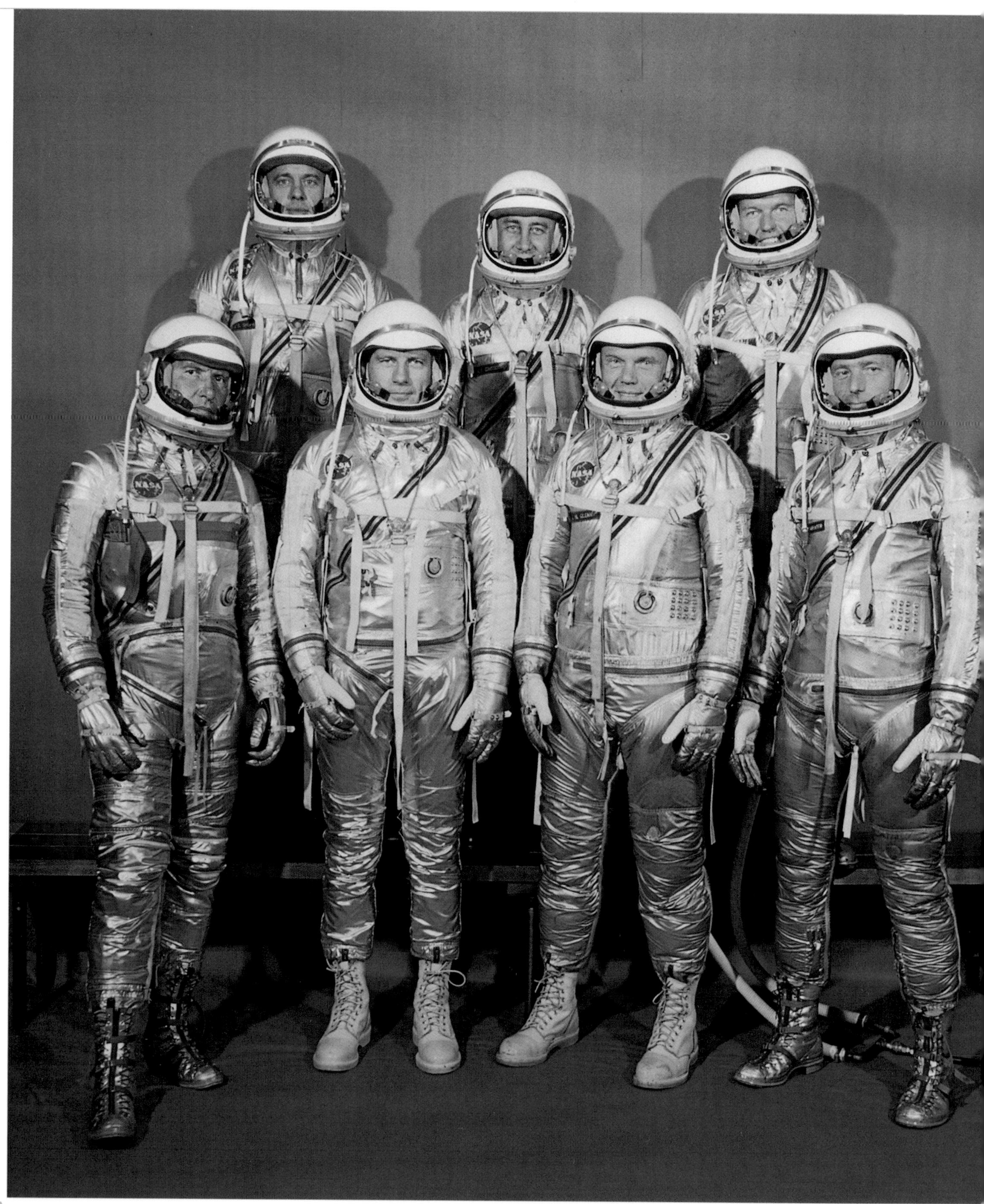

Heroes before they flew in space, the Original 7 became the point men in the nation's Cold War bid to conquer the heavens. From left: Wally Schirra, Alan Shepard, Deke Slayton, Gus Grissom, John Glenn, Gordo Cooper, and Scott Carpenter.

competent astronauts available for any flight. As the missions progressed he would just keep rotating crews, until one of them accomplished Apollo's ultimate goal. His message this day was simple: Any crew could fly any mission. In his mind they were all equal. The real selection, he said, was made the day each of them joined the Astronaut Office. It wasn't the first time the astronauts had heard Slayton make such a statement; it would not be the last.

For some of the men in this room, like Pete Conrad, the journey to the moon had offered few obstacles so far. For the rookies, like Bill Anders, it had been laden with delay and frustration. But none of these astronauts understood precisely how they had been selected to fly these missions. And if you had asked any of them, especially the rookies, whether Slayton's statement was really true—he might as well have said that the moon doesn't go around the earth.

One morning early in 1963, six shiny Corvettes tooled through the "space suburbs" of Timber Cove and Friendswood, and converged on Interstate 45, heading for Houston. Moving onto the freeway, the six drivers encountered each other and took off. They roared past startled motorists at 100 miles an hour, passing them two at a time, one on the left, one on the right. They jockeyed for position, hunting for clear lanes, weaving through the rush-hour traffic. As quickly as they had come, the six exited the freeway and sprinted through city streets to the Farnsworth and Chambers building, where the astronauts had temporary offices. The cars pulled into the parking lot, still crackling with the heat of the race. Doors opened and out came Gus Grissom, Wally Schirra, Gordo Cooper, Deke Slayton, Jim Lovell, and Pete Conrad. The men laughed as they locked up and headed for their offices. Deke Slayton, clenching the stub of a cigar in his teeth, muttered, "Goddammit, we gotta knock this shit off. We'll get ourselves arrested."

In 1963, Pete Conrad's first full year as an astronaut, an early morning drag race with the Original 7 was the perfect expression of the best that life had to offer. Never mind the fact that he was still a rookie waiting for his chance to fly in space; even now, he could see himself headed for the moon. Not that the moon itself held any particular fascination for Conrad. He was here because John Kennedy had directed that America was going to the moon. If Kennedy had ordered NASA to build a space station instead, he would have been just as happy to do that. But the moon was NASA's goal, and when he joined the astronaut corps the previous October, it became his. He was a member of the second astronaut group—or the New 9, as they were

IN 1963

Life publishes its first cover photo of the Vietnam War.

Lawrence of Arabia wins best picture Oscar.

Pope Paul VI is elected.

The U.S. Postal Service introduces ZIP codes.

President Kennedy is assassinated in Dallas, Texas. Police arrest the suspected perpetrator, Lee Harvey Oswald, the same day.

dubbed by *Life* magazine—and they were key players in the Apollo effort. But to the world, the only real astronauts were the Original 7.

One of Conrad's first assignments was to accompany John Glenn on visits to Apollo contractors around the country. Glenn was likable enough, and Conrad could see that as the first American to orbit the earth he had risen to a hero status no pilot had attained since Charles Lindbergh. At every airport an excited crowd would quickly surround Glenn, asking for autographs, and Glenn would smile and put down his bag and oblige them. In order to keep moving, Conrad ended up carrying Glenn's bags as well as his own. A few weeks of this were enough to shatter any illusions Conrad may have had about his status as one of the nation's astronauts. ☾

In fact, Pete Conrad was the last person anybody would think to ask for an autograph. A wiry five feet six, his blond hair nearly gone, his features were best described as gnome-like, with mischievous blue eyes, a prominent and pointed nose, and a grin that exposed a wide gap between his two front teeth. It wasn't unusual for tourists to come through Timber Cove and the other space suburbs, and when that happened Conrad would stand out on his lawn wearing a baseball cap, scratching his balls. Getting recognized was the last thing Conrad worried about.

What mattered to every one of the Nine was flying in space as soon and as often as possible. Conrad was well aware that as a group, the Nine's credentials were even more impressive than those of the Original 7. Not only were the Nine test pilots, they had advanced degrees in aeronautics or the new field of astronautics. Some, like Frank Borman and Tom Stafford, had been instructors at the air force's test pilot school at Edwards Air Force Base. There were those, like John Young, who had set a record or two. And there was even a rocket pilot among them: Neil Armstrong, a NASA man out of Edwards, who had flown the X-15 rocket plane to the edge of space. In short, the Nine were ready and able to compete on the same field as the Seven.

But that wasn't going to be easy. Conrad had the distinct feeling that the Original 7 were not at all happy to see newcomers invading their ranks. Mercury wasn't even over yet, and already the title of astronaut was something they had to share. It came out in subtle ways—the goodies, for example. The Nine came in already knowing about the lucrative contract for the astronauts' personal stories with *Life* and Field Enterprises, which meant an extra $16,000 per year for each of the now sixteen astronauts. After years of raising a family on a military pilot's pay, Conrad was glad to hear about the money. ☾

But there were a host of other perks the Original 7 had scored, great deals on homes, Corvettes on loan from dealers, you name it—that the Nine knew nothing about. Conrad's group had to pry information out of the Original 7. When one of the newcomers stumbled across a deal such as motel rooms for a dollar, or free tickets to Houston Oilers games, one of the Seven would say, "Oh, yeah, we forgot to tell you about that."

Of course, the cars and the other goodies didn't matter to the Nine, not compared to the real issue. The only one of the Seven Conrad knew well was his Pax River classmate, Wally Schirra. Now that he was an astronaut who had flown in space, Schirra was still the same breezy, fun-loving fighter jock he'd always been, but there was a subtle difference about the way he related to the Nine. He was like a pro ball player visiting with the farm team. But Schirra never put it into words. It was Gus Grissom who came right out and said it: You're not an astronaut until you fly.

But that hardly mattered to *Life* magazine. In September 1963, *Life* ran a cover story on the Nine: "The New Astronauts: They go head over heels into training." Inside, the Nine were introduced to the American public with a

Conrad had the distinct feeling that the Original 7 were not at all happy to see newcomers invading their ranks.

fanfare that, if it didn't measure up to the Original 7's, was still good coverage. There were pictures of the Nine in training, climbing into a Gemini mockup, trying out a Gemini life raft, eating space food while floating around in a training aircraft. Then came their personal stories, with each of them talking about the role he was playing in the assault on the moon. And pictures from the home front: there was Ed White going for a bike ride with his family, Jim McDivitt out for a weekend spin with the wife and kids, Conrad and his clan seated in front of the fireplace, and so on. Even Deke Slayton had an article, with a shot of the Nine sitting around a conference table in his office, dressed in business suits, looking like the young executives of some rising corporation. Slayton told the *Life* people, "There will be plenty of flights for everybody." And he meant it.

Around this time, there was a definite change in the Original 7's attitude: they began to warm up to the Nine. With Mercury over, Gemini was on the near horizon, and it was clear that even sixteen astronauts would

Three members of the Nine, Elliot See *(left)*, Pete Conrad, and Neil Armstrong, try out an early version of an Apollo life raft in 1963. The inflated arches would support a shelter against foul weather.

have their hands full planning and flying those missions. Meanwhile, Apollo was gaining momentum. NASA was already in the process of selecting more astronauts.

Of course, the Seven could afford to be friendly, because none of the Nine were going to take a mission away from any of them. That was due to an arrangement called the pecking order, a holdover from the military. The Astronaut Office was not a military outfit, but the Nine came to understand that the pecking order was a reality just the same. Here it was based not on military rank but how long you had been an astronaut; the Original 7 were permanently first in line.

But for the most part, the Nine didn't have much time to think about when they would fly in space. After ground school with the Original 7 that included jungle and desert survival training and courses on all facets of spaceflight from orbital mechanics to computers, they were swept up in the Apollo whirlwind. Each man was assigned a slice of the massive engineering effort. Conrad drew the enviable task of helping to design the controls and instrument displays for the lunar module, the spacecraft that would actually land on the moon. In the months that followed Conrad found himself standing in a plywood mockup of the lander, surrounded by painted switches and dials, imagining himself flying over a silent, cratered world. He wriggled down rope ladders, descending to a simulated moonscape. He hung suspended from a Peter Pan rig, to familiarize himself with the trampoline-bounce of walking in the moon's one-sixth gravity. Always, in the back of his mind, Conrad was certain that some day he would be doing these things for real.

Conrad knew that when some people

Face to face with dinner, Apollo astronauts watch an instructor display the evening's main course during jungle survival training in Panama. Bill Anders turned down his portion of iguana, explaining, "It just depends on your appetite. I've already eaten once this week."

thought of astronauts they imagined the thrill of exploration, but that wasn't what he felt. It was his *job* to learn how to walk in lunar gravity, to figure out what kind of instruments a lunar lander ought to have. He wasn't here because he had an insatiable desire to set foot on another world. For him it was not the destination but the journey, one that offered the most challenging test flying anyone could imagine. He was hungry for the chance to take control of a moonship and fly as no one had ever flown before, to take it all the way down to the surface of the moon, and when he was there he would do

some good, useful work, and then he would fly it home again. The lure for Pete Conrad was not to walk on the moon, but to stand on the summit of his profession. If people didn't quite understand that, if they wondered why astronauts didn't sound like explorers when they talked, he had a simple answer. He wasn't an explorer; he was a test pilot.

As the months passed, the Nine came into their own, and at the same time the Original 7 waned. Alan Shepard, the first American in space, considered by many to be the most skilled pilot of the Seven, had been taken out of the running by an inner ear disorder and, like Deke Slayton, was grounded indefinitely. John Glenn was spending so much time making public appearances and so little time in the Astronaut Office that Wally Schirra had criticized him in a television interview for shirking his duties to the program. But Glenn was flying off on a new trajectory; there were noises about him considering a bid for the Senate in the 1964 election. Scott Carpenter's spaceflight career had also come to a premature end; unlike Glenn, he had no choice. Many at NASA felt he had botched his Mercury mission by wasting precious maneuvering fuel, misaligning his spacecraft for reentry, and firing his retrorockets late, which caused him to splash down some 250 miles off target. Though no one ever said so publicly, it seemed clear Carpenter would never be assigned to another space mission. By the end of the year, only three of the Original 7 were available to fly, Gus Grissom, Wally Schirra, and Gordo Cooper. What that meant for the Nine was that more seats would be available on the two-man Gemini flights and the three-man Apollo missions.

And so Pete Conrad found himself engaged in an unspoken but real competition for a seat on a Gemini mission. There were no stated rules, and not even any tangible measures of progress. During ground school, for example, there had been no grades; the Original 7 would never have put up with that. Besides, how would Slayton weigh a less-than-perfect showing in the classroom against golden hands in an airplane? There were some hot pilots in the second group, particularly Tom Stafford and Jim McDivitt, who had a rare combination of qualities: good stick work and good head work. They could fly with a skill and grace that made even the most complex maneuvers look like a work of art; at the same time, they had instincts that in an emergency made the difference between buying the farm, as the saying went, or living to tell about it. The others, Conrad thought, were a mixed bag; some matched him as pilots, others didn't.

During desert survival training in Nevada, the Original 7 fashioned burnooses from strips of parachute cloth to keep from broiling during the day, when the temperature could soar to 120°F.

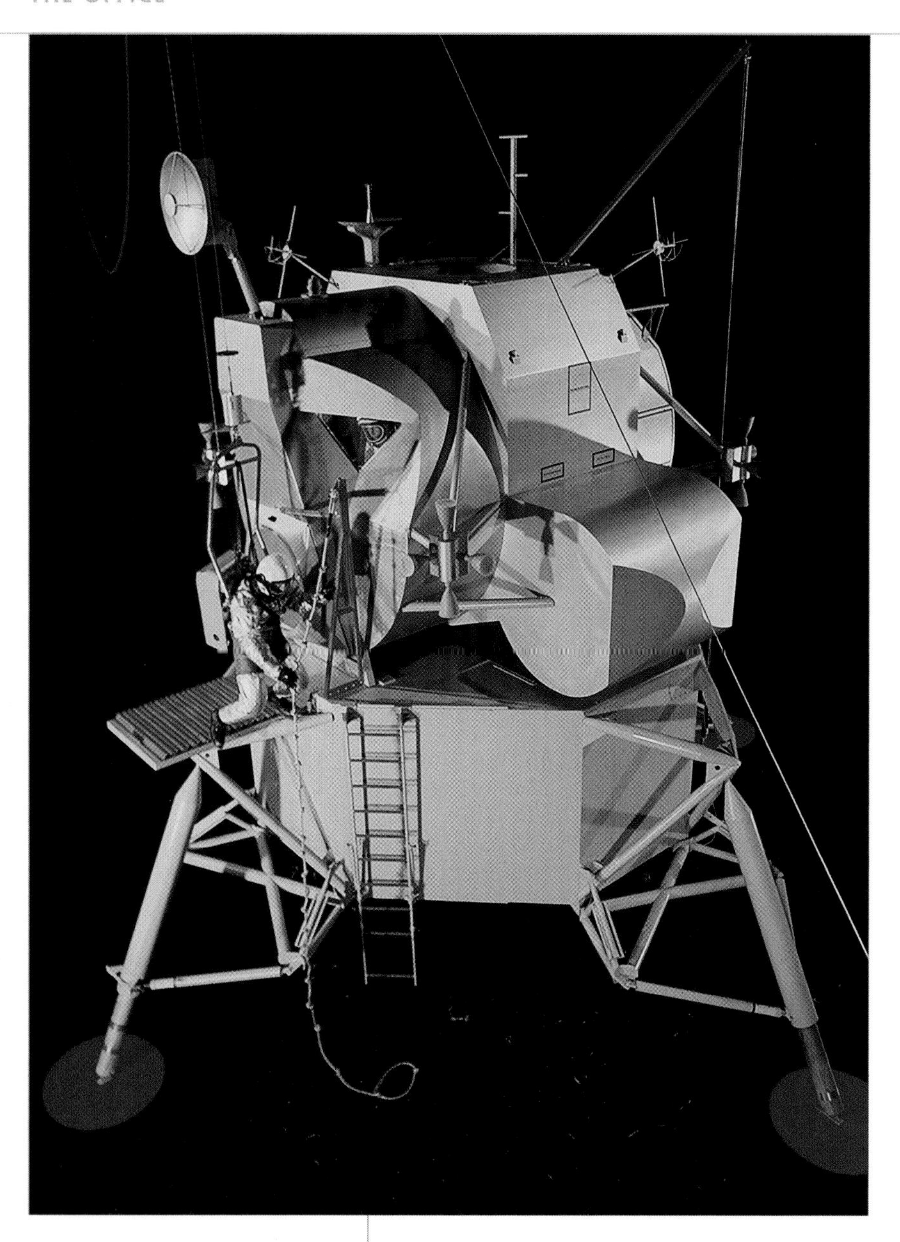

Hanging from a "Peter Pan" rig to simulate lunar gravity, Pete Conrad tests a rope ladder attached to an early mockup of the lunar module. In the end, engineers abandoned the rope ladder in favor of a rigid one attached to the lander's front leg.

Soon it was clear that the competition embraced everything they did. In the summer of 1963, Conrad and the other astronauts made trips to Johnsville, Pennsylvania, to the navy's Acceleration Laboratory, to ride the huge centrifuge, which was known simply as The Wheel. Inside the passenger gondola of The Wheel was a mockup of the Gemini cockpit, in which Conrad was put through a series of simulated reentries from earth orbit. During the exercise the doctors could monitor him on closed-circuit television. One run simulated a fast, steep reentry. Conrad felt the force of tremendous deceleration, just as he would if his spacecraft were slamming into the dense layers of atmosphere. When he hit 8 g's, Conrad could not move his arms, and still the g-forces mounted. At 12 g's, his eyeballs flattened out of focus, causing his vision to blur. Still the load kept building. When it peaked at 15 g's, Conrad felt as if his chest were being torn open. Thankfully, it didn't last long—but he came out with a punishing hangover and ruptured capillaries all over his back, as a souvenir.

Next, the centrifuge was programmed for a long, shallow reentry. This time the acceleration built up to only 5 g's—but it stayed there for five solid minutes. At 5 g's, with his chest nearly pressed against his backbone, Conrad had to force his rib cage open just to take in breath. Five minutes became an eternity. Halfway through the run, Conrad was about to stop breathing. He was sure he was going to black out. He was trying to figure out how he might manage to do that without closing his eyes so that as the centrifuge kept whirling and the doctors watched him, no one would know he'd lost consciousness. To his relief, Conrad managed to get through the run fully conscious, and thank God, because if he hadn't, Slayton would have heard about it, and it wouldn't have been good for his career.

When Conrad was back in Houston, he and the other astronauts breezily queried each other:

"Hey, how'd you do on the five-g run?"

"Oh, super! Piece of cake."

"You didn't black out, did you?"

"Me? Naaah, not at all."

And Conrad couldn't help but smile, certain that they'd probably gone through the same thing he had and that they were all lying through their teeth. And it was always that way, after a practice session in the docking simulator, or a parachute drill, or anything they did. When they were in Houston, astronauts rarely socialized together, but on those occasions when Conrad and Jim Lovell or Tom Stafford or any of the Nine gathered on a day off, and the air filled with the smell of steaks on the patio barbecue, the one thing they didn't talk about was who might get picked to fly, and when. None of the Nine was about to show a weakness like worrying about his career, not in front of another astronaut.

In 1963, while Pete Conrad and the rest of the Nine were drag racing with the Original 7 and searching for their paths into space, Bill Anders was a twenty-nine-year-old fighter pilot with one goal, and that was to become a test pilot. He hadn't grown up with dreams of flying; his childhood ambition had been to become a career naval officer like his father, who had commanded the gunboat *Panay* when it was under attack by the Japanese in 1937. He was raised to be practical and competitive; he didn't smoke or drink, and he had always sought to push the limits of whatever he was doing. There had never been any question about where he would go to college; he applied to the Naval Academy at Annapolis and was accepted. It was there that Anders gravitated toward flying. But during his first cruise on an aircraft carrier as a midshipman, it seemed as if there were an accident every other day. He saw a plane hurtle toward the carrier deck on final approach, miss the arresting wire, strike a line of parked jets, careen off the deck and plummet into the sea. Two or three other pilots died on the same cruise, all of them during takeoff or landing. Anders was willing to accept the risk of flying jets, but he wanted to take that risk in a dogfight at 30,000 feet, or evading enemy guns—not trying to land. Later, when a young air force colonel came to Annapolis

The navy's centrifuge at Johnsville, Pennsylvania, gave astronauts a taste of the g-forces they would experience during launch and reentry. The ride could be punishing, as shown by the strain apparent on Wally Schirra's face during a high-g run *(inset)*.

to tell the cadets that his branch of the service offered a better path for advancement, Anders made up his mind. After graduation, he was off to the air force, to earn his wings.

By 1959, fresh from a tour of duty with an interceptor squadron in Iceland, Anders knew he had become a hot pilot. But he also knew he wasn't like the other squadron pilots, who lived only to fly. Flying didn't give Anders the intellectual stimulation he craved; he wanted something more. Test pilots have to be engineers as much as pilots, and that appealed to Anders. Most of all, for any hot fighter pilot, test pilot school was the next step up the ladder. Anders had already amassed the 1,500 hours of flying time in high-performance jets required by the air force's Experimental Flight Test Pilot School at Edwards Air Force Base, in the high desert of California. But when he went to Edwards to make his case, things didn't go as he'd hoped. One of the admissions people told him, "If only you had an advanced degree, you'd be a prime candidate."

"Advanced degree?"

"Yeah, we're pushing academics."

Along with his mild disappointment, Anders felt a sense of irony; he wondered what Chuck Yeager—the most famous test pilot in the world—would have thought of this new requirement. In 1947, Yeager had become the first human being to shatter the sound barrier, flying a rocket plane over the desert at Edwards. The world of test flight had changed since then, but not Yeager; he was still the prototypical rocket ace, an alloy of split-second reflexes, go-for-broke daring, and unflappable calm. There was hardly a fighter pilot in the country who didn't long to be molded in his image. But Yeager had not had so much as a college diploma.

But this wasn't Yeager's school, and Anders knew he had no choice but to accept the new direction. In fact, graduate school had been on his list of things he hoped he might accomplish. He enrolled at the Air Force Institute of Technology at Wright-Patterson Air Force Base in Dayton, where he planned to steep himself in the intricacies of astronautical engineering, then return to Edwards. It wasn't until Anders packed up his three children and his wife, Valerie (who was expecting their fourth), and moved to Dayton that he was informed that the astronautical engineering program was full. Because of his excellent math grades at the Naval Academy, the school assigned him to nuclear engineering. Maybe that wasn't so bad; the air force was planning to develop a nuclear-powered airplane, and this kind of experience would make him a natural for the project. But Anders wasn't taking any chances. Even as he labored through the two-year degree program at Wright-Pat, he went to night school at Ohio State University, taking courses in aeronautics. It was a wise decision: the nuclear airplane program was soon canceled.

And something else happened while Anders was in school. One day in May 1961, John Kennedy announced that the United States should put a man on the moon. Valerie couldn't imagine how anyone could accomplish such a feat, and she was even more stunned when her husband said to her, "That's what I'd like to do." But it wasn't the flying that was uppermost in Bill Anders's mind, or the benefit to his career; it was the chance to be an explorer. As a boy he'd devoured tales of exploration: Magellan's circumnavigation of the globe, Lewis and Clark's ventures into the American wilderness. The *terra incognita* of the twentieth century was a quarter of a million miles away, and Kennedy was saying men would go there. If there was flying involved, so much the better, but Anders would have signed up to make the trip by barge. What he wanted most of all was the chance to walk on another world.

But the astronaut program, he knew, was beyond his reach as long as he

The first man to break the sound barrier, Chuck Yeager in 1963 ran the air force's Aerospace Research Pilots School—the portal, at that time, for Air Force pilots into the space program.

lacked test pilot credentials, and after graduation in 1962, he was ready to try again at Edwards. By now, the air force had developed an interest in space, and the test pilot school had been replaced by the Aerospace Research Pilots School, otherwise known as ARPS, to help air force pilots gain the qualifications to be astronauts.

When Anders contacted ARPS he learned the school wasn't taking any new students for a while. Discouraged, he accepted an assignment as an engineer and instructor pilot at Kirtland Air Force Base in Albuquerque, where the air force was conducting a program of nuclear research. And he bided his time, waiting for the hiatus at ARPS to end. That happened the following year, 1963. Anders readied another application, sure that his graduate work would impress the selection committee.

But ARPS was now under the command of none other than Chuck Yeager. He told Anders, "We've changed the requirements. We're looking for flying time." Anders realized that if he'd stayed in his squadron for the last three years flying an F-101, he'd be in—some of his squadron mates were being accepted. Undaunted, Anders applied anyway. As often as he could, he took a break from his duties at Kirtland to fly to Edwards, demonstrating his skills by duplicating the flying maneuvers required for the entrance exam. Then it was back to Albuquerque, to wait for some word.

He was still waiting on a hot Wednesday early in June 1963. That day he was driving home from work in his Volkswagen microbus and listening to the news when there was an announcement that NASA was looking for new astronauts. As the announcer read the requirements, Anders checked them off in his mind: Age, thirty-five years or less. Two thousand hours flying time in high-performance jets. Advanced degree in engineering or physical science. And then—most important of all: Test pilot experience preferred, but not mandatory. Anders met every requirement on the list. He had never given serious thought to the astronaut program before now, because he knew he wouldn't get in without test-pilot credentials. But if NASA would take him, he didn't need Yeager. By the time he reached the next traffic light, he had decided to apply.

No one was more surprised than Anders when he found himself still in the running after each cut. When it came time for the interview with Deke Slayton's selection committee in Houston, Anders tried hard to anticipate what they would want. He stressed his fighter-pilot experience, of course, but he also played up his work at Kirtland on how to shield spacecraft from space radiation. NASA was concerned about the hazards from intense solar flares to moonbound astronauts, who would be the first to venture outside the earth's protective Van Allen belts. Anders didn't think there was too much of a risk—you'd get dosed, and maybe some of your hair would fall out, but it wouldn't kill you—but he wasn't about to tell the committee that. Radiation was a real problem, he told them, and he could help them solve it. Anders returned to Albuquerque satisfied he had done his best but entirely unsure of his chances.

On October 14, three days before his thirtieth birthday, Anders got a call from Deke Slayton inviting him to fly for NASA. Not long afterward, the space school turned him down, but Anders couldn't have cared less. By then he was already getting his family ready for the move to Houston. One day, on a trip to the Pentagon to check on some paperwork, Anders ran into Yeager.

"Too bad you didn't make it," Yeager said.

"That's okay," Anders said wryly; "I got a better offer."

"A better offer?"

"Yeah, I'm going to NASA. I made it into the program."

Anders still remembered how Yeager fumed—"NASA took you, and you haven't been through the school? I'll see about that . . ." But, Yeager could bitch all he wanted, and it wouldn't make a difference. Anders had been accepted into the most elite flying fraternity in the world. Of course, "fraternity" was best used loosely in this case, as Anders would find out.

●◐◐○○○◑●

By the time fourteen new astronauts came to Houston at the end of 1963, the Astronaut Office had moved to Ellington Air Force Base, just down the road from the construction site of the new Manned Spacecraft Center. The astronauts had their offices in some rehabbed World War II barracks near the flight line, which made it easy to hop into a T-33 jet trainer and head out to a design review halfway across the country. To Pete Conrad, flying was a great escape from the barrage of paperwork—of which he'd never seen so much in his life. Fortunately, the new astronauts had arrived just in time to take up the slack, fourteen new rookies whose place on the totem pole was even lower than the Nine's. Conrad's group saw no threat from the newcomers; the first of the Fourteen probably wouldn't lift off until after the last of the Nine had splashed down. And Conrad had some good friends in the new group, including Dick Gordon, who had been his roommate on the carrier *Ranger*, and Al Bean, who had been one of his brightest students at Pax River. Conrad wished them well. And if the Fourteen came in with a nose for the goodies, the Nine weren't keeping any secrets. Neither were the Original 7. Wally Schirra even sat down with the newcomers and spelled out, clearly and in detail, how the goodies worked, the cars and the *Life* contract and the rest of it.

Before long the Fourteen were ready to take on engineering assignments. Conrad handed off his job on Apollo controls and displays to his buddy, Dick Gordon. Then he and the rest of the Nine left the new rookies to their own devices and headed off to work on Gemini. Never mind the moon, for the time being; Gemini was upon them, and every one of the Nine was itching to fly it. The first manned flight was slated for early 1965. Nine more missions were planned after that, brimming with techniques vital to the moon program like space rendezvous, dockings, and space walks. Gemini would be a test pilot's dream.

Grissom, Schirra, and Cooper were first in line for Gemini, and the Nine could only guess at how they might be jockeying for position. The two real piloting plums would be the first manned flight, Gemini 3, and the first rendezvous mission, Gemini 6. By early 1964 it was getting to be time for Slayton to fill the seats for the coveted first flight. It went without saying that Grissom, who was at the top of the rotation and had a key role in Gemini's development, would command the mission. But no member of the Original 7 would fly right seat; that was a job reserved for the rookies.

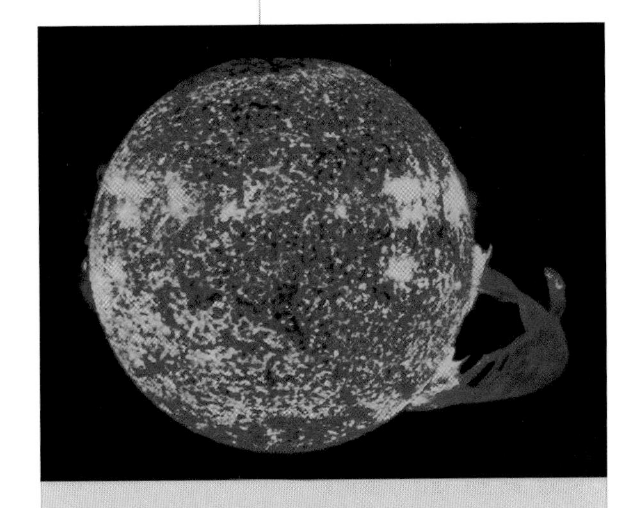

HERE MAY BE DRAGONS

Giant solar flares were among the phenomena seen as potential threats to the Apollo missions; moon-bound astronauts, engineers feared, would be vulnerable to bursts of solar and cosmic radiation. Even a small meteorite, traveling at 10 miles per second, could strike a spacecraft with devastating impact. One scientist predicted a thick blanket of moon dust that could swallow the descending astronauts as they tried to land. Another warned that lunar grime on the astronauts' space suits might burst into flame as soon as it was exposed to oxygen inside the lander's cabin.

One day in April, Slayton called all the astronauts together in the briefing room at Ellington and announced that Gus Grissom was going to command Gemini 3 and that his copilot would be John Young. Backups to Grissom and Young, Slayton announced, were Wally Schirra and Tom Stafford. ☾

Conrad didn't dwell on his disappointment. But in July, Slayton named two more crews, and they were all from the Nine. Jim McDivitt and Ed White would fly Gemini 4; their backups were Frank Borman and Conrad's buddy, Jim Lovell. Conrad was beginning to lose his patience. McDivitt, White, and Borman were air force, like Slayton. There was a lot of good-natured ribbing between the air force and navy, but it was meaningless—or was it? Was Slayton taking care of the air force?

By the end of 1964, signs of order emerged. Wally Schirra and Tom Stafford, who were still training as Grissom and Young's backups on Gemini 3, were unofficially named as the prime crew for Gemini 6, the first rendezvous mission. That was why none of the Original 7 had showed up on Gemini 4; compared to the first rendezvous, it was a much less desirable flight; there wasn't a single new objective, except staying in orbit for four days, and not much real piloting. Furthermore, the Nine realized, Slayton was setting up a rotation: You serve on a backup crew, skip two flights, and then fly as the prime crew. Presumably, as long as you didn't screw up, you could

Gus Grissom and John Young, depicted in a Norman Rockwell painting, suit up for the coveted first flight of the new spacecraft.

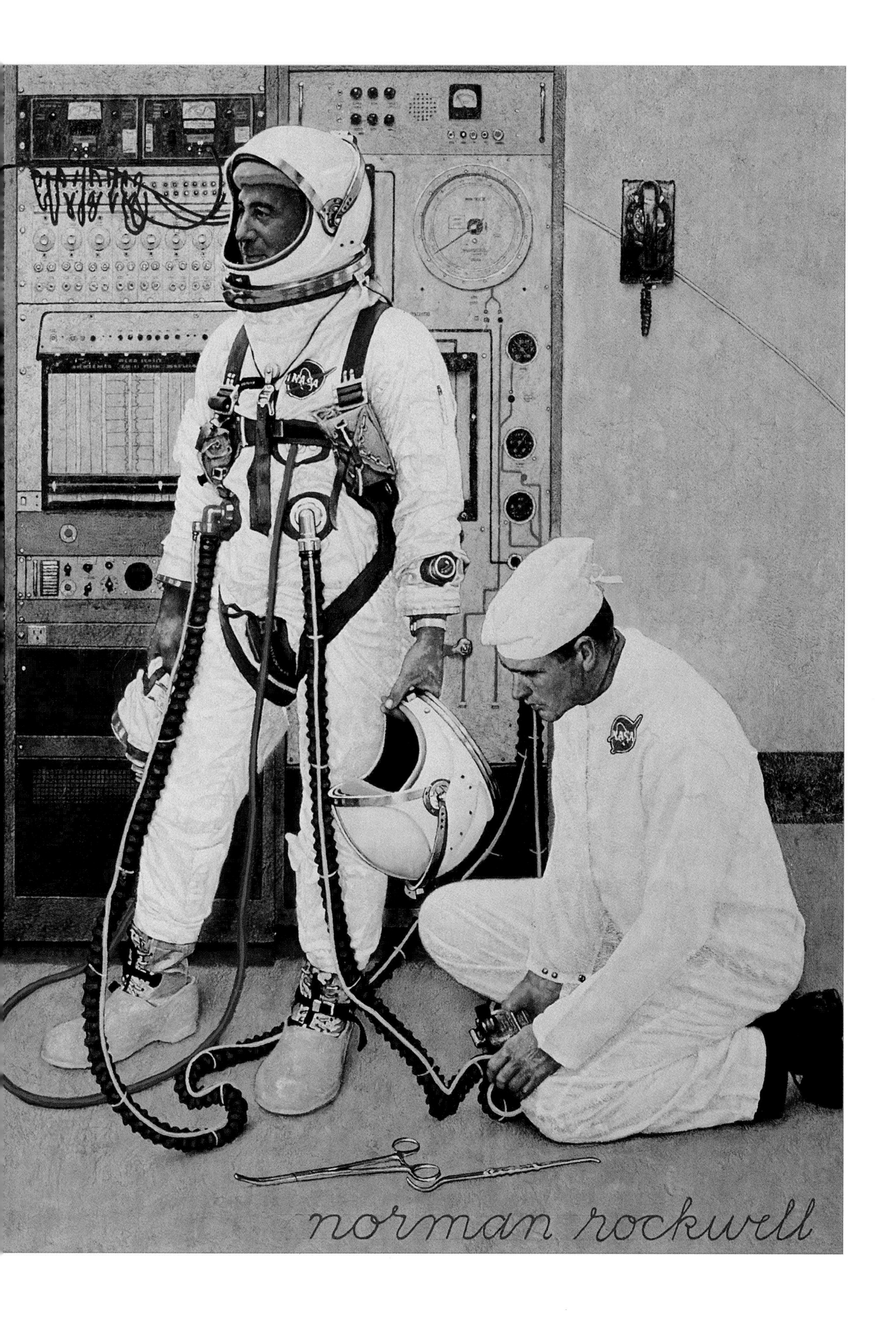
NASA
NASA
norman rockwell

stay on the rotation indefinitely. At that point, you were "in the pipeline."

But getting into the pipeline to begin with—that was the difficult part. Conrad reassured himself that the longer he waited to fly, the better the missions got, since the last several Gemini flights would all be rendezvous flights. And Slayton had said that rendezvous experience was essential to getting a seat on a lunar mission.

In February 1965, Conrad's wait came to an end, without warning or explanation, when Slayton told him he would be assigned as Gordon Cooper's copilot on Gemini 5. It was anything but a plum assignment: a week in orbit, in a cabin no bigger than the front seat of a Volkswagen. But it was fine with Conrad. And this was just the beginning. If he didn't screw up, he'd stay on Slayton's crew rotation and keep flying. He'd have a shot at one of the rendezvous and docking missions, maybe even as mission commander. And then it would be on to the moon. As far as Conrad could see, he had it made.

IN 1964

The Federal Trade Commission requires health warnings on cigarette packages.

President Lyndon Johnson signs into law the Civil Rights Act of 1964.

Martin Luther King Jr. wins the Nobel Peace Prize.

Mod fashion takes off—hot pants, miniskirts, and vinyl boots.

Looking back on it all years later, Bill Anders would admit that he'd been unprepared for what he found when he joined the astronaut corps. He'd expected to be in pretty fast company—you don't find a collection of overachievers like the astronaut corps every day—and Anders knew that he was lucky to be here. But he never had any doubts about his own abilities, and he'd come in fully expecting that with hard work and perseverance, he would find the way into space open to him. He would never have imagined that he would still be waiting for his chance to fly more than three years later, in April 1967, as Deke Slayton and the other astronauts were sitting down to help plan the lunar missions. Looking back, he would wish he had been a little smarter about the crew selection game.

When the Fourteen reported for work at the end of 1963, the Manned Spacecraft Center was surging with activity. There were seven years left to meet John Kennedy's lunar landing deadline, and the men and women of NASA were giving their all to Gemini and Apollo without a thought for the long hours. These were the most dedicated bunch of people Anders had ever seen. And Anders fit right in; before long he was working as hard as he had in grad school. He soon came to understand, however, that the hardest part of the job wasn't the long hours, or the demands of keeping in shape, or the flying—it was simply finding a way to stand out and get selected for a space crew. From the beginning, the whole process was shrouded in total mystery. The only thing anyone knew was that Deke Slayton made the choices, and that Al Shepard, who had been appointed chief of the Astronaut Office after being grounded, was in on it too. Unlike Slayton, of whom the astronauts saw relatively little, Shepard was their day-to-day boss. Around the office they wryly referred to him as "Big Al." ☾

Shepard wasn't like the rest of the Original 7. He had charisma to fill a room, but he lacked the good-natured bluster of Wally Schirra, or the easygoing friendliness of Gordo Cooper. He was no Father Slayton. You didn't have to be around Shepard too long to find out how well he thought of himself; he was dripping with arrogant self-confidence. Everyone knew his ready wit, his toothy grin, his appetite for a party, his penchant for obscene jokes; they also knew his icy, penetrating, blue-eyed stare. Either incarnation could surface without notice.

You didn't have to be around Shepard too long to find out how well he thought of himself; he was dripping with arrogant self-confidence.

From what the other astronauts could tell, Shepard's influence extended well beyond the walls of the Astronaut Office. Since becoming grounded, he'd built up something of a small empire in business. He was part owner of the Baytown Bank near downtown Houston, and he had made money in hotels, shopping malls, and a variety of other ventures. He was the only astronaut who lived apart from the space suburbs, in a downtown apartment.

In the Astronaut Office, Shepard usually kept a chilly distance from his troops; a hello might be returned by a grunt, or nothing at all. And he had a way of intimidating the other pilots, especially some of the younger ones, by staring right through them and asking questions on obscure technical details, to throw them off balance. It was all very subtle; Shepard was too smart to be obvious; but many astronauts felt an inner alarm go off when he entered a room. To be sure, there were astronauts who had no fear of Shepard and some who even seemed to get along with him—though they would never use the word "friends" to describe their association. But even they knew that crossing Big Al was bad head work. Years later, one astronaut would say, "You had the feeling that if it came down to you or him, frankly, he would cut your balls off so fast you wouldn't know they were gone for a little while." More than one of the Fourteen tried to avoid him. ☾

Al Shepard, feared by some astronauts, respected by others, confers here with Gemini 7 commander Frank Borman. At times, Shepard could display a charming wit and engaging warmth; more often, he kept an icy distance from his subordinates.

But there was no way around the fact that along with Slayton, Shepard had ultimate power over an astronaut's career. But what did they look for? Late in 1964, near the end of basic training, Shepard gave the Fourteen a homework assignment called a peer rating. Each man was to evaluate his colleagues' abilities and rank them in the order they ought to fly in space, leaving out himself. The peer rating gave Anders a chance to think about the competition. For example, Mike Collins and Charlie Bassett, both of whom came to NASA from coveted test-flying jobs in Fighter Operations at Edwards. And Dave Scott, who not only had been through ARPS like Collins and Bassett, but had a couple of advanced degrees in aeronautics and astronautics from MIT. And here was Dick Gordon, a Pax River grad who had been the project test pilot on the F4 Phantom and had used it to win the Bendix Trophy transcontinental air race. Gordon had been on a first-name basis

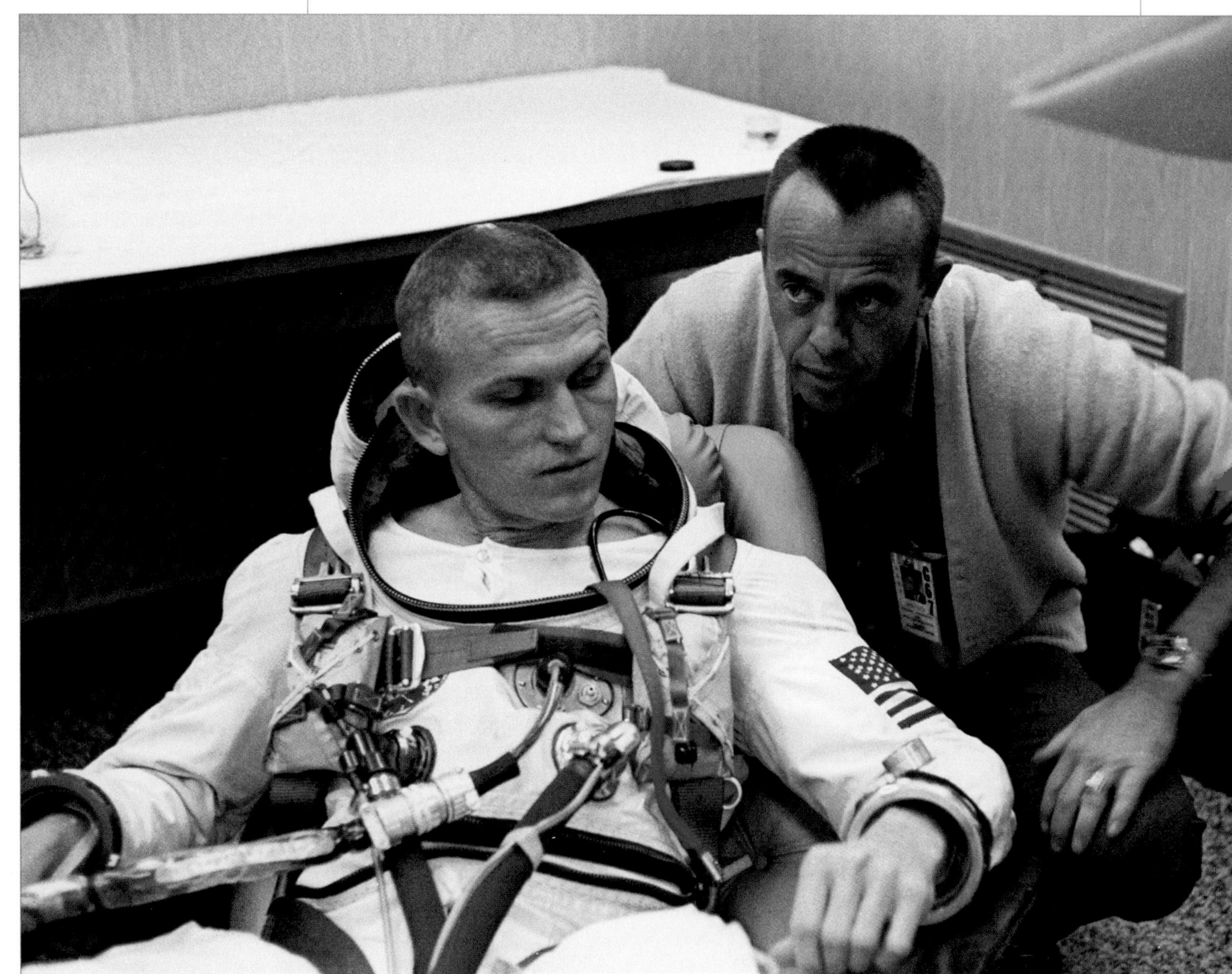

with Al Shepard, Deke Slayton, and Wally Schirra before he even came to NASA. It wasn't hard to figure out that Collins, Bassett, Scott, and Gordon ended up near the top of many of the Fourteen's ratings.

Where Bill Anders fit in, he had no idea. If the measure of an astronaut was his flying skills, then Anders was ready to face off against anybody in the office. Because he'd been a T-33 instructor at Kirtland, Anders was given the job of teaching his colleagues in the Fourteen how to fly it, and he'd already encountered astronauts whose flying abilities left something to be desired. One came close to crashing until Anders grabbed the controls away from him, much to the man's chagrin. Presumably Shepard and Slayton were flying with the Fourteen enough to make their own assessments, but as far as Anders could see, flying wasn't the basis of this competition. And if not, then what was? Anders's motto was simple: Work your tail off, and someone will notice. ☾

When it came time for the third group to divvy up pieces of the Apollo design effort, Anders chose the environmental control system, the complex assemblage of pumps, pipes, sensors, and valves that would maintain the spacecraft's breathing oxygen, proper cabin temperature, and so on. Getting all that plumbing to work was a challenge that appealed to Anders; it fit well with his Naval Academy training. But it didn't take him long to realize that he would've been better off if he'd landed an assignment that had to do with actually flying the spacecraft. A good example was the spacecraft's hand controller, the equivalent of the control stick in a jet fighter. The hand controller was important to the pilots, and the astronaut who got it could, in the course of his efforts, circulate among the Old Heads' offices and solicit opinions. They would try out a model of the controller while wearing a space suit glove, then engage in a thoughtful discussion of the best angle at which it ought to be mounted, the "breakout forces" required to move it in a given direction, and so on. If the advice was valuable, so was the chance to be seen. The hand controller was a plum assignment.

No one asked Anders's advice on the hand controller; he was too far down on the totem pole for any of the Seven or the Nine to care what he thought. And his own assignment—the environmental control system—wasn't the best way of getting noticed. Walking into Wally Schirra's office with a section of coolant pipe just didn't produce the same effect. Most of the time, though, Schirra and the rest of the Old Heads weren't even around; in fact, the only time all the astronauts saw each other was at the Pilots Meetings that took place in the Astronaut Office conference room every Monday morning. These meetings, which

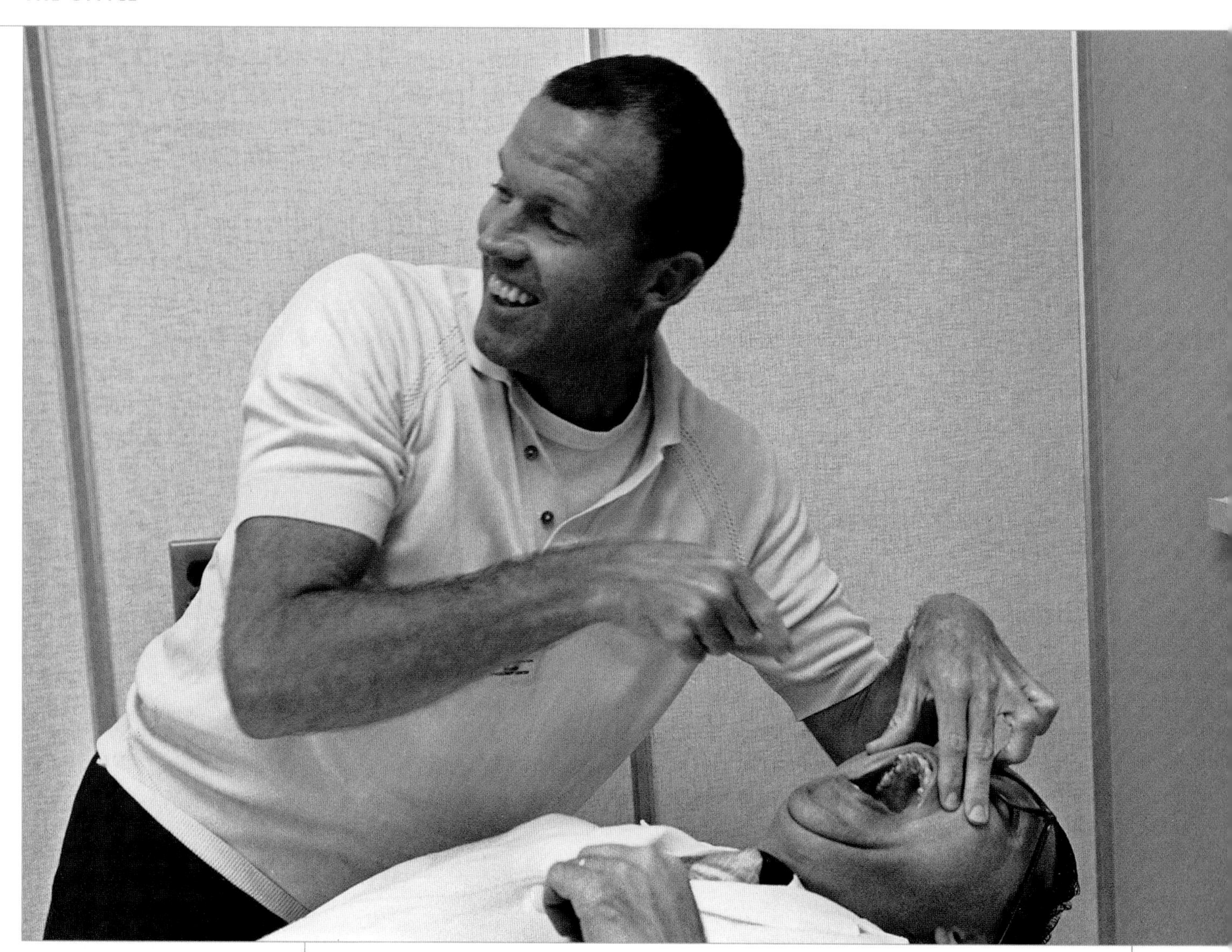

In a moment of levity, Gordo Cooper turns the tables on NASA's chief physician, Chuck Berry, who oversaw the poking and prodding that astronauts endured. Many of the astronauts regarded doctors as the enemy and sometimes cooperated only grudgingly.

were run by Shepard, gave the men a chance to coordinate their activities.

It was at the Pilots Meetings that the Old Heads would band together against a common enemy—the doctors. Astronauts and doctors were natural adversaries; every pilot knew there are only two ways you can walk out of a doctor's office, fine or grounded. The Original 7 had to put up with the doctors' alarmist predictions about spaceflight, none of which came true. They talked about having astronauts wired with heart catheters and rectal thermometers, until the men put their foot down. Now they were irritating the Old Heads by coming up with medical experiments that cluttered up the Gemini flight plans. At the Pilots Meetings, when the doctors cooked up some new plan to have the next Gemini crew wired with sensors from stem to stern, the Old Heads' rallying cry was, "Look what those bastards are trying to do to us now!" Joining the protest was probably good politics, but the truth was that Anders didn't mind the doctors or their experiments. To send people into space and not try to understand the effects was just plain stupid. But Anders kept these thoughts to himself.

Gradually Anders understood that he was different from the other pilots, and that it wasn't helping him. The most obvious difference was on his résumé: He wasn't a test pilot. There were five others like him among the Fourteen, including a wiry marine named Walt Cunningham and a tall, red-headed air force pilot named Rusty Schweickart. Like Anders, they were highly educated, but they didn't have diplomas from ARPS or Pax River. Now that they were here, the seven of them, the unspoken question in the minds of the Old Heads—Anders could sense it—was, What the hell are *they* doing here? ☾

The answer was that the astronaut corps was evolving. NASA no longer deemed test pilot experience essential to flying in space. Personally, Anders felt the test pilot distinction was overrated. Besides, he and Cunningham and Schweickart had been selected because they had scientific expertise that made them valuable players in their own right, or so Anders had thought. In reality, it was a liability. Most of the Old Heads seemed to regard the scientists who developed experiments for flights as nuisances who thought the program revolved around them and who simply didn't understand the demands of flying in space. The fact that Cunningham had done doctorate-level research in upper atmospheric physics and Schweickart had studied astronomy at MIT didn't boost their standing in the Astronaut Office. Schweickart's assignment was to coordinate the scientific experiments, and Anders wondered if someone put him there to get him out of the way.

It was easy to see why Schweickart irritated a few of the Old Heads. Some detected an air of intellectual superiority. Not all of them appreciated his barbed wit. Schweickart didn't defer to them on technical matters; he didn't hesitate to offer his opinions, even when the Old Heads weren't interested. Rusty Schweickart was an unusual commodity in the Astronaut Office; he was a free spirit. He shared an office with Walt Cunningham, and despite some superficial similarities—Cunningham was also very bright, even more outspoken, and just as superior in his attitude—the two could not have been more different. While Cunningham was decidedly right of center, Schweickart had left-wing leanings on most of the day's social issues. It would be years before Schweickart would let his hair grow and sport a beard, prompting Pete Conrad to label him the Astronaut Office's "token hippie." But even now, under that orange crew cut, there was enough about him to break the fighter pilot mold. A few years younger than most of his colleagues, Schweickart seemed curious about everything that was going on in the world. He and his wife, Claire, hosted a literary discussion group. He showed an openness to new things, and listened in fascination to the stories of a friend who had witnessed San Francisco's drug culture. In short, there was much more to him than flying.

Anders, Cunningham, and Schweickart were odd men out. What did not make sense was that they were joined by a thin, quiet navy test pilot named Al Bean who'd been near the top of his class at Pax River and had flown in a couple of attack squadrons before coming to NASA. He'd barely missed getting selected with the second astronaut group. Bean was different from the other test pilots in the office. He didn't project the same macho image. Most astronauts went hunting or worked on cars in their spare time; Bean liked to paint. He seemed more inclined to hang out with the likes of Anders, Cunningham, and Schweickart than with the other test pilots. They became a kind of Four Musketeers, the closest thing to a pal group in the Astronaut Office. Whenever they had a free moment together, they talked about one thing more than any other: Who would be the first of their group to fly?

Al Bean *(left)*, Walt Cunningham, Bill Anders, and Rusty Schweickart visit Arizona's Kitt Peak National Observatory in 1964. Whenever the four found themselves together, they would talk about one thing: how to get selected for a space mission.

The suspense ended one day in the fall of 1965 when many of the Fourteen were on a geology trip in Oregon. Slayton got them all together in his motel room and announced that some of them would be moved into the Gemini flight rotation. Dave Scott would fly right seat on Gemini 8; Charlie Bassett would copilot Gemini 9. Mike Collins, meanwhile, was going to be Jim Lovell's backup on the upcoming Gemini 7 mission, and Dick Gordon would back up Scott on 8. According to Slayton's rotation, that meant Collins would fly right seat on Gemini 10, Gordon on 11. In one fell swoop, four of them were in the pipeline. No sooner had the group returned to Houston than Collins and the others soared off into the world of Gemini flight crews. They began simulator training, and got fitted for space suits, and went to mission planning meetings, while the rest of the Fourteen were left to their comparatively routine lives.

Late in the year most of the astronauts went to Boston for an introductory course at MIT on the Apollo computer and guidance system. Between lectures, Anders, Cunningham, Schweickart, and Bean took in the sights, venturing to the Concord Bridge to stand where the Shot Heard 'Round the World was fired, and then back to Cambridge for a stroll among the shops and cafés of Harvard Square. Everywhere they went, they talked not about the historic landmarks they were visiting, but about the crew selection game. And Anders had had enough.

"It doesn't work," Anders told the other three.

Cunningham responded, "What doesn't?"

"Working hard and hoping someone notices. Nobody does."

In December two more Gemini crews reached orbit. The world's first space rendezvous brought Wally Schirra and Tom Stafford in Gemini 6 face-to-face with their two bearded colleagues in Gemini 7, Frank Borman and Jim Lovell, who were midway through their own two-week mission. Deke Slayton had suggested, optimistically, that each of the Fourteen might fly a couple of space missions. But the Gemini program was already half over.

●◐◐○○○◑◑●

Nineteen sixty-six was America's calm before the storm. The war in Vietnam was in full swing, with 184,000 American troops in southeast Asia and more on the way, but political support for the war at home was still strong. Lyndon Johnson's Great Society was at the top of the political agenda. The struggle for civil rights was in transition; in the inner cities, hope was giving way to frustration and rage. In the Astronaut Office, Bill Anders paid little attention to these events. He would admit to feeling a little guilty about the fact that he'd made it into the astronaut corps but wasn't flying in space; he might as well be in Vietnam, doing his duty. But like the rest of the Astronaut Office, he was caught up in Gemini's spectacular progress, which was closing the gap in the race to the moon. The Soviets had not flown a manned space mission since 1965, when they had scored the first walk in space. In the meantime, Ed White had made his own space walk, and he'd topped his Russian counterpart by staying out twice as long and by using a hand-held maneuvering gun. With the completion of Borman and Lovell's fourteen-day marathon, Americans had taken a commanding lead in man-hours in space. The rendezvous and docking missions were next, something the Soviets hadn't even attempted. The flight schedule was proceeding at a manic pace, a new launch every other month, each one a totally different mission. Gemini had the flavor of the old barnstorming days of flight, transferred to the high-tech world of space. Wing walking had given way to spacewalking; aerobatics yielded to rendezvous and docking. Gemini's star was rising, and with it, the Fourteen's. Dave Scott was about to become the first of the third group to fly, on the Gemini 8 mission, and there was a member of the Fourteen on every Gemini crew from then on. None of them, however, included Anders, Cunningham, Schweickart, or Bean. ☾

Ironically, Anders noticed, the outside world treated even rookie astronauts as if they had been anointed. He did his share of NASA PR activities—Rotary Club appearances, mayors' luncheons, and so on—and he felt a little strange; what could he talk about except that he was training to go into

Gemini had the flavor of the old barnstorming days of flight, transferred to the high-tech world of space.

space—someday? But it was possible to give a speech on astronaut training and get a standing ovation. Occasionally he and Valerie took in the Houston social life, in which no party or charity ball was complete without an astronaut. But every Monday morning at the pilots' meetings, surrounded by men who had been in space, some not once but twice, Anders had no illusions.

Try as he might, Anders could not piece together an understanding of the crew selection game. The problem with Cunningham, Schweickart, and himself went beyond the fact that they weren't test pilots. Background alone wasn't the answer. Otherwise, why was Al Bean, a Pax River grad like Dick Gordon, still waiting? And look at Gene Cernan, another of the non-test pilots. Gregarious, easy-going, and well liked, Cernan fit in well with the Old Heads, and he was named as Tom Stafford's copilot on one of the later Gemini missions. Years later, Anders would decide that there weren't any selection criteria. Maybe Shepard and Slayton threw darts. Or maybe this whole thing was based on how well you fit the mold. But Anders couldn't be a chameleon. He couldn't turn himself into a hard-charging fraternity brother. And he didn't know what else he could do.

In February the Astronaut Office got a reminder that the pecking order could

Capping a string of Soviet coups in space, on March 18, 1965, cosmonaut Aleksey Leonov becomes the first human to venture outside an orbiting spacecraft. For much of the 12 minutes he spent outside Voskhod II, Leonov struggled to reenter the craft in his stiff, pressurized space suit.

be overruled at any time, by death. The prime crew of Gemini 9, Elliot See and Charlie Bassett, were killed when their T-38 struck the roof of the building housing their spacecraft at the McDonnell plant in St. Louis. Backups Tom Stafford and Gene Cernan were named to replace Bassett and See. At the same time, Jim Lovell and Buzz Aldrin, who up to now had tended a dead-end assignment as backups for Gemini 10—there would be no Gemini 13—became the Gemini 9 backups, putting them in line to become the prime crew of the final mission, Gemini 12. The last Gemini seat had been taken.

It was around that time that somebody finally noticed: Anders was assigned as a Capsule Communicator on the Gemini 8 mission. Being a Capcom put him a step closer to getting on a crew. On the first day of the mission, March 3, 1966, Anders walked into the Mission Control

Center just minutes after Neil Armstrong and Dave Scott had successfully docked with their unmanned Agena target. The craft was out of radio contact, somewhere over China. Jim Lovell was just ending his shift at the Capcom mike.

"You'll pick him up in about five minutes," Lovell told Anders. "Pretty boring." Lovell left the control center, leaving Anders to his Capcom debut. But that debut coincided with one of the most harrowing moments in the history of spaceflight. When Gemini 8 came back in radio range, Anders was startled to hear Armstrong report, "We've got serious problems, we're tumbling end over end. . . ." Gemini 8 was spinning out of control, and it wasn't until many anxious minutes had passed that Armstrong stabilized the craft. By that time, flight director Gene Kranz was preparing to abort the mission. Anders radioed the countdown for retrofire to Armstrong and Scott. Half an hour later, Gemini 8 was floating in the Pacific, and almost before it had started, Anders's first real job as an astronaut was over.

Still, he must have impressed somebody, because he was soon named as Armstrong's copilot on the backup crew of Gemini 11. As glad as he was to get the assignment, he wasn't happy about working for the Gemini 11 prime crew, Pete Conrad and his old navy buddy Dick Gordon. Anders thought they were a couple of obnoxious loudmouths—until he realized that they were two of the most competent astronauts in the office. He'd never seen anyone work so hard and have such a great time doing it. After a couple of months Anders liked Conrad and Gordon so much that he would've been happy to fly in space with either of them.

Anders knew from the start that backing up Gemini 11 was a dead-end assignment, but he dug in anyway. He learned the ins and outs of flying in Gemini's right seat, including his part in a space rendezvous, until it was second nature. And in a special training airplane that could produce brief intervals of weightlessness, he sweated through hours of practice space walks, hot and nauseated in his pressure suit. There was always the chance that the prime copilot, Dick Gordon, would break his leg, and if that happened, Anders would be ready to step in.

The most amazing thing was that suddenly he mattered. People listened to what he had to say. Conrad and Gordon were hot to set a new altitude record during their flight, but they had to prove to the doctors that there wasn't an undue hazard from radiation. Anders used his nuclear expertise to help make the calculations that won the managers' okay. And in September 1966, when Conrad and Gordon soared to 850 miles, Anders listened from mission control with not only envy but a sense of accomplishment. One more

IN 1966

Sound of Music wins an Oscar for best picture.

The National and American Football Leagues agree to merge.

Richard Speck kills eight Chicago nurses in their dormitory.

Feminist leaders form the National Organization for Women (NOW).

The first episode of _StarTrek_ airs.

Gemini flight came and went in November, and then the program was over.

By that time, the Manned Spacecraft Center was teeming with astronauts. Nineteen new pilots had come in, joining a handful of scientist-astronauts selected the previous year and swelling the ranks of the astronaut corps to forty-six. Slayton had been wary of taking on so many pilots; even with plans for perhaps a dozen lunar landings and a space station in earth orbit, he suspected that he would end up with nineteen astronauts he couldn't use. But the NASA management had pressed him to get "manned up" for the future flights, and he'd gone along. In the meantime, those flights were in danger of being cut from NASA's budget. By the time the Nineteen arrived in Houston, they sensed that they were in for a long wait. In fact, they seemed to be ignored, as if they weren't really there. They wryly called themselves the "Original 19." ☾

But for Bill Anders, the wait was finally over. Slayton assigned him as the lunar module pilot for an earth-orbit test of the Apollo spacecraft, with Frank Borman in command and fellow Gemini veteran Mike Collins as the command module pilot. Meanwhile, Walt Cunningham and Rusty Schweickart had made it into the pipeline too. But not all of the Four Musketeers had found deliverance. Al Bean still wasn't on a crew; in fact, he wasn't even working on Apollo any longer. He'd been transferred to the Apollo Applications Project, the earth-orbit space station planned for the 1970s. Anders could only shake his head when he thought of Bean; if he wasn't going to the moon, none of this game made much sense.

It was funny; Anders never heard any of the Old Heads talk about what it would be like to go to the moon. But sometimes, flying cross-country at 45,000 feet, he could see it through the clear canopy of his T-38, and he would think about how much he wanted to explore it. Now that he was on a crew, it was becoming a real possibility. He'd stay in the pipeline, and eventually, he'd find himself on the crew of a lunar landing mission. In the meantime, as 1967 opened, Anders shared the soaring confidence that filled the space center as Gus Grissom and his crew readied for their Apollo 1 flight.

NOVEMBER 9, 1967
MERRITT ISLAND LAUNCH COMPLEX, KENNEDY SPACE CENTER

On a chilly November dawn, a crowd of spectators gazed across the still waters of a tidal basin to Pad 39-A, 3½ miles away. There, the first Saturn V moon rocket was minutes from its maiden voyage, and most of the astronaut corps was here to witness it. In the crowd, Pete Conrad gazed admiringly at the mammoth white booster as it spewed trails of vapor into the morning air.

By this time, Pete Conrad was one of the most seasoned astronauts in the office. In the summer of 1965 he'd copiloted Gemini 5, an eight-day marathon in earth orbit. Then, on Gemini 11, he and Dick Gordon had made the first one-orbit space rendezvous. And for much of this year, Conrad had been immersed in training for Apollo. Like all the astronauts, Conrad thought it was a waste to launch a Saturn V unmanned, but NASA would never have considered putting astronauts on a new and untried rocket, especially the most powerful one ever created. And a successful launch, even an unmanned one, would clear the way for the flights to come and give Apollo badly needed momentum.

Suddenly a torrent of fire erupted at Pad 39-A. For many long seconds it continued, and still the great rocket sat motionless, as if it might never leave the earth.

During the fall, NASA planners had firmed up the Apollo flight schedule into a series of methodical, incremental steps that are the hallmark of test flying. On Vu-graphs and blackboards around the center they were conveyed in a series of letters. Today's Saturn V launch was one of the unmanned test flights called the "A" and "B" missions. The command module would make its manned, earth-orbit debut on the "C" missions. Then, on the "D" and "E" missions, the lunar module (LM) would be added for a test of the full Apollo spacecraft in earth orbit. If they were successful, the astronauts of the "F" mission would make the first flight to the moon, for a "dress rehearsal" of the lunar landing. Then, at last, would come the first attempt to land on the moon: the "G" mission. NASA would follow these milestones, each mission bolder than the one before, closing the distance one step at a time, like a ladder to the moon.

By November 1967, Pete Conrad was beginning to think he might find himself at the top of that ladder in 1969. He and his crew were backups for the D-mission, and by Slayton's crew rotation, if everything went according to plan, they would become the prime crew of the G-mission. To Conrad, and all the astronauts, the first landing mission was considered the ultimate test flight. No assignment equaled the chance to be the crew that actually carried out Apollo's mission.

Conrad knew the lunar landing was a task of enormous complexity, the details of which had yet to be worked out. Even if everything went perfectly, would the first attempt at the actual landing be successful? It might take several tries before they succeeded. And the time left to meet Kennedy's deadline was steadily dwindling. A catastrophic failure could place the entire moon program in jeopardy.

Over the public address system, Conrad could hear the public affairs commentator calmly giving status reports as the Saturn was fully pressurized with fuel. Then the count entered its final minute.

"T minus fifty seconds and counting. We have transferred to internal power and the transfer is satisfactory. . . . T minus thirty seconds and counting. . . . T minus twenty, nineteen, eighteen . . ."

Conrad wished he were on that rocket.

". . . eleven, ten, nine, ignition sequence start. . . ." Suddenly a torrent of fire erupted at Pad 39-A. For many long seconds it continued, and still the great rocket sat motionless, as if it might never leave the earth. At zero, the Saturn slowly rose from the pad, trailing a tail of incandescent flame. It crept upward past the towering launch structure while the crowd cheered. Suddenly their cries were drowned out by an onslaught of sound. Witnesses felt the shock waves pound against their chests. The Saturn arced upward, gaining speed, and Conrad's eyes followed it into the clear morning sky, until it was gone. Nine hours later NASA had chalked up a flawless test flight of the booster and the Apollo command module, and for the first time in NASA's worst year, the moon came a little closer.

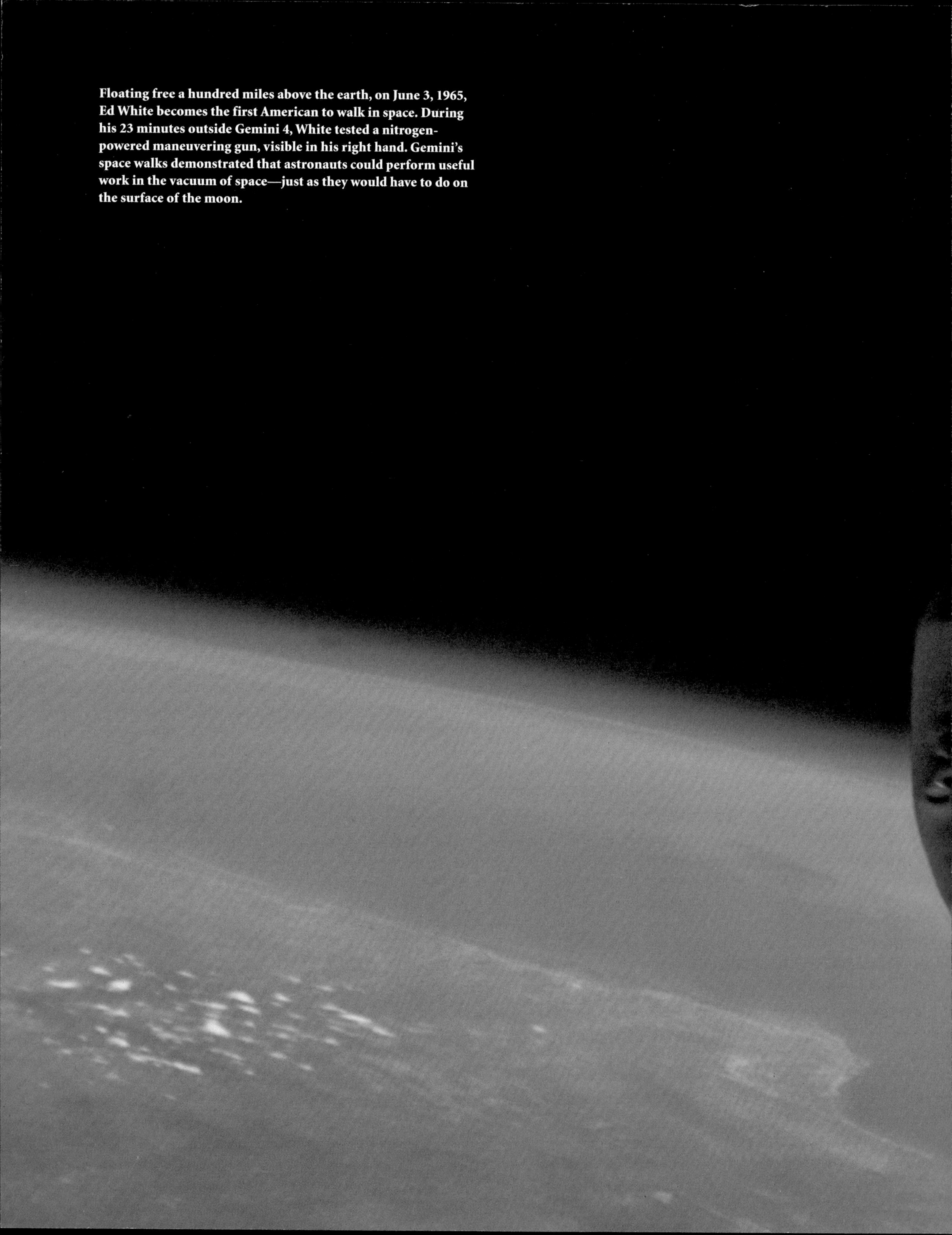

Floating free a hundred miles above the earth, on June 3, 1965, Ed White becomes the first American to walk in space. During his 23 minutes outside Gemini 4, White tested a nitrogen-powered maneuvering gun, visible in his right hand. Gemini's space walks demonstrated that astronauts could perform useful work in the vacuum of space—just as they would have to do on the surface of the moon.

At the climax of an orbital ballet, Gemini 6 closes within a few feet of Gemini 7 on December 15, 1965, in history's first space rendezvous. Astronauts returning from the surface of the moon would have to match this feat in order to link up with the orbiting command module for the trip home.

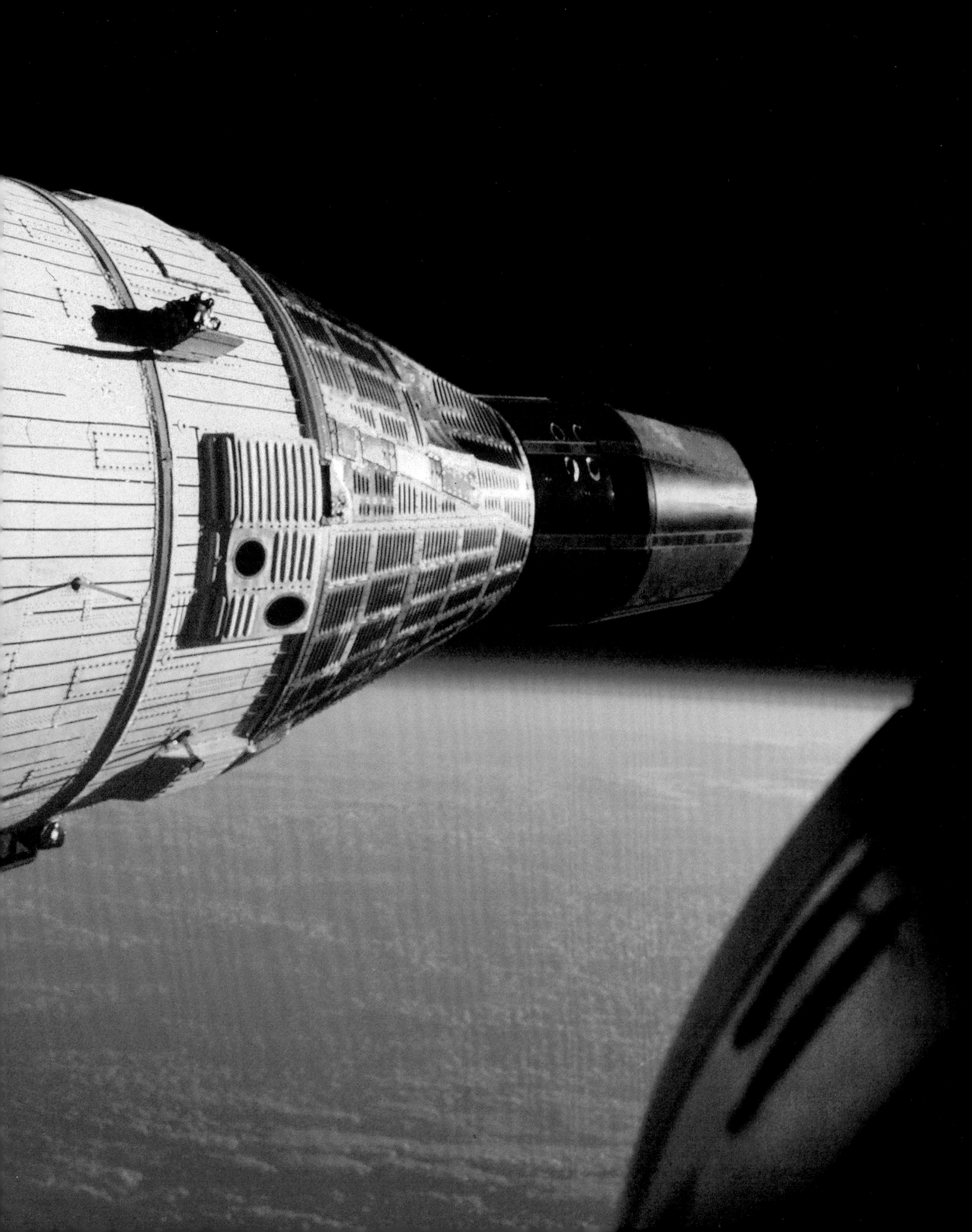

UNITED STATES
USA
USA

FIRST AROUND THE MOON

APOLLO 8

I. THE DECISION

The news was bad in the summer of 1968. A nation reeling from the assassinations of Martin Luther King and Robert Kennedy now confronted more images of violence in its living rooms: the blood of young soldiers in Vietnam, the blood of demonstrators outside the Democratic National Convention in Chicago. And even within NASA's world, where these events were overshadowed by the race with the decade, there was bad news. The second unmanned test of the Saturn V moon rocket had been a near disaster. Minutes into the launch the booster began to vibrate badly. Then two of the second-stage engines shut down prematurely. Later, a third engine refused to reignite in space. And if that weren't enough, there were ongoing headaches with the Apollo spacecraft. The redesigned command module was coming along well at North American, and the craft slated for Apollo 7, the command module's manned, earth orbit debut, was already at the Cape being readied for an October launch. But the lunar module was facing one technical problem after another. From the beginning, engineers at the Grumman Corporation in Bethpage, Long Island, had struggled to keep the lander's weight from exceeding forbidden limits.

Perched atop a giant Saturn V rocket four days before launch, the Apollo 8 spacecraft stands ready for its historic voyage around the moon. Towering 363 feet above its launch platform, the entire "stack" weighed more than six million pounds when fully fueled.

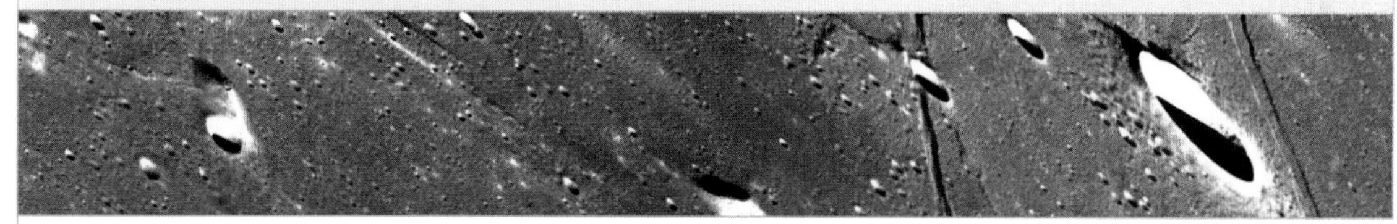

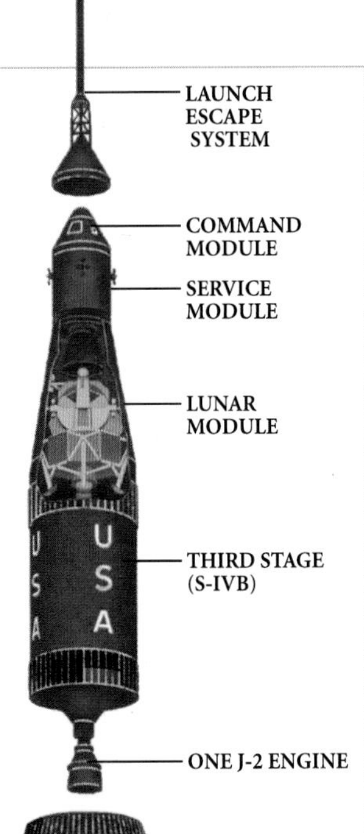

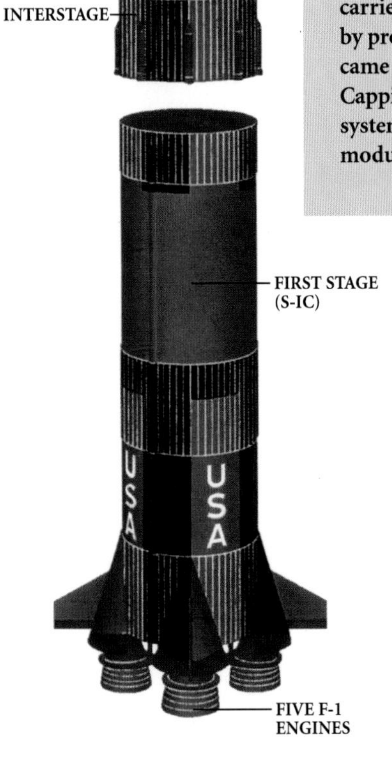

And there were other woes: faulty wiring, corroded metal, and most serious of all, troubles with the LM's ascent rocket. And when the first manned lunar module was shipped to the Cape in June, quality control inspectors found 100 separate defects. At NASA, no one who heard the reports on the lander was happy with the situation. Apollo 8, the LM's first manned flight, would almost certainly be delayed beyond the end of the year, throwing the whole sequence of Apollo missions into jeopardy. The end-of-the-decade deadline for the lunar landing was slipping out of reach. ☾

All that began to change in early August. A plan emerged, elegant in its simplicity, astounding in its boldness, that altered the course of the moon program. It was the brainchild of George Low, the quiet engineering genius who oversaw the development of the Apollo spacecraft from Houston. If Apollo 7 went well in October, Low reasoned, why keep Apollo 8 in earth orbit? Even if the LM wasn't going to be ready for its debut, the second command ship could go to the moon by itself in December. Already, during the spring, Low had quietly raised the possibility of a circumlunar flight in which the joined command module/lunar module pair would execute a figure 8 loop around the moon and then come home. His new plan was even more ambitious. Low wanted to send the command module to the moon by itself, not to fly a figure 8 loop, but to go into lunar orbit. Even without a lunar module, that would let NASA practice the elements of a basic lunar mission: navigating across the vast translunar gulf, executing the precise rocket firings to get into and out of lunar orbit, communicating across a quarter-million miles, and the critical reentry into the earth's atmosphere at hypersonic speeds. Then, by the time the LM was ready—estimates said February—Apollo would have taken a giant step forward. ☾

ROCKET TO THE MOON

The first stage of the Saturn V rocket that launched Apollo spacecraft moonward had five F-1 engines producing a total of 7.5 million pounds of thrust. An equal number of J-2 engines on the second stage generated just over one million pounds of thrust. The third stage, powered by a single J-2 engine, contained the booster's guidance system and propelled the spacecraft first into earth orbit and then across the 240,000-mile void between earth and the moon. Atop the third stage rested the lunar module (not carried on Apollo 8), which was concealed by protective panels during launch. Next came the service and command modules. Capping the assembly was a launch escape system intended to whisk the command module to safety if the booster failed.

But there was another reason for urgency. Reports from the Central Intelligence Agency said the Soviet Union was about to resume flying its new Soyuz spacecraft—after the first Soyuz crashed, killing its lone cosmonaut, Vladimir Komarov, in April 1967—and were on the verge of sending one around the moon. Most experts doubted the Soviets had the capability to land on the moon before the end of the decade; for one thing, they had yet to

In 1963 Valentina Tereshkova *(right)* became the first woman in space. Her flight was one of a string of Soviet pioneering feats dating from Sputnik 1, suggesting that the Soviets might succeed in sending a pair of cosmonauts around the moon in late 1968.

test a rocket, like the Saturn V, powerful enough to propel the necessary payload to the lunar surface. Even a lunar orbit flight was probably beyond them. But with the booster they already had, they could fire a Soyuz, with one or two cosmonauts aboard, on a trip around the moon.

From the beginning, without warning, the Soviets had upstaged the United States in space with their own spectacular firsts. In 1957 it had been the first earth satellite, Sputnik I. More than three years later it was Yuri Gagarin's one-orbit flight that stunned the world and sparked John Kennedy's decision to go to the moon. Then came the first woman in space, the first multiperson space crew, the first spacewalk. If the Soviets got to the moon first—even if they did nothing more than loop around it—the world would hardly notice the difference between that accomplishment and NASA's more difficult lunar orbit mission.

There is no way of knowing what would have happened to Low's plan if NASA Administrator James Webb had been in Washington, but the fact was he was not; he and his deputy George Mueller were in Vienna attending a conference. In their absence Associate Administrator Thomas Paine, a bright, young engineer with a penchant for the visionary, was in charge. When Paine's deputy, Apollo program director Sam Phillips, told him about Low's

Webb wasn't ready for what he heard. He yelled over the transatlantic phone line, "Are you out of your mind?"

idea Paine immediately saw the logic in it. But it remained to convince Webb, and that might not be easy. ☾

Webb was not an engineer. He was, however, a canny bulldog of a politician. When Kennedy said "Go to the moon" it was up to Webb to keep Congress from having second thoughts, which he did by any means of persuasion he found necessary—including a knack for knowing where congressional skeletons were hidden. Year after year, he was Apollo's champion on the Hill, where it counted most. He had persevered even as the war in Vietnam claimed more and more of Lyndon Johnson's attention and Apollo became a target of congressional opposition. If Americans reached the moon by the end of the decade, it would be due in large measure to Jim Webb. But Webb would not be at NASA to see it; he already knew his tenure would end when Johnson left office. ☾

Webb took Paine's call at the American embassy, where there was a secure phone, and then Sam Phillips got on the line. Webb wasn't ready for what he heard. He yelled over the transatlantic phone line, "Are you out of your *mind*?" Webb reviewed what was apparent to any sane person: They hadn't even flown a manned Apollo spacecraft, and here they were with a scheme to send the second flight to the moon. And with no lunar module! All along, the LM had been thought of as a measure of safety, a lifeboat in case something happened to disable the command ship's rocket engines. Sending the command module by itself only increased the risk of what was already a risky mission. With the Fire still fresh in the memory of the public and the

Congress, Webb could only imagine the effect of another space tragedy. He warned his two deputies, "You're putting the agency and the whole program at risk."

Webb was right. For all its logic, Low's plan was audacious. Many would look back on it as the boldest decision NASA ever made. Still, by the time Paine and Phillips hung up the phone to Vienna, Webb had agreed to give the idea a chance.

In Houston, they were already working on it. Low had asked Chris Kraft, director of Flight Operations, to find out whether his people could be ready to send Apollo 8 to the moon in December. They went ahead with their study in secret. When their office mates asked—"What's all this lunar stuff you're working on?"—they replied coolly, "Oh, it's just a what-if type of study . . ."

Within a week Kraft's team had an answer. By the summer of 1968, after years of intensive effort, the basics of sending a manned spacecraft to the moon were all but perfected. The biggest hurdle: finishing the computer software that mission control would need to help Apollo 8 navigate to and from the moon. Making the December launch date would be tight, but Kraft's people were confident they could do it. Meanwhile, at the Marshall Space Flight Center in Huntsville, Alabama, Wernher von Braun's rocket team reported the problems with the Saturn V were being ironed out. It remained for the Apollo spacecraft to prove itself. If all went well on Apollo 7, slated for October, there wouldn't be anything to stand in the way.

SATURDAY, AUGUST 10, 1968
NORTH AMERICAN AVIATION,
DOWNEY, CALIFORNIA

It wasn't unusual for Frank Borman to be working on Saturday; the past nineteen months had been among the busiest of his life. After the Fire he had plunged into the recovery effort, beginning with the accident review board at the Cape and then many long months here, at the North American plant in Downey, California. Until a few months ago Borman had all but lived here at the factory, helping to redesign the command module. Now that effort was behind him, and the Block II command module was on the assembly line, with several of the cone-shaped craft taking form in the sterile whiteness of North American's clean-room. One of those spacecraft, command module number 104—the official designation for the fourth Block II command module—was his. There were problems with 104, but that was to be expected, and Borman was here with his crew, Jim Lovell and Bill Anders, to nurse his ship along and help ready it for delivery to Cape Kennedy.

CLEAR CREEK
60
CLEAR CREEK
85

Frank Borman gets a lift from his sons Frederick *(left)* and Edwin. A key player in NASA's recovery from the Apollo 1 fire, Borman was determined to vindicate the Apollo spacecraft. When Deke Slayton offered him a chance to fly it around the moon, he accepted without hesitation.

When Frank Borman walked into a room, you knew that he was in charge. Looking at him—he was a sturdy man with a square, slightly oversized head—you could still see a tough, scrappy kid. He'd been molded at West Point; at age forty he still wore his dirty-blond hair as short as a cadet's, and he still lived by the Point's simple motto: Duty, Honor, Country. The mission came first. It was always that simple for Borman. When he applied for the second astronaut group, the psychologists on the selection board said in amazement, "Nobody's that uncomplicated!" He did not deal in nuance, he did not make small talk. His hearing in one ear had been bad ever since he ruptured an eardrum early in his air force career, but there were times when he seemed not to hear very well with the other one; in meetings he'd listen to a number of views and then make a decision as if no one had spoken. When it came to his crew Borman was a model of the old-school military commander. If there was a decision to be made Borman made it; then he told his crew about it.

He hated few things more than wasting time. During meetings if he got bored or too impatient he would get up and walk out. He made decisions quickly—so fast that if you didn't know Borman you would have thought him impulsive—and once he convinced himself he would do something a certain way it was all but impossible to dissuade him. But he was usually right. And if he seemed mostly gruff and unyielding, Borman understood people, and in addition to his talents as a pilot and engineer he was a capable manager. He was also a favorite of the NASA management. After Borman had finished his work on the Accident Board—and after appearing before Congress to say, "Let's stop the witch hunt and get on with it"—Bob Gilruth, the head of the Manned Spacecraft Center, personally asked him to oversee the recovery efforts at North American. Borman would always look back on that time, working with the engineers here in Downey, as the most productive episode of his astronaut career. Getting Block II ready took much longer than people expected, but it was a superb spacecraft, free of most of the problems of its predecessor. It had a new, one-piece hatch that could be opened in as little as three seconds. It was fireproof, and soon it would be spaceworthy. And now that Block II was on the assembly line, what Borman wanted most of all was to vindicate that spacecraft: he wanted to *fly* it. ☾

Borman had his own command module and his own crew, but he wasn't very enthusiastic about his mission. Deke Slayton had given him command of the third manned Apollo flight, scheduled for sometime early in 1969. Eventually it would be called Apollo 9, but for now it was known by its letter designation, the E-mission. True, Borman's crew would be the first to ride the huge Saturn V booster. And during the mission they would change the high point of their orbit to a record 4,000 miles. For a few days, while they traveled that lofty ellipse, they would see the entire earth at a glance. But aside from those frills, the E-mission was a repeat of the previous flight, which belonged to Jim McDivitt. McDivitt's D-mission would be the first manned test of the entire Apollo spacecraft in earth orbit. It would include tests of the rocket engines on both the command ship and the lunar lander, and the first Apollo

A successful lunar orbit mission would lift NASA out of the shadow of the Fire, restore its shaken confidence, propel it to a lunar landing.

rendezvous maneuvers. With all those firsts, the D-mission was a test-piloting bonanza, and Borman would have gladly traded places with McDivitt. But this wasn't the first time Frank Borman was going to fly a mission he did not want.

In December 1965, after waiting three years for his chance to fly in space, Borman lifted off on what was considered the "dregs" mission of the Gemini series: Gemini 7, a two-week marathon in earth orbit. Borman and his copilot, Jim Lovell, spent fourteen days in Gemini 7's tiny cabin wired from stem to stern with medical sensors. Two weeks in a flying men's room. It wasn't an enviable assignment, but Borman took it without complaint. Dregs or no, he wanted to fly. And in his mind, he'd accepted whatever assignments might come his way on the day he joined the astronaut corps—including the E-mission.

On Saturday, August 10, Borman was knee-deep in testing command module 104 when he was called away for a phone call from Deke Slayton.

"Frank, get back to Houston right away. I need to talk to you," Slayton said.

"So talk to me now, Deke; I'm busy."

"I can't do this over the phone. Grab an airplane and get back here."

Borman was irritated at the interruption—the test was far from over—but when he reached Slayton's office and his boss asked him to close the door, he understood the urgency. Borman listened as Slayton described the CIA report on the Soviets and George Low's plan to send Apollo 8 to the moon.

Borman immediately saw the logic. A successful lunar orbit mission would lift NASA out of the shadow of the Fire, restore its shaken confidence, propel it to a lunar landing. Slayton was proposing that Borman and McDivitt swap places and command modules, that Borman's crew fly in December. One thing Slayton did not say was that he had already offered the lunar mission to Jim McDivitt, who had turned it down. As for himself and his crew, Borman needed no time to weigh his decision. To him Apollo was like a war; nothing less than the nation's prestige was at stake. He wanted to see the war won. When Slayton asked him if he wanted to go to the moon, Borman said yes.

MONDAY, AUGUST 12
TIMBER COVE, TEXAS

Marilyn Lovell was feeling pleased with herself. She had just finished a long, hot day of hunting for bargains on vacation clothing in the August department-store sales. She could barely remember the last time the family had taken a vacation. Was it before Jim became an astronaut? But just a few weeks ago Jim had agreed to take the family to Acapulco during his week off between Christmas and New Year's. Marilyn was jubilant, and today, in the Houston swelter, she'd found some great buys. She would tell her news to Jim, who would be home tonight from a trip to California.

Marilyn and Jim had met and dated in the mid-1940s, when they were both attending high school in Milwaukee. Even then, Jim had his mind on the stars; sometimes, on clear nights, he would take her to the roof of his apartment building to give her a tour of the heavens. And during World War II, Jim had been fascinated by the reports of German V-2 rockets. Marilyn witnessed his experiments with rockets made from mailing tubes and fueled by gunpowder mixed with airplane glue. While he and a classmate were busy in the open field across the street Marilyn would sit in the apartment with Jim's mother, both of them hoping the two boys didn't blow their heads off. But this was no reckless stunt. The young experimenters took every precaution, wearing welder's goggles and gloves. And it was a good thing—one rocket exploded, sending the nose cone about 80 feet into the air. No one was injured—in fact, Jim proudly proclaimed the "flight" a qualified success. ☾

After high school Jim went to Annapolis, where he had his sights on becoming a rocket engineer. For his senior thesis, he wrote about interplanetary rocket travel, and Marilyn, now his fiancée, typed the paper for him. When she came to the end, she read with disbelief Jim's statement that someday people might ride rockets to the moon. But in 1952 there weren't any jobs for

In good spirits after a two-week space marathon on Gemini 7, Jim Lovell and Frank Borman meet the press aboard the carrier U.S.S. *Wasp*. The two men became friends during the mission, which paved the way for Apollo's lunar voyages.

rocket engineers to speak of, and after Annapolis, Jim found himself flying airplanes for the navy. His life had taken a different path, and to Marilyn's joy and amazement it had led to Houston and into space.

Like all the astronaut wives, Marilyn accepted the constant demands the space program placed on her marriage. Her husband's competition for flights spilled over into her own life, just as it did with all the wives. In 1965, while Jim was training for his first Gemini mission, Marilyn discovered she was pregnant with their fourth child. She kept the news to herself for weeks, because she was afraid Jim might be taken off the crew. And for the next three and a half years, as Jim plunged happily into the maw of training, going from backup crew to prime crew to backup and prime again, Marilyn was head of the household, financial manager, mother and father to their kids. During the week Jim would always call in from wherever he was, and somehow they managed to work through the day-to-day crises over the phone. But seeing him in person—that was something she enjoyed only on weekends.

Now Jim was working with Frank Borman again, and his lunar module pilot Bill Anders. But with the flight set for late next winter, the most intensive training was still ahead, and Marilyn was glad that her husband had finally made plans to take some time off at the end of the year.

When Jim came home, Marilyn proudly reported the success of the day's shopping expedition. In the midst of her excitement, she noticed Jim had a peculiar expression, and when she asked him what was wrong, he took her into the privacy of his study, where pictures from his space flights adorned dark wood paneling, and closed the door. "I hate to tell you this," he said, "but we're not going to Acapulco for Christmas."

Marilyn didn't try to hide her disappointment. "What do you mean we're not going? Where on earth do you think you're going to be if you're not going to be with the family for Christmas?" Years later, Marilyn would still get goose bumps thinking about what happened next. Jim paused for just a moment, his eyes bright, and he said, "Would you believe, the moon?"

Some people—other astronauts—had the mistaken impression that Jim Lovell flew through life on luck. Tall, relaxed, and outgoing, Lovell got along with everybody. Frank Borman would later say that he had never worked with anyone who faced life with such consistent, good-natured optimism. But Lovell's easy-going manner masked his competitive energy and sharp mind. He'd graduated first in his test pilot class at Patuxent River, ahead of

Wally Schirra and Pete Conrad. Like Conrad, he had been turned down for the Mercury selection, then made it with the second astronaut group.

Lovell had to wonder at the twist of fate that had brought him and Frank Borman together again. Lovell knew Borman well; after two weeks cooped up in Gemini 7, he probably knew him better than anyone except his wife. The two men were the same age (Lovell was eleven days younger), and although they had chosen different branches of the service (Borman the air force, Lovell the navy), they were the same equivalent rank. In Lovell's mind they had been equally qualified to command a space mission. When Slayton named him as Borman's copilot on Gemini 7, Lovell took the assignment without ever revealing his disappointment. He was a rookie, and he wanted to fly.

Jim Lovell designed this insignia for the Apollo 8 mission. The patch features a figure 8 with the earth in one loop and the moon in the other.

Two weeks in space only whetted Lovell's appetite for more, but Slayton handed him a dead-end assignment as backup commander for the final Gemini mission. That changed when Elliot See and Charlie Bassett were killed; Lovell and Buzz Aldrin reached orbit for Gemini's finale. Lovell expected to move into Apollo with his own crew, but instead found himself as backup command module pilot for Frank Borman's E-mission. But in July, fate again changed Jim Lovell's plans. Borman's command module pilot, Mike Collins, was forced off the crew by a bone spur on his spine that threatened his body and his career. Collins would have corrective surgery, but even if it went perfectly he would be out of action for months. Lovell stepped in to take his place. And now, he found himself flying right-seat to Borman once more.

When Borman came back from Houston with the news that they were going to the moon, Lovell was electrified. He could still remember how, as a boy, he had devoured such science fiction classics as Jules Verne's *From the Earth to the Moon,* in which a trio of adventurers ride a craft fired out of an enormous cannon on a circumlunar voyage. It had been beyond his wildest teenage dreams that he might see human beings live out the adventure Jules Verne had imagined, let alone that he would be one of them. Even now, as an astronaut who had made the moon his goal, he found the news almost too exciting to believe. Later, under the clear canopy of a T-38 as he and Borman flew back to Houston, he had sketched a design on his knee-pad that would become the mission emblem of Apollo 8: a figure 8 with the earth in one loop and the moon in the other.

Lovell never had any second thoughts about the lunar mission. To him, it was worth the risk for the adventure alone, never mind the potential for scientific discovery. And then, to be pathfinders for those who would follow, the benefits Apollo 8 would bring to the program—who could question the logic of it? Once Lovell had a chance to get used to the idea, he wondered why that hadn't been the plan all along.

●●◐○○○○◑●

In the summer of 1968 Bill Anders's fortunes seemed to be improving all the time. He had spent much of the spring as one of a handful of astronauts who were learning how to fly the dangerous and unwieldy trainer called the Lunar Landing Research Vehicle. Neil Armstrong was flying it too, and he and Anders had a friendly competition to see who could make the better simulated lunar touchdown. No one had said, "This means you're in line for a landing mission," but you didn't get an assignment like that for no reason.

Meanwhile, he was hard at work as Frank Borman's lunar module pilot for the E-mission. As a lunar module pilot, he was developing an intimate knowledge of the lander, and the experience he'd get on the E-mission would surely make him an ideal candidate for a landing mission down the line. It looked as though Shepard and Slayton had finally recognized what he could do and were going to give him a chance to prove it. Anders figured he had an 80 percent chance of walking on the moon in the near future.

When Frank Borman came back to Downey with the extraordinary news that their mission had been changed, Anders felt disappointment underneath his excitement. He was losing his lunar module, and with it his chance to one day land on the moon. Borman tried to give him a fatherly talk, saying it wasn't so bad, that he'd still get his chance later. But Anders wasn't convinced. The good news was he was going to the moon in four months; the bad news was he would probably never walk on it.

●●◐○○○○◑●

With only four months to go until a December launch, no one had time for second-guessing, least of all Frank Borman. As mission commander he had to be intimately involved in working out the flight plan for his mission. Normally that took months, and since the circumlunar mission was an entirely new creation it should have taken even longer. But to Borman's great satisfaction—he would look back on this as the space program at its best—the basic design for Apollo 8 was hammered out in a single meeting one August afternoon in the office of Chris Kraft. ☾

Christopher Columbus Kraft, Jr., was the engine that powered mission control. In the infancy of manned space flight Kraft had created the persona of a flight director as an actor creates a timeless role, setting the standard for all who would follow. Now in his mid-forties, the Virginia-born Kraft possessed the special brand of grace under pressure required to direct a space mission in which human lives are at stake. To his team of flight controllers, the men who manned the trenches of mission control and kept watch on every bit of telemetry from a spaceborne machine, he was stern, generous, perceptive, and inspiring, sometimes all at once. They idolized him the way an infantry division does a beloved commander. In addition to his forthrightness—you always knew where you stood with Kraft—he had the ability to sort through differences of opinion and get to what really mattered. To Frank Borman, Kraft was one of NASA's giants.

Kraft's style, honed during a score of manned and unmanned space missions, was an almost inexplicable blend of caution and boldness. Borman

Chris Kraft *(left)*, the master of mission control, watches the progress of an unmanned Apollo test flight in 1968. With him is Bob Gilruth, director of the Manned Spacecraft Center.

knew it well, for it had been Kraft who had saved Gemini 7. For Borman the last three days in earth orbit were the worst three days of his life. Thrusters went bad; fuel cells were threatening to quit; they were low on maneuvering fuel. He and Lovell were just drifting around the earth, trying to hold out long enough to fly the full fourteen days. Had it been up to him, they never would have made it, because he was ready to come down. It was Kraft's persistence and deft handling of the crises that kept Borman going to complete his mission.

On the afternoon of August 19, Borman and several of Kraft's best people, including such stalwarts as Bill Tindall—the tireless and ebullient mission planner who spearheaded the effort to work out detailed techniques for all phases of the lunar landing—gathered in Kraft's office to design the first flight to the moon.

The six-day mission was slated to begin on December 21, just after the new moon. The timing was chosen so that when Apollo 8 arrived the sun would just be rising across the Sea of Tranquillity, throwing the landscape into relief, allowing Borman's crew to reconnoiter a potential touchdown spot for the first landing.

Borman's crew would become the first to ride the three-stage Saturn V booster into space. Reaching orbit, they would leave the rocket's third stage attached. For about three hours, while they circled the earth twice, Borman, Lovell, and Anders would check the systems aboard their spacecraft, while mission control scrutinized Apollo 8 via telemetry. A failure in any of thousands of components might prompt the controllers to cancel the lunar mission and direct Apollo 8 to stay in earth orbit for a ten-day alternate mission. But if everything was in order they would send Borman's crew a historic message: Go for Translunar Injection. In that maneuver—the astronauts called it "TLI"—Borman's crew would relight the Saturn's third stage and accelerate out of earth orbit. When the engine shut down Apollo 8 would cast off the spent booster and continue on a course for the moon, 240,000 miles away. From then on, Borman, Lovell, and Anders would be in uncharted territory.

Consider the sheer scale of the voyage. Up to now, human beings had barely strayed from their home planet; the world's altitude record, set by Gemini 11 astronauts Pete Conrad and Dick Gordon, was a mere 850 miles. If the earth were a basketball, that would amount to just one inch from the surface. But in the same scale model the moon, 2,160 miles in diameter, would be a baseball 23 feet away. Getting to the moon and back would require acts of precision more demanding than any previous space flight.

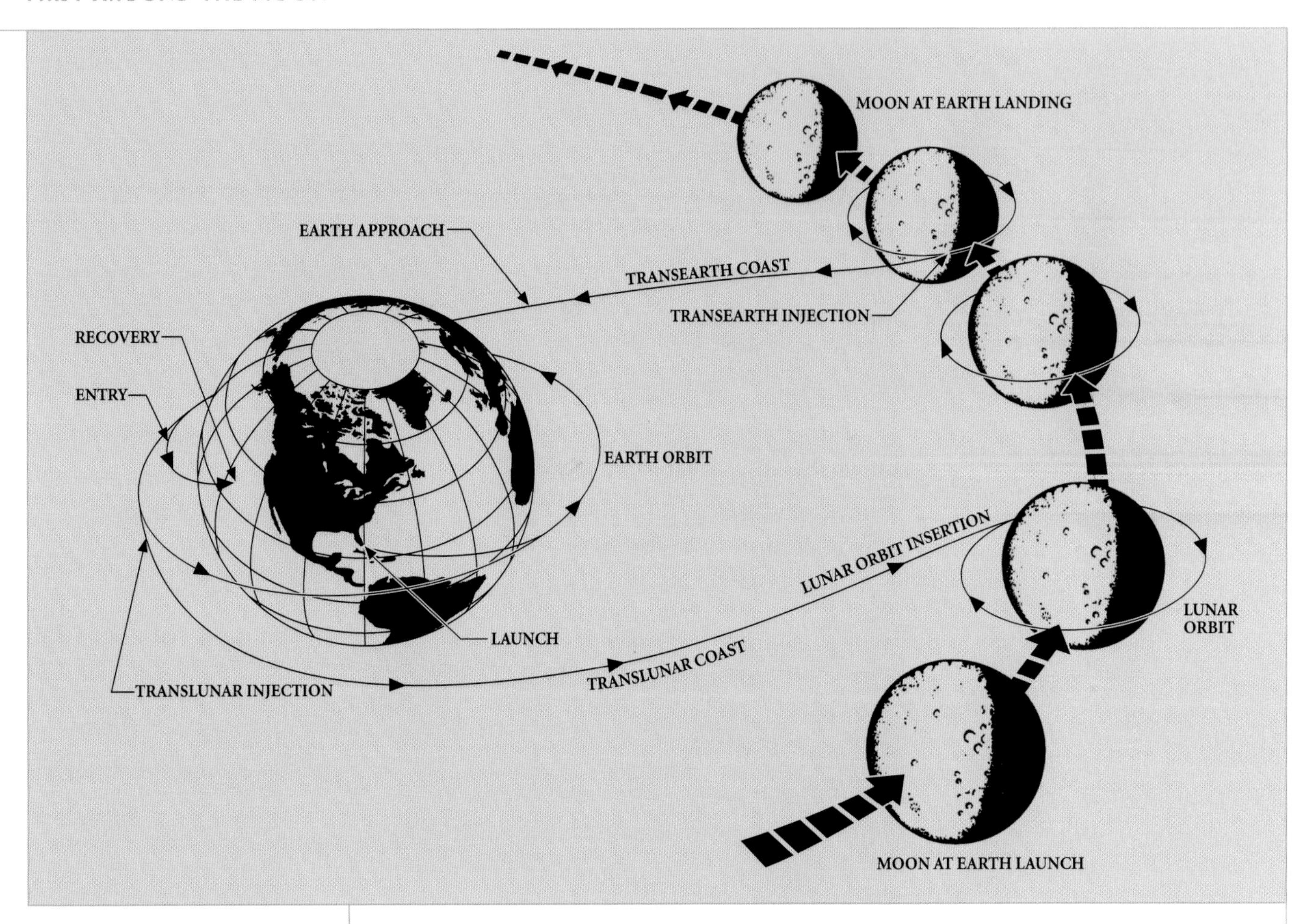

Intercepting a moving target—the moon—was the goal for Apollo 8 and every lunar mission thereafter. To accomplish this feat, the third-stage rocket had to propel the astronauts from earth orbit to the point where the moon would be approximately 66 hours later.

To make matters more difficult, the moon is a moving target, barreling along in its orbit at a speed of 2,300 miles an hour. Apollo 8 would have to reach the moon's orbit just as the moon was arriving. Then, like a car racing a locomotive at a crossing, the spacecraft would zip in front of the moon's leading edge. After speeding behind the moon, Borman's crew would fire the spacecraft's main rocket engine and go into an orbit with a low point of 69 miles above the lunar surface—eight one-hundredths of an inch from the skin of the baseball.

But that was only if everything went as planned. Suppose Borman's crew suffered a serious malfunction, an engine failure, for example, before they reached the moon? Without some other way to turn around, would they simply speed past the moon, condemned to a lonely death when their oxygen supply ran out? Fortunately, the laws of celestial mechanics—whose strictness gave space flight an extraordinary predictability—made it possible to give Borman's crew a built-in ticket home. The trajectory specialists had taken advantage of this in an elegant creation called the free return. By aiming Apollo 8 at just the right distance from the moon, they could use the lunar "gravity well" like a curve in a toboggan course, to bend its path around the moon and send it back toward earth. In theory, it was possible to fire Apollo 8 out of the starting gate so precisely that it would fly a perfect figure 8 around the moon even if Borman's crew never touched the controls. There wasn't a man in Kraft's trajectory division who wasn't so confident of this that he would have offered to go in their place.

Borman took them at their word. He wanted the trajectory people to pledge they'd keep Apollo 8 on an essentially perfect free-return path, but that was a promise they could not make. The problem, they stressed, was that it would simply not be possible to keep perfect track of Apollo 8's position in space at any given moment. There would always be some error in the measurement—but how big an error, no one could say until Apollo 8 was actually moonward bound—and without that precise knowledge, a perfect free return was impossible to guarantee.

Furthermore, there was a crucial, built-in safety measure. Unlike a rifle bullet—or, for that matter, Jules Verne's manned cannon shell—Borman's crew would be able to correct Apollo 8's path along the way with bursts from its small maneuvering thrusters. If the spacecraft drifted off course, mission control would be able to tell, and they would radio Borman's crew instructions for a so-called midcourse correction. As long as the maneuvering thrusters were functioning, there would be plenty of chances along the way to keep Apollo 8 on course.

The translunar crossing of some 234,000 miles would take 66 hours, roughly half the time it takes an ocean liner to cross the Atlantic. At last, on the morning of December 24, Apollo 8 would reach its destination. If everything checked out onboard mission control would give Borman's crew the go-ahead for Lunar Orbit Insertion (LOI). Over the lunar far side, out of radio contact with earth, they would fire Apollo 8's big Service Propulsion System (SPS) rocket engine for some four minutes—just long enough to slow to the speed necessary to go into lunar orbit, about 3,700 miles per hour. When the engine shut down there would be no more free ticket home. Apollo 8 would be a satellite of the moon, and Borman's crew would have to trust the SPS to work perfectly when the time came to return to earth. ☾

While millions on earth would be among friends and loved ones on Christmas Eve, Borman, Lovell, and Anders would spend it circling the moon, taking pictures and observing its pockmarked surface at close range. From their orbiting platform they would scout landing sites, and take navigation sightings on lunar landmarks. Meanwhile, Kraft's trajectory people would track Apollo 8, amassing data on its orbit and keeping close watch for any changes. Already, from some of the unmanned Lunar Orbiter probes, scientists knew that the moon's gravitational pull was uneven, probably due to buried masses of relatively dense rock that they called mascons (short for "mass concentrations"). Mascons had caused small, unexpected shifts in the probes' orbits, and they would surely do the same thing to an Apollo spacecraft. Before anyone could commit a crew to a lunar landing mission, mascons had to be understood. ☾

But here was the makings of a dilemma. Some of Kraft's people wanted Apollo 8 to circle the moon for as long as possible to accumulate the maximum amount of data on its orbit. Frank Borman, on the other hand, didn't want to spend a minute longer in lunar orbit than necessary; the longer they stayed, the greater the chance of something going wrong. In the end, with Kraft mediating, they settled on 10 orbits—a total of 20 hours circling the moon.

The first minutes of Christmas Day would bring the moment of truth for Borman's crew: the final blast from their big rocket engine, called Transearth Injection (TEI), to free them from the bonds of lunar gravity and send them toward a small and distant earth. On December 27, in the last hour of their homeward voyage, Borman's crew would cast off the spent service module and steer their command ship through one final act of precision, reentry into the earth's atmosphere at 25,000 miles per hour. The angle of approach would be critical: too shallow and the command module would bounce off

Someone anxiously pointed out that the timing for a 10-orbit mission meant Apollo 8 would splash down in darkness.

the atmosphere like a stone skipping across the waters of a pond; too steep and it would be torn to pieces by the forces of deceleration. The zone of safety was a perilously narrow cone just 2 degrees wide. Think once more of the earth as a basketball; reentry was like trying to hit the edge of a thin sheet of cardboard balanced atop it.

These were the weighty matters under discussion that afternoon in Kraft's office—the boldest venture in the history of space exploration, the safety of the three men who would make the journey. And if Borman had shown some unease, he also displayed characteristic pragmatism and bluntness. Someone anxiously pointed out that the timing for a 10-orbit mission meant Apollo 8 would splash down in darkness.

IN 1968

Dr. Benjamin Spock, the Rev. William Sloane Coffin, and three other antiwar activists are indicted for conspiracy to aid and abet draft evasion.

Dr. Norman Shumway performs the first successful heart transplant operation in the U.S.

American navy intelligence ship *Pueblo* is seized by North Korean forces in the Sea of Japan.

Peggy Fleming wins the only U.S. gold medal of the 10th Winter Olympics in Grenoble, France.

"What the hell difference does it make?" asked Borman. "If the parachutes don't open, we're dead anyway, whether it's day or night." Some risks, Borman understood, weren't worth worrying about.

For now, the new mission was still kept secret, even from the other astronauts. There were still too many what-if's unanswered, the biggest being whether the command module would perform well on its maiden voyage, Apollo 7. But for Borman, Lovell, and Anders, there could be no waiting, and as the summer of 1968 neared its end they plunged into training for the first flight around the moon.

For Borman and his crew, Apollo 8 began on September 9, inside the command module simulator, a jumble of angular shapes that resembled high-tech sculpture. A carpeted stairway led to a small, square hatchway and an exact replica of the command module cabin in which every switch, readout, and control—from engine gauges to circuit breakers to hand controllers—was simulated to function like the real thing. Without ever leaving earth, Borman's crew practiced the complex task of piloting a moonship. They confronted untried maneuvers like the Lunar Orbit Insertion burn and the hypersonic reentry. Here is where Borman's crew earned their own self-confidence, practicing every phase of the mission from liftoff to splashdown until it became second nature.

Outside, seated at consoles, were the simulation instructors, young men with sharp minds and mischievous spirits who understood the command module as well as the astronauts did, perhaps even better. Much to the chagrin of any astronaut within earshot, they liked to say that they could take any reasonably bright individual off the street and, in a year's time, teach them to fly to the moon. It would have been an empty boast if not for the command module's onboard computer. Utterly primitive by today's standards, this mid-sixties vintage had only 33,000 words of memory, a fraction of any modern desktop model. But in that memory lay coded instructions for a flight to the moon, permanently written on bundles of magnetized wire encased in plastic, so that not even a total power failure would erase them. Although the computer was incapable of adding two numbers, when it came to getting to the moon and back it was as good as putting Isaac Newton himself to work on the problem. It could calculate the command module's position and path through space with the same equations of motion used by Newton to study the orbits of celestial bodies. And it was so essential to flying the command module that the astronauts

An astronaut in training hurtles down an escape slide. To be used if the Saturn V threatened to explode before launch, the slide led to an underground shelter, where the astronauts would wait until danger passed.

thought of it as a fourth crew member. Normally it was the computer that would align the command module's gyroscopic navigation platform with the stars, fire its rocket engine with precision, and keep its antenna aimed at the earth. Thanks to the computer, the spacecraft would almost fly itself—that is, if everything worked. The premise of simulation, borne out by hundreds of hours of manned space flights, was that not everything would work.

Again and again, under the watchful eyes of the simulation instructors, Borman's crew rode the rumbling Saturn booster off the earth. Before long the instructors were throwing in simulated malfunctions—engine failures, errant trajectories, electrical trouble—and in each case the astronauts had to respond, quickly and correctly. Sometimes, for example, when the Saturn strayed dangerously off course, the only thing to do was abort. Early in the launch, Borman would simply take hold of the abort handle in his left hand and twist it, setting off an automatic chain of events to fire a rocket, called the escape tower, perched on the spacecraft's nose. It would yank the command module away from the careening booster and then depart, leaving Borman's crew to a high-speed, roller-coaster ride through the atmosphere before parachuting into the Atlantic. But if the mishap struck later, when the escape tower was already cast off and the Saturn had propelled them to the fringes of space, Borman's crew would have to cut loose from the booster and then, racing the clock, fire up the service module's big rocket engine to kick them onto the right trajectory for a safe splashdown. Emergencies like that test the mettle of a space crew, and there were plenty of them in all phases of their simulated lunar journeys: An engine malfunction during Lunar Orbit Insertion. Communications trouble on the way home. Computer failure just before reentry. Yes, the simulation instructors would readily admit: if things went wrong it would take a highly skilled and experienced test pilot to escape disaster.

But none of them could have mastered it all, not with only four months to train, and so each man had to specialize. As commander Borman had the overall mission responsibility; he also trained to steer the command module through its fiery reentry in case the computer went out. Lovell, meanwhile, concentrated on navigation; he would use the command module's sextant to make star sightings throughout the voyage to verify the craft's trajectory. In lunar orbit, he would take bearings on lunar landmarks. And if the radio went out at any time, it would be up to Lovell to get them home. And to Bill Anders fell the role of command module systems expert. Then there was the service module, the 16-foot cylinder that would be joined to the command

Naval Academy graduate Bill Anders, seen here sailing with his family, was the sole rookie on the Apollo 8 crew. But as he noted at a pre-flight press conference, "When it comes to going to the moon, everyone's a rookie."

module's base throughout the voyage; it contained the rocket engines, propellants, oxygen, and electric-power fuel cells. All of it was Anders's responsibility, and Borman had told him in no uncertain terms, "I'm expecting you to make sure it works."

The first time Anders climbed into the simulator, sometime in 1967, he was greeted by an almost bewildering array of switches, dials, displays, and controls. Over time they grew familiar, as Anders learned to navigate the miles of wiring, the intricate array of pipes, valves, tanks, antennas, and relays. Subjected to one simulated malfunction after another, he came to play the spacecraft like a virtuoso. If plumbing to the big rocket engine became blocked he rerouted the flow of propellants. When one of the power-producing fuel cells conked out he rearranged electrical connections to keep current flowing. In time, these sessions in the simulator seemed less like spaceflight than a ride in a game box. Even when he wasn't in the simulator, Anders was learning his machine, immersing himself in a world of fuel cells and evaporators, relays and accelerometers. Seeming bits of minutia that Borman simply didn't have time for became Anders's turf. To Borman, Anders was like a precocious younger brother, and it rankled him that the junior man sometimes knew more about the hardware than he did—especially when Anders realized why one of Borman's ideas wouldn't work: "Goddammit Anders, *show* me!" ☾

In his intimacy with the machine, Anders found reassurance. He couldn't help but be impressed with the beauty of the Apollo design. Nothing exemplified this better than the SPS rocket engine, rated at 20,500 pounds of thrust, that would get Borman, Lovell, and himself into and out of lunar orbit. The beauty of the SPS was its simplicity: It was a no-frills rocket engine. There were no fuel pumps because pumps have moving parts that could break down. Instead, pressurized helium forced the propellants into the combustion chamber. It had no ignition system; none was required because the SPS burned hypergols (in this case hydrazine and nitrogen tetroxide), chemicals so reactive that they need only come in contact with each other to explode into hot gases. As long as the valves opened, that engine would *fire.* And for all its simplicity, the SPS was an extremely high performance rocket engine.

And the SPS also exemplified Apollo's other watchword, redundancy. The fuel tanks, pipes, valves, quantity sensors, everything except the combustion chamber and the huge engine nozzle itself, came in duplicate sets. If one component malfunctioned, there was a backup standing by.

But would it work? In the moment of truth would it blast them out of lunar orbit on a course for home, or would there be only silence? According to its designers, Apollo was "three nines" reliable: The odds of an astronaut's survival were .999, or only one chance in a thousand of being killed. Anders, for one, didn't believe it; the odds couldn't possibly be that good. Soon after he found out about the lunar mission he took time to ponder his chances of coming back from Apollo 8, and he made a mental tabulation of risk and reward in an effort to come to terms with what he was about to do. Frank Borman would never have conducted such an exercise. ☾

Danger—like speed—is an experience that follows a logarithmic, ever-shallowing curve. Once you've been exposed to a lot of risk (and Anders had been, as a fighter pilot) then *a hell of a lot of risk* doesn't seem like that much more. So it wasn't just his own welfare that he considered. He thought of his wife, Valerie, and of their five young children. He had barely any life insurance; no one would sell it to an astronaut about to go to the moon. Was it irresponsible to hazard leaving them alone so prematurely? Was a seat on the first flight around the moon worth that chance?

In Anders's mind there were three factors. First, and most important, it was the greatest opportunity for adventure and exploration a man could have in the twentieth century. He would see the hidden face of the moon with his own eyes, and he would be first.

Second, as an American and as a military man he felt a sense of duty, and Apollo was his country's most important mission. As a pilot, he could not deny Apollo 8's lure: the first manned test flight of the Saturn V, the first voyage of the Apollo spacecraft to lunar distance.

Finally, Apollo 8 would put his name in the history books. Somehow—Anders couldn't say exactly why—all those things combined were worth it. If he had two chances in three of coming back—and he figured the odds were probably a good bit better than that—he was ready.

●◐◐○○○○◑●

As the summer ended NASA lifted the veil of secrecy surrounding the circumlunar mission. As if to heighten the sense of urgency, the Soviets achieved a notable advance in mid-September: an unmanned spacecraft called Zond 5 flew a figure 8 loop around the moon and splashed down in the Indian Ocean, all under automatic control. This was the same type of free-return trajectory being planned for Apollo 8. Before long, many speculated, a man would follow.

Meanwhile, in Houston, the flurry of activity and anticipation surrounding Apollo 8 almost eclipsed Apollo 7, the first manned Apollo flight. And yet, Apollo 7's outcome would make or break the decision for a moon flight. Everyone knew the odds were small that a craft as complex as the command module would sail through its maiden voyage with no major problems—but that is just what happened. On October 11, Wally Schirra, Donn Eisele, and Walt Cunningham rode a Saturn 1B booster into orbit for an eleven-day shakedown cruise. Schirra's crew met every engineering objective and performed every scientific experiment in the flight plan, and then some. Only minor problems turned up in the command and service modules. The big SPS engine passed every test.

Perhaps the biggest problem of Apollo 7 was its irascible commander. Schirra's performances in Mercury and Gemini had won him admiration, but this time around he surprised observers by acting ornery and short-tempered. He caught a monster head cold early in the mission and sneezed and sniffled his way through the eleven days. Along the way, he battled with

Kraft's flight controllers, canceling a scheduled television broadcast from space, then refusing to accept changes in the flight plan. At times Schirra's bluster seemed to rub off on his crew. After one test, Eisele quipped with obvious annoyance that he wanted to talk to "the man, or whoever it was, that thought up that little gem." The man was Glynn Lunney, the flight director. By the time Schirra and his crew were ready to come home, Kraft's men were so angry that one of them offered, only half-jokingly, to bring Apollo 7 down into a typhoon—and Kraft was half-ready to have him do it. Years later Schirra would pin his bad mood on some broken promises, including a decision to launch Apollo 7 under less than ideal conditions. Other astronauts felt there was no excuse; to make such outbursts with the world listening, they would say, was plain unprofessional. Cunningham would publicly sur-

Wally Schirra *(left)* and his Apollo 7 crewmates Donn Eisele and Walt Cunningham spoof themselves before their mission with open-cockpit-era silk scarves. "Levity," Schirra once said, "is appropriate in a dangerous trade."

Apollo 7 roars heavenward against a backdrop of the giant Vehicle Assembly Building, where Saturn V moon rockets were built. On board the spacecraft in earth orbit *(inset)*, mission commander Wally Schirra shows the strain of the eleven-day earth-orbit mission.

mise that the grind of training for almost a decade had simply worn the man out. But for Chris Kraft, there was another explanation for Schirra: The Fire—which had claimed his next-door neighbor Gus Grissom—had scared the daylights out of him. ☾

IN 1968

Bobby Gentry wins best female vocalist Grammy for "Ode to Billie Joe."

Troops under Lt. William Calley murder 22 South Vietnamese civilians at My Lai.

Pierre Trudeau succeeds Lester Pearson as prime minister of Canada.

President Johnson announces diminished bombing of North Vietnam.

Discord aside, Apollo 7 was better than anyone could have hoped for. Now NASA made it official: Apollo 8 was going to the moon. Within weeks, however, the Soviets made their next move. Eighteen months after the Komarov tragedy, they lofted Soyuz 3 with Georgi Beregovoy aboard and returned him four days later. Whether another Soyuz was being readied for a circumlunar mission, no one knew. ☾

In Houston the space center geared up to support Apollo 8, and at the center of it all, Frank Borman tried to keep his mission from growing beyond bounds. If someone brought up an idea that wasn't strictly essential to the mission, Borman cut him off. It didn't matter how small the extra—even an improved type of in-flight meal—Borman's reaction was always swift and negative. Some of his reluctance was understandable, but to Jim Lovell, his behavior seemed entirely in character. In the last days of Gemini 7, Kraft wasn't the only one who tried to keep Borman's spirits up. When Borman talked about reentering early, Lovell encouraged him to keep going. When Borman worried they might have to come down in some remote stretch of ocean Lovell reassured him, "The navy would find us, Frank. They know what they're doing." Not that Borman's fretting affected his performance. But as far as Lovell could tell, behind that macho, take-charge exterior was a very apprehensive astronaut. ☾

Was Borman more uncomfortable with the lunar mission than he let on? Years later, he would say that to his mind, his worry was simple: he wanted to make sure he and his crew didn't get handed a mission they couldn't perform. He'd seen Gemini commanders accept flight plans so crowded with tests and experiments that nobody could have done them all; he was determined that wasn't going to happen to him. In Borman's mind anything that wasn't essential to the mission—circle the moon ten times and come home—was inviting trouble. They were going to the moon—that was enough!

●●◐○○○◑●

By late November, with a month to go until launch, Borman and his crew were working seven days a week, spending three or four days in the simulators at the Cape and coming back to Houston for meetings with mission planners. Now there was no time for weekend parties or dinners with friends. Sunday existed only to wade through the piles of mail on their desks. So demanding was the pace that they had less time than they wanted for exercise, and they were a somewhat tired crew by the time LBJ threw a bon voyage party for them in Washington.

As the fall of 1968 wore on, apprehension surfaced once more within NASA. Would the Soviets try to beat Apollo 8 to the moon? Due to the latitude of the Baikonur launch site in central Asia, their lunar launch window opened in early December—well before the Americans'. The weeks passed with no news from the Soviet Union. By the beginning of December, everyone at NASA realized, with great relief, that there would be no Soviet circumlunar attempt in 1968. The field was clear; now it remained to send Apollo 8 on its way.

On December 10, with only eleven days to go until launch, Borman, Lovell, and Anders flew their T-38's to the Cape for the last time. From now on, they would live there, in the Spartan crew quarters. Each man had a small bedroom, and shared a living room, a conference room with maps of the moon and the stars on the walls, and a dining room that was something like a ship's mess hall. The decor was strictly Holiday Inn, but there were compensations—not the least of which were the high-calorie meals served up by a former tugboat cook named Lew Hartzell, brimming with steak and potatoes and mile-high sandwiches. Access was highly restricted, to protect the men from last-minute illness. The sign at the entrance to the crew quarters read, "No one with a cold, or symptoms of a cold, may pass beyond this point."

In this comfortable monastery, the three men settled in for their last days on earth. Occasionally—less than they would have liked—they had a chance to go for a run or to work out in the nearby exercise room. They reviewed the flight plan, and received briefings on the readiness of their spacecraft and Saturn booster. And above all, there were the daily sessions in the simulator; at times, Apollo 8 seemed to be an exercise in switches and valves and maneuvers, not the first flight away from the earth. But on December 20, the day before launch, Borman's crew had a visitor who brought home the historic impact of what they were about to attempt.

Charles Lindbergh, one of the most enigmatic figures of the twentieth century, emerged from his retreat to visit Borman, Lovell, and Anders in the crew quarters. Forty-one years after flying solo across the Atlantic, Lindbergh appeared tall, tanned, and surprisingly fit for his sixty-six years. Accompanied by his wife, Anne, herself an accomplished pilot and author, Lindbergh arrived to have lunch with three fellow fliers about to navigate an ocean far more vast and untraveled. ☾

To most of the astronauts Lindbergh had been a boyhood hero, and Borman

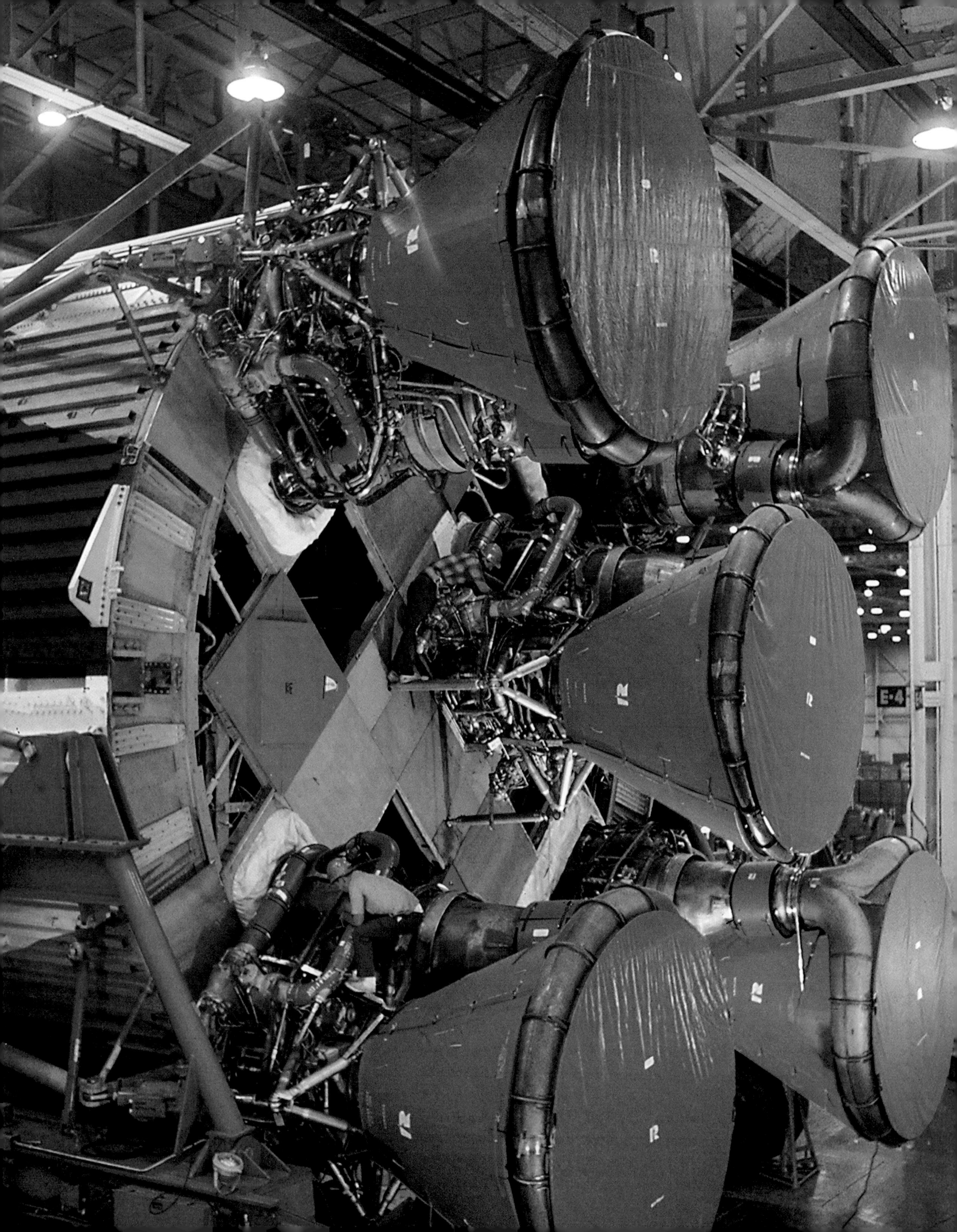
E-4

The five F-1 engines of the Saturn V's first stage dwarf workers at NASA's cavernous Michoud Assembly Facility in New Orleans. Each F-1 produced 1.5 million pounds of thrust during the opening 2½ minutes of flight.

was no exception. Now, in the quiet of the crew quarters, it was just one flier talking to other fliers. Gathered around the table with Lindbergh and his wife, Borman's crew and their backups shared questions, recollections, and humor. They were fascinated by his accounts of meetings with Robert Goddard, whose experiments with liquid-fueled rockets in the New Mexico desert had foretold the space age (and fired the imagination of a teenage Jim Lovell). Goddard had conceived of flights to the moon, Lindbergh said, but was daunted by the fantastic cost of the venture—he had mused, "it might cost a million dollars." With that, the room exploded in laughter.

The great flier asked Borman's crew about the navigation system that would take them to the moon. Then he told the astronauts how before his own trip, he and a friend had gone to the library, found a globe, and measured, with a piece of string, the distance from New York to Paris; from that he had figured out how much fuel he would need for the flight. Lindbergh asked how much fuel the Saturn V rocket would consume during its climb into space; one of the astronauts did a quick calculation: 20 tons per second. Lindbergh smiled. "In the first second of your flight tomorrow," he said, "you'll burn ten times more fuel than I did all the way to Paris."

●◐◐○○○◑◑●

Eight miles away, the Saturn V towered above Pad 39-A, looking more like a skyscraper than a rocket. Some 363 feet tall—about six stories higher than the Statue of Liberty—the Saturn was more than three times the size of the Titan missile Borman and Lovell rode in Gemini 7. It was, far and away, the most powerful thrust machine ever flown, the crowning achievement of Wernher von Braun and his team of rocket engineers at NASA's Marshall Space Flight Center in Huntsville, Alabama. Through it, the sheer difficulty of reaching the moon was made visible. It was a monument to human audacity. ☾

For now, the Saturn stood empty. But overnight, even while Borman's crew slept, technicians would ready it for departure. By morning its enormous fuel tanks would be filled with super-cold propellants, until the rocket would contain the explosive energy of an atomic bomb. This engineering masterpiece was designed to tame that energy and liberate it in a sustained, fiery release of power. Public relations people for the contractors that built the Saturn were always coming up with new analogies to convey its incredible might. Someone estimated that the thrust from the booster's first stage engines at liftoff would equal more than twice the hydroelectric power that would be obtained if all the rivers and streams in North America were chan-

Too big to be transported by road or rail, the second stage of the Saturn V traveled by ship from the North American Aviation plant in Redondo Beach, California, through the Panama Canal, and then to the Kennedy Space Center in Florida.

neled through turbines. Everything about the Saturn V was grossly out of scale with the rest of the world. For example, each of the five F-1 engines that powered the first stage had an engine bell measuring 12 feet in diameter. At liftoff, those engines would deliver a combined thrust of 7.5 million pounds—about 160 million horsepower.

Three separate stages would do the work of pushing Apollo 8 off the earth and toward the moon. The first stage alone, nearly half a football field long, would burn half a million gallons of kerosene and liquid oxygen in just two and a half minutes, cutting off at a height of 40 miles, and then it would fall away. The second stage would fire for just over six minutes until its supply of liquid hydrogen and liquid oxygen was spent. By then Borman's crew would be 120 miles up, and all that would remain would be a three-minute push from the third stage to place Apollo 8 in orbit around the earth. The third stage would do its best work three hours later, when Borman's crew would execute the Translunar Injection maneuver to break the bonds of their home planet.

At nightfall the floodlights at Pad 39-A came on, turning the Saturn into a huge glowing monument and reaching past it into the Florida night.

For now, the Saturn stood in the embrace of its steel launch tower, waiting for the launch window to the moon to open. Evening twilight revealed the moon's thin crescent glowing briefly in the west—Anders went out in the parking lot with a couple of visitors to look at it—then slipping beneath the horizon of a turning earth. At nightfall the floodlights at Pad 39-A came on, turning the Saturn into a huge glowing monument and reaching past it into the Florida night.

In the crew quarters, Frank Borman lay awake in his room, confronting his darkest fear. It wasn't blowing up on the Saturn, being stranded in lunar orbit, or being burned to a cinder in reentry. Some people had wanted him to make a tape in case he didn't come back; he'd scoffed at the idea. Sure, he had some anxiety—who wouldn't have, preparing to go to the moon for the first time? But if he thought he wasn't coming back, he wouldn't be going.

Borman was afraid of one thing: that they would be in earth orbit and some malfunction, however small, would arise, and the managers in Houston would cancel the lunar mission right there. He would be stuck with the alternate mission, ten long days in earth orbit doing nothing but keeping the spacecraft going. He hated the thought of it. And as the night dragged on, while technicians worked under the floodlights of Pad 39-A, readying his booster for an early-morning launch, Borman prayed not that he would come back from the moon, but that he would have the chance to go.

II: A HOLE IN THE STARS

SATURDAY, DECEMBER 21, 1968

Frank Borman's sleepless night came to an end at a few minutes past 2:30 A.M. when Deke Slayton came to the door to awaken him. The night was clear, Slayton told him, and the weather at liftoff—set for 7:51 A.M.—was expected to be good. Minutes later, after undergoing a final medical exam, Borman, Lovell, and Anders sat down to their last meal on earth, the traditional astronaut's breakfast of steak and eggs. Deke Slayton was there, and Al Shepard, along with backup crewmen Neil Armstrong and Buzz Aldrin, scientist-astronaut Jack Schmitt, and the man who had envisioned this mission, George Low. Years later Anders would remember the conversation as decidedly unremarkable, his mood as matter-of-fact. And if some of the support people seemed extra careful, extra serious, even a little nervous this morning, that was in striking contrast to the three of them who were about to leave the planet.

When this brief, earthly ritual ended the three men headed for the suiting room. There technicians, wearing surgical masks as part of the pre-flight health quarantine, were waiting to help them into their space suits. Someday suits like these would protect astronauts on the surface of the moon, but on this flight they were merely a precaution against a loss of cabin pressure during launch. They were also fireproof, thanks to a pristine white covering of glass-fiber Beta cloth coated with Teflon. Hidden from view were layers of insulation and pressure restraints, special joints and cables to facilitate motion when the suit was pressurized, all of which made these suits seem less like garments than wearable machines. Each of the three astronauts, clad in long johns, climbed into his modern-day suit of armor. It was all familiar from tests and practice runs, but this morning the technicians had little tension breakers—a tiny stocking hanging from a paper Christmas tree for Frank Borman, a clean white handkerchief for the pocket of Jim Lovell's space suit.

Next the men donned communications hats resembling the headgear of

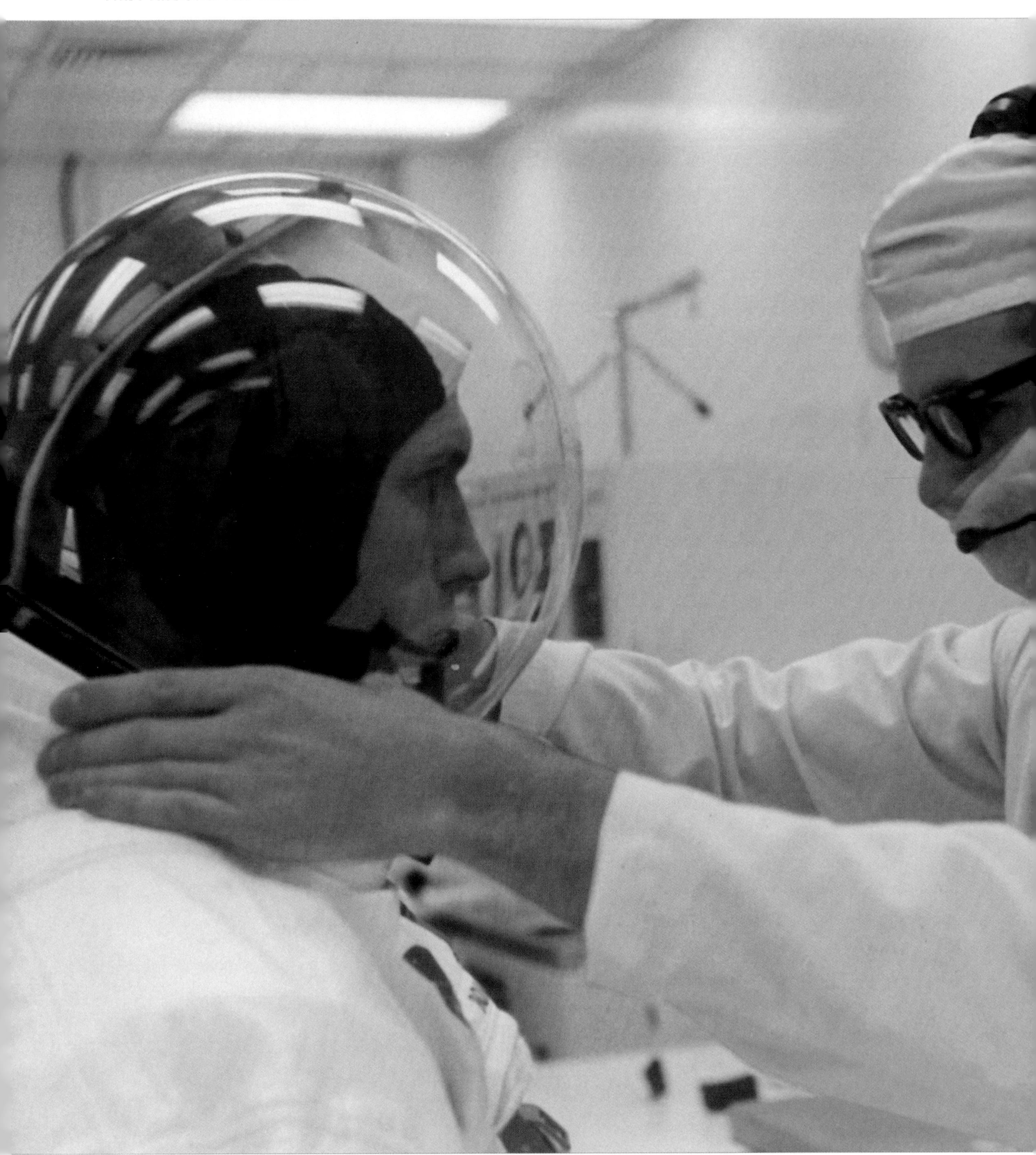

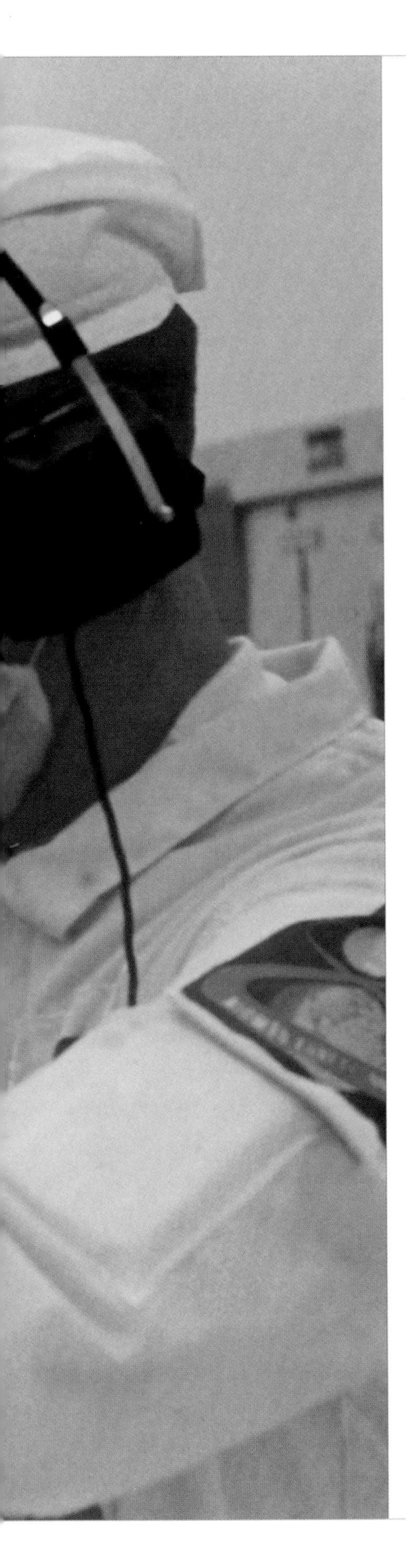

Less than four hours before launch, a technician secures Frank Borman's helmet to his space suit. Both wear radio headsets, allowing them to communicate with each other as the work progresses.

a World War I flying ace. Oxygen hoses were mated to metallic blue and red connectors on the chest. Then came black rubber pressure gloves, joined to rotating rings at the wrists of the suit. Finally, a clear bubble helmet was lowered into place and snapped onto a metal neck ring. At that moment, each man was a self-contained universe. Aside from the occasional voice of a technician in their headsets, they heard only the sound of their own breathing; they felt cool oxygen flowing past their faces. For a time they rested, letting the pure oxygen purge their bloodstreams of nitrogen. Then, at last, it was time to go. Toting portable oxygen units they headed down a long corridor to the outside with stiff-legged strides. At the entrance to the Manned Spacecraft Operations Building they were greeted by the glare of television lights and a small crowd of well-wishers. Only a hint of applause penetrated their bubble helmets as they boarded a special transfer van for the 8-mile ride to Pad 39-A.

Brilliantly lit, the Saturn V stood naked next to its launch tower, its tanks full of cryogenic propellants, spewing plumes of vapor into the predawn darkness. The sight of it filled Borman with awe. Accompanied by a suit technician, the three men entered into the service tower, and within that complex of steel girders, massive fuel pipes, and machinery, they boarded a small elevator and ascended past the Saturn's huge first stage, then the second and third stages, to the 320-foot level. There they strode across an access arm to the small White Room. First Borman, then Anders, climbed into the spacecraft, assisted by the closeout crew. Jim Lovell, meanwhile, waited alone for several minutes outside the crowded White Room, within the metal gridwork of the access arm. From this lofty perch he gazed down at the most powerful rocket in existence, and all at once the realization came to him—*My God, they're serious!*—that the very thing they had all been talking about and practicing for four months was about to happen, that NASA was really going to seal him up in that command module and fire him off to the moon. In the far distance Lovell could see the headlights of cars making their way to viewing sites: Thousands of spectators had descended on the Cape, eager to witness the departure of the first moon voyagers. For a time, sealed within his private universe, Lovell savored this communion with awe.

Then Lovell joined his crewmates, sliding on his back into the center couch. He lay still while the closeout crew hooked up oxygen hoses and communications lines. Strapped in, fully suited, he had almost no room to move;

With 3 hours, 43 minutes, and 54 seconds to go, engineers and technicians in the Saturn V Launch Control Center monitor launch preparations. Hundreds of displays relay temperatures, pressures, and other critical data from the rocket as it takes on fuel.

020
3:43:54
3:43:54
3:43:54

he was literally rubbing elbows with his crewmates. To his left, Borman lay before gauges and readouts for the Saturn V; he would keep a watchful eye on the booster's performance during the ascent into space. To Lovell's right, Anders manned the controls for the spacecraft's electrical and communications systems. From the center couch, Lovell would operate the command module's onboard computer and monitor their trajectory into space.

At last it was time to close the hatch. Rookie astronaut Fred Haise, who had been inside the spacecraft checking switch positions when Borman's crew arrived, now wriggled underneath the couches and through the open hatchway, then offered his hand in farewell. The technicians swung the massive hatch closed and locked it, sealing the three astronauts inside. It was 5:34 A.M.—T minus 2 hours, 17 minutes and counting.

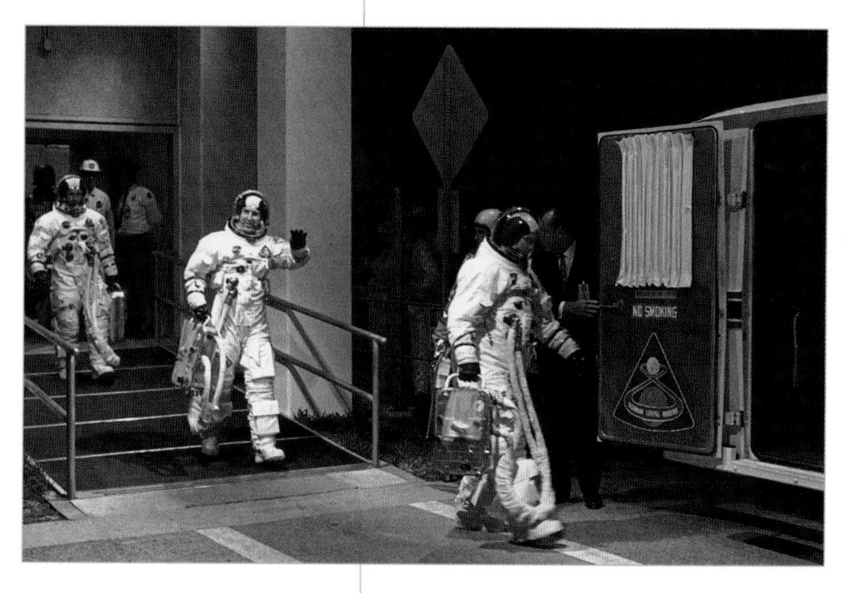

Borman, followed by Lovell and Anders, boards the transfer van outside the Manned Spacecraft Operations Building for the trip to Pad 39-A. Once settled into the spacecraft, the crew will spend the next 2¼ hours setting switches and making other preparations for the flight.

Inside Apollo 8 all was quiet. Within their helmets Borman's crew heard the voice of test conductor Dick Proffitt talking to them from the Launch Control Complex 3½ miles away, where hundreds of engineers monitored data from the spacecraft and booster. For most of the next hour they followed his instructions, setting switches, as part of the complex process of readying the command module for flight. Over in the right couch Anders was amazed at how calm he felt. It was just like a simulation; he was almost bored.

Around Pad 39-A the glare of floodlights yielded to a clear dawn. During a break in the switch-settings, out of the corner of his eye Anders noticed something moving; he looked over at a window in the protective heat shield, called the boost protective cover, that covered the command module. A hornet buzzed around and landed, worked for a short while, then flew off and returned. She's building a nest, Anders thought, and did she pick the wrong place to build it!

Far below, the great rocket was filling with fuel. If disaster threatened now, while the Saturn was still earthbound, the men would scurry out of their craft and into a small gondola attached to a slide wire stretching from the launch tower to a concrete bunker. And if there wasn't time for that, Borman could, with the twist of a handle, fire the escape rocket poised above the command module's nose, whisking them up and away from the disaster. But an abort from the pad involved considerable risk of injury, and that was something everyone hoped would not be necessary. Meanwhile, 1,000 yards from the pad, armored tanks stood at the ready in case the men had to be rescued. Inside Apollo 8, Borman asked Dick Proffitt, "How is the booster doing?" Proffitt assured him that all was well.

Up to now Borman, Lovell, and Anders had known, in the back of their minds, that there was the possibility that a malfunction would turn this countdown into just another practice run and they would have to climb out and try again another day. But now, as the count reached T minus 15 minutes, there seemed no doubt: they were really going. In the control center the flight surgeons monitoring telemetry from the spacecraft saw Borman's heart rate start to climb. With 7 minutes left Dick Proffitt took his final status check, and in the middle of his lengthy poll he called out, "Spacecraft" and Borman, Lovell, and Anders answered together, in full voice: "Go!"

Now the pace quickened. With just 5 minutes to go the White Room and its access arm swung away. At 3 minutes, 7 seconds, the launch pad's automatic sequencer took over, monitoring the last influx of propellants and controlling the final events before liftoff. By T minus 60 seconds all three stages were fully pressurized; 10 seconds after that the booster went on its own power. With 45 seconds to go Borman confirmed the last switch settings to ready the command module for launch. Thirty seconds, now 20. And now test conductor Proffitt began to count: "*Nine, eight, seven . . .*" For a moment the men heard faint sounds of fuel pouring through manifolds to the five huge F-1 engines. ☾

"*. . . Ignition.*"

Suddenly the base of the Saturn spawned a cauldron of smoke and flame that gave way to a river of golden-white fire, spilling out from both sides of the launch platform. Yellow smoke billowed into the chill morning. For long seconds the behemoth strained against a set of enormous hold-down clamps, while the first-stage engines built up to full power. Up in the command module there was no noise or vibration, but somehow Borman and his crew could sense the growing power far beneath them. At T minus 3 seconds there came a distant rumbling, like thunder on the horizon, that swelled into a roar. Finally, in the midst of the heightening commotion came a sudden, mild jolt, and Borman's crew heard Proffitt cry, "*Liftoff!*"

Borman glanced at the mission clock on the instrument panel. "Liftoff," he called, his voice charged with adrenaline. "The clock is running." The Saturn ascended, seemingly wracked by spasms of uncertainty, steering nervously past the launch tower, its engines correcting and recorrecting in quick, spasmodic jerks. Up in the command module these corrections translated into sudden, jarring motions that threw the men from side to side against their harnesses. No simulation had even hinted at the violence of this ride. In the post-flight debriefing Anders would say only that he was "impressed" by the Saturn's "very positive control," but in reality, he felt as if he were helpless prey in the mouth of a giant, angry dog. After all those simulations—if the first 10 *seconds* were this different, what would the rest of the flight be like?

Long seconds passed in thunder while the rocket climbed its own length and still higher. Borman's crew barely heard Proffitt shout, "Tower clear!" The danger of collision with the launch tower past, they kept climbing. Now the rocket turned and headed onto its programmed flight path. "Roll and pitch program!" called Borman, his voice shaking with the vibrations of the ride. Meanwhile, the Cape launch center yielded command to mission control in Houston, where Mike Collins was serving as Capcom. If the booster suddenly went berserk and mission control ordered an abort, it would be Collins who would relay the command, but inside Apollo 8 the Saturn's roar was so loud that Borman's crew would not have heard him. And they could no longer hear each other; they were no longer a crew but three passengers riding in a fury of sound.

Just 40 seconds after liftoff the Saturn went supersonic and the ride smoothed out. Now there was quiet again. In Apollo 8's left seat, Borman kept a watchful eye on the trajectory readouts. If at any point the Saturn

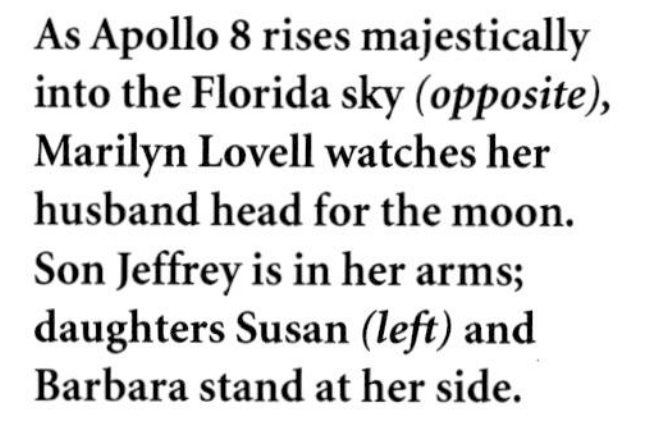
As Apollo 8 rises majestically into the Florida sky *(opposite)*, Marilyn Lovell watches her husband head for the moon. Son Jeffrey is in her arms; daughters Susan *(left)* and Barbara stand at her side.

should turn angry he would be able to whisk them away from it by twisting the abort handle, setting off the escape rocket, but the beast was behaving itself beautifully. And from Houston, Mike Collins's message of reassurance came through loud and clear: "Apollo 8, you're looking good."

The Saturn tore through the atmosphere on a great bonfire column of light hundreds of feet long. Under the commands of its own gyroscopic brain the booster arced slowly over until it was almost horizontal, following the curve of the earth, picking up speed and receding into a deep blue sky. As the rocket penetrated the rarefied upper atmosphere the exhaust fanned out into a broad plume of golden flame.

This NASA illustration shows the separation of the now unneeded launch escape system and its attached boost protective cover, following the ignition of the Saturn V's second stage.

In the command module Borman and his crew scanned the instruments and felt the mounting force of acceleration as the massive load of fuel in the first stage was consumed. Soon their chests begin to flatten. The g-meter registered three times the force of gravity, now four, and was still climbing. Their arms were leaden. Then, just as the g-meter hit 4½, the forces of acceleration abruptly vanished as the first stage shut down on schedule. At that moment, the men could have been sitting on a catapult. They flew forward against their straps with tremendous momentum—Anders was sure he would go right through the instrument panel—but their harnesses held them firmly. Borman felt the sudden jarring and thought warily of the stress placed on the booster.

Suddenly, right on schedule, there was a muffled bang as jets of smoke and flame heralded the departure of the now unneeded emergency escape rocket, taking the boost protective cover along with it. Daylight streamed into the cabin as the command module's windows were uncovered. For a stolen moment rookie Anders glimpsed a view available only to the space traveler, a vivid bright arc of ocean and clouds against a darkening sky.

Five minutes into the flight now; it was amazingly quiet. The g-forces had lessened. And there were more welcome words from Mike Collins: "Apollo 8, your trajectory and guidance are Go."

"Thank you, Michael," said Borman, sounding pleased.

Apollo 8 sped out of the last fringes of the atmosphere, picking up speed. Now 7 minutes. At about 8 minutes a rapid vibration set in, the same kind that had rattled the previous Saturn V to the point of malfunction. Thankfully, it did not build beyond a mild shaking, but Borman was relieved when the second stage shut down and fell earthward, its work done. At 8 minutes, 45 seconds the third stage kicked in with a mild jolt and chugged along, getting up the last bit of velocity until, 11½ minutes after liftoff, it too fell silent. Apollo 8 was in orbit.

●◐◐○○○◑◑

A hundred and fifteen miles up, circling the earth at a speed of more than 17,400 miles per hour, Apollo 8 moved in that exquisite balance between gravity and momentum called orbit. Had they done nothing, Borman, Lovell, and Anders would have remained there for days, slowed only by friction with the scant upper atmosphere. The bold difference in this flight was set for just 2½ hours from now, late in their second orbit. At that time they would reignite the third-stage engine for a little over five minutes and tip the balance between gravity and momentum enough for Apollo 8 to leave earth orbit and reach the moon's gravitational sphere of influence. And that was what Borman lay awake worrying about.

"I don't want to see you looking out the window!" Borman's voice. They had a timeline to stick to; Anders knew that as well as his commander. But to be in orbit for the first time and not look outside! That was easy for Borman to say; he and Lovell had been here before. Once or twice, when Borman wasn't watching, Anders stole glimpses of the earth, a magnificent panorama of color and bright clarity that filled his window. Brilliant white plumes and swirls of cloud crisscrossed land and ocean. Entire continents swept past in minutes. Somewhere over the midnight earth—was it New Zealand?—lightning glowed in the clouds far below like flash bulbs going off under wads of cotton. And when they came over the coast of California he spotted San Diego, the scene of his childhood explorations of hills and rabbit trails. He wanted to linger here, taking in the ever-changing beauty of his home. Anders wished they weren't going to the moon—not yet.

But the time flew by. There was a small mishap when Lovell, reaching under one of the couches to adjust a valve, accidentally inflated the life vest attached to his space suit; he would always remember the disgusted look on Borman's face. But aside from that, everything went like clockwork. What so

Apollo 8's third-stage engine propels the astronauts out of earth orbit toward the moon. The maneuver, captured in this view by a telescopic camera in Hawaii, was visible to the unaided eye as a point of light moving rapidly across the night sky.

many had doubted, including Borman, was actually happening: Apollo 8 was checking out perfectly. And at last came the word the three astronauts had been waiting for, and ironically, it came from the man originally slated for Apollo 8's center seat, Mike Collins. One of the most momentous directives ever given, it was spoken with remarkable calm and in the coded language of space flight: "Apollo 8, you are Go for TLI."

As Apollo 8 drifted through darkness over the Pacific the last minutes ticked by until the scheduled ignition of the third stage. If everything went as planned, Borman and his crew would be mere passengers while the computer did the work. With 10 seconds to go until ignition the computer gave a coded message to the astronauts, a flashing number 99. Translated, it said, "Are you sure you want to do this?" Lovell answered by pushing the button marked "PROCEED," and moments later, at Mission Elapsed Time 2 hours, 47 minutes, and 37 seconds, the third-stage rocket came to life with a long, gentle push. This time, the ride really did feel like the simulator; the men sank into their couches with barely more than the force of normal gravity.

Immediately, they sensed the rocket veering to one side as it headed out of earth orbit and onto a course for the moon. Trajectory specialists in Houston hawkeyed the moonship's path and sent word, via Mike Collins: "You're looking good here, right down the old center line." Borman kept his eye on the attitude indicator, ready to take over steering if the booster's automatic system failed. Anders monitored the pressures and temperatures in the fuel tanks. And Lovell called out their ever increasing speed from the computer readout. The numbers galloped upward: 30,000 feet per second . . . now 33,000 . . . and finally, 35,532 feet per second, some 24,226 miles per hour, the speed necessary to reach the moon on a free-return path. At that instant, 5 minutes and 18 seconds after ignition, the computer shut down the engine automatically. Apollo 8 was on its way to the moon. ☾

From mission control Collins had good news for the departing moon voyagers: "We have a whole room full of people that say you look good." And one of those people was Chris Kraft, sitting in the back row of the control room. Kraft rarely came on the radio during a mission, and Borman was surprised to hear his exultant sendoff: "You're on your way—you're really on your way now!"

Still, inside the command module there was nothing to convey this departure to the senses, no sensation of speed whatsoever, just numbers on the computer. That changed dramatically when Borman cut loose from the spent third-stage booster, pulled away with a burst from the service module's small maneuvering thrusters, and spun Apollo 8 around. At first, the sight of the third stage itself—a hulking cylinder aglow in the unfiltered sunlight of space—caught their attention. But then, as the spacecraft turned, Borman's crew could see the place they left behind, not a landscape but a *planet,* a luminous sphere whose roundness was apparent to the eye. Apollo 8 was departing at such fantastic speed that the men could see their world receding from them almost as they watched. Already the entire globe fit neatly within the round window of the command module's side hatch.

Whatever names humans gave their earth, it deserved to be called the Blue Planet, for its dominant aspect was the vivid, deep blue of oceans. In striking contrast were the clouds, brilliant white flecks and streamers that embraced the globe, swirling along coastlines and across oceans. Where land masses peeked through, the vivid oranges and tans of the deserts were easy to spot. More elusive were the jungles and temperate zones; because their verdant hues did not easily penetrate the atmosphere, they showed up as a bluish gray with only a hint of green. And everywhere, beyond the planet's bright, curved edge, a blackness so deep as to be unimaginable.

Right now, though, it wasn't time to look at the earth; Borman was more concerned about the cast-off third stage. As the flight plan called for, Borman had pulled up within a few dozen yards of the booster, to demonstrate the maneuvers that future crews would use to extract a lunar module from its berth. But Borman, anxious to save fuel and to avoid any maneuvers that would affect their trajectory, did not want to prolong the exercise. Furthermore, he knew the booster was scheduled to blow off its excess fuel sometime in the near future—and when that happened it would be better not to be anywhere nearby. All he wanted to do was get away from it. After conferring with mission control, Borman pulsed the hand controller and fired the maneuvering thrusters to pull away.

But the third stage seemed to be following them. Already it was spewing fans of brilliant ice particles into space, reminding Borman of a huge lawn sprinkler. For the better part of an hour Borman made anxious queries to Houston on how to get away without disturbing the free-return trajectory. Lovell's attempts to realign the command module's navigation platform were to no avail; the sky was full of "false stars" from the booster, and it was impossible to find any real ones. And right now the best landmark in this dark, sunlit ocean—the earth—was out of view. When Collins in mission control outlined a small evasive maneuver, Borman replied, "Okay, as soon as we find the earth, we'll do it." In mission control Borman's words triggered brief, amazed laughter.

Finally, after more than an hour, Anders saw the world drift into his right-hand window, and after more deliberations with Houston Borman fired the maneuvering thrusters once more. Slowly, the third stage dwindled until it was just a bright star, and Apollo 8 was alone in the translunar void.

●◐◐○○○◑◑●

"Would you pass me the flight plan, Bill?"

Anders reached for the three-ring book floating in the air next to him. He gave the book a gentle push and it drifted across the cabin into Lovell's open hand. Apollo 8 was coasting moonward like a baseball fleeing the strike of the bat, and everything inside it—including the three astronauts—was weightless. There was a moment, back in earth orbit when he unbuckled his harness and his body hung literally in midair, suspended above his couch. For years he'd heard other astronauts talk about zero g, but there was no way to anticipate it. No simulation could have prepared him.

In the first few hours of the mission there was no time to enjoy this

A space-age derelict, the Saturn's third stage looms before the command module's windows after separating from the spacecraft. Droplets of escaped fuel, which instantly froze in the cold of space, glint in the brilliant sunlight.

CRUISING TO THE MOON

Apollo crews sat side-by-side in the command module *(blue)* of the spacecraft that ferried them to lunar orbit and back. Because Apollo 8 did not land on the moon, the command module had no docking probe in the nose, which, on later missions, linked the lunar lander to the command module. Behind the crew, the service module *(green)* contains fuel and oxygen tanks, fuel cells to provide electricity, and a high-gain antenna for earth communications. At the rear of the package is the nozzle of the rocket engine used to place the spacecraft in lunar orbit and to send the crew home.

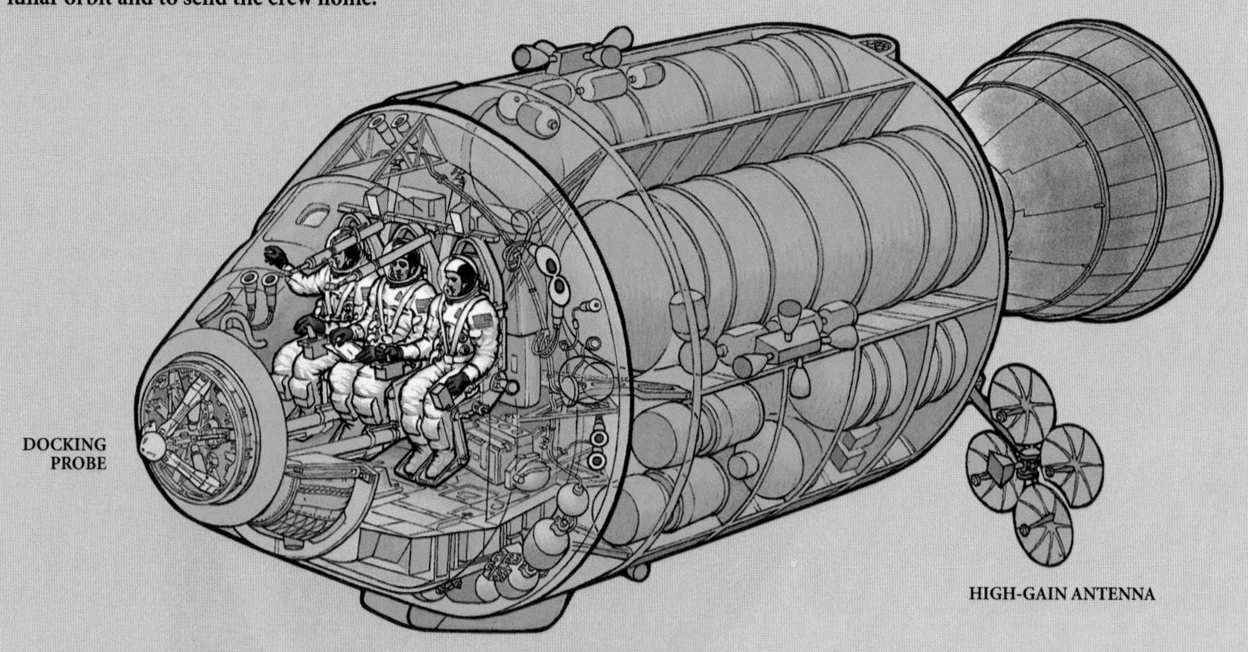

strange new world. But sometime after TLI, well on the way to the moon, Anders climbed out of his space suit and found a freedom unlike any he had ever experienced. Wearing only a pair of Beta-cloth coveralls over his long johns, he floated unencumbered. Suddenly the cramped cabin seemed to grow roomy. With a push of a fingertip against his couch he propelled himself slowly past the instrument panel into the open area they called the lower equipment bay, which housed storage lockers and Lovell's navigation telescopes. There he found enough room to stretch out, or to hang inverted, with his feet up by the top hatch and his head pointing at the floor. "Up" and "down" were whatever he wanted them to be. He could even float underneath the seats, among coolant pipes and storage compartments. The command module seemed to have suddenly doubled in size. And in zero g it became a wonderland. Water formed perfect shimmering, dancing spheres. Cameras twirled and tumbled with the touch of a fingertip, or lingered in midair when not in use. This was a world of action and reaction, a three-dimensional ice rink. There wasn't room for gymnastics, but in the lower equipment bay Anders had enough space to tuck his body into a ball; a nudge against the wall set him tumbling, like an acrobat magically suspended at the top of his arc. It would have been great fun but for one thing: it was making him ill. Suddenly, in the midst of his acrobatics, Anders felt a wave of nausea come over him.

For years the NASA doctors had worried about motion sickness in space, fearing that zero g would confuse the inner ear, which gives the body its sense of up and down. But no astronaut had ever returned from orbit with anything but glowing enthusiasm for weightlessness. Borman and Lovell, for example, had spent two weeks in free fall with no ill effects. But the command module had something that the phone-booth-sized Gemini didn't—room to move. The doctors feared that simply by floating around, an astronaut would push his vestibular system over the edge. ☾

The doctors feared that simply by floating around, an astronaut would push his vestibular system over the edge.

Still, Schirra's crew came back from Apollo 7 with no complaints. Maybe it came down to the individual, and there was no way of knowing who would be sick and who wouldn't. The only thing Anders knew was that he needed to be still for a while; soon he felt better. Several hours later, Anders not only wasn't sick, he was so comfortable that he felt as if he had always been weightless.

But one aspect of weightlessness was so unpleasant that even the thrill of exploration didn't make up for it. If this marvel of engineering called Apollo had one major design flaw, it was the "Waste Management System," perhaps the most euphemistic use of English ever recorded. For urine collection there was a hose with a condom-like fitting at one end which led, by way of a valve, to a vent on the side of the spacecraft. On paper, at least, it seemed like a reasonable if low-tech way to handle urinating in zero g, assuming you got over your anxiety about connecting your private parts to the vacuum of space. You roll on the condom, open the valve, and it all goes into the void where it freezes into droplets of ice that are iridescent in the sunlight. One astronaut answered the question "What's the most beautiful sight you saw in space?" with "Urine dump at sunset."

In reality, using the urine collector didn't work out so well. For one thing, it could be painful. If you opened the valve too soon, some part of the mechanism was liable to poke into the end of your penis, which tended to prevent you from urinating. And at that point, as if to confirm your worst fears, the suction began to pull you in. Now you were being jabbed and pulled at the same time, so you shut off the valve, and as the mechanism resealed itself it caught a little piece of you in it. It only took one episode like that to convince you not to let it happen again. Next time you had a strategy: start flowing a split-second before you turn on the valve. But once you began to urinate the condom popped off and out came a flurry of little golden droplets at play in the wonderland, floating around and making your misfortune everyone's misfortune. And in no time the whole device reeked; it was an affront to the senses just sitting there.

One of the Apollo astronauts said the smell was so bad it woke him up out of a deep sleep.

Anders got used to the urine collector, though, and he got used to mopping up afterwards. But there was no getting used to the other part of the Waste Management System. Tucked away in a storage locker was a supply of special plastic bags, each of which resembled a top hat with an adhesive coating on the brim. Each bag had a kind of finger-shaped pocket built into the side of it. When the call came you had to flypaper this thing to your rear end, and then you were supposed to reach in there with your finger—after all, nothing *falls*—and suddenly you were wishing you'd never left home. And after you had it in the bag, so to speak, you had one last, delightful task: Break open a capsule of blue germicide, seal it up in the bag, and *knead the contents* to make sure they were fully mixed. At best, the whole operation was an ordeal. In the confined space of the command module, your crewmates suffered too. One of the Apollo 7 astronauts said the smell was so bad it woke him up out of a deep sleep. When Schirra's crew came back they wrote a memo about it: "Get naked, allow an hour, have plenty of tissues handy . . ." Anders saw the memo and heard the stories, and before the mission he decided he was going to do everything in his power to avoid it. The food on Apollo 8 was specially formulated to produce as little residue as possible, but

Anders wasn't taking any chances. He started his own low-residue diet a few days before launch. Six days was a long time, but he was determined. He'd go all the way to the moon and back on Lomotil, if he had to.

The hours passed in steady activity. By noon, Houston time, some five hours into the flight, all three men had doffed their bulky space suits and stowed them underneath the couches. Around 1 P.M. Lovell began taking star sightings for navigation. And there were more tasks into the afternoon—replace an air-filtration canister, look after a battery, service a power-producing fuel cell. At 6 P.M. the astronauts made the first, brief firing of the service module's SPS engine. Though it lasted only two seconds—the engine slammed Borman's crew back into their couches, then released them—it was enough to correct Apollo 8's path after Borman's earlier maneuvers to get away from the third stage. Just as important for engineers in Houston, the firing gave a crucial look at the engine's performance in space. The SPS passed its first brief test with flying colors.

By then, more than eleven hours had passed since launch, and aboard Apollo 8 it was getting to be a long day. It was time for Borman to get some sleep. The flight plan called for at least one man to be awake at all times to keep an eye on the spacecraft and maintain contact with Houston. For now, Lovell and Anders would stand watch while Borman slept. As he floated into the sleeping bag attached to the underside of his couch, Borman was more than ready for a rest, but his mind would not cooperate. It wasn't easy to just turn off the mission and fall asleep. Two hours later, still keyed up, he called down to Houston and got permission to dig out the medical kit and take a Seconal. He hated pills, but it was more important that he rest.

9 P.M., HOUSTON TIME
14 HOURS MISSION ELAPSED TIME
77,000 MILES OUT

Already, a bit more than eleven hours after Translunar Injection, Apollo 8 was a third of the way to the moon. But even as Borman, Lovell, and Anders sped moonward, the earth tried to pull them back, slowing their flight. It was as if the moonship were coasting up a hill, one that became less and less steep as it went along. About two days from now, on the afternoon of December 23, Apollo 8 would reach the gentle crest of that hill, the place where the earth's gravitational influence gave way to the moon's. From then on it would begin falling toward its destination.

For now, though, there was no sense of speed—or for that matter, any normal sense of time. To Borman's crew time was told by the mission clock on the instrument panel. Their wristwatches were still set to Houston time, but all vestiges of day and night had vanished. They moved in the unrelenting glare of an unfiltered sun in a black sky. To keep the sun's heat and the frigid cold of space evenly distributed on the hull, Borman had set the spacecraft rotating slowly on its axis, making one full turn in an hour. The astronauts nicknamed this the "barbecue mode." Every once in a while, as the craft turned, the men caught sight of the earth. With each passing hour it dwindled. They couldn't see the change as they watched, but if they turned away from it and looked again later, they noticed that it was a little smaller and more distant. Presently it was about the size of a baseball held at arm's length.

At least now Anders could tell what he was looking at. It hadn't been so easy a few hours ago, when the planet still loomed big and bright. Back then, fresh out of his space suit, Anders had his first chance to savor the view, and to his embarrassment, he couldn't tell what in the world he was looking at. As a kid, he'd prided himself on being something of a geography expert. He knew every country and major city on the schoolroom globe. But the real earth wasn't like the schoolroom globe. It had *clouds,* for one thing, and the countries weren't different colors—no small detail! A large land mass peeked from beneath the clouds. Was that Africa? He could just about make out the bulge of the Sahara, the point of Cape Town. Wait a minute, he thought, that doesn't make any sense. If that's Africa, then where is South America? And what is *that* thing out in the middle of the Pacific?

It was time to go back to basics. There was a big white patch near the edge; it had to be ice or clouds. Anders thought it looked more like ice. Winter in the northern hemisphere, and there's a big patch of ice in sunlight. That had to be Antarctica. But how could that be right when it was at the top? Then he realized: Because *we're* upside down. Anders turned himself until the white patch was at the bottom, and suddenly everything fell into place. There was the great south polar ice cap. Above it, not the Horn of Africa but the coast of Chile, and all of South America, from rain forest to coastal desert, wrapped in clouds. North America hid beneath a winter overcast, but he could spot the Florida peninsula under clear skies. In the Maritimes a cyclone's brilliant white pinwheel sprawled across the Atlantic. And in the Caribbean, the shallow waters of the Bahamas gleamed like a turquoise jewel lit from within; he would be able to spot it all the way out to the moon. He wished he could spin the planet around on its axis and see the rest of it, but it turned at its own pace, just as it always had.

Most of the Western Hemisphere can be seen in this view of earth taken early in the outbound leg of the Apollo 8 mission. Much of the United States is hidden by clouds, but the shape of South America is clearly visible in the upper center of the photograph.

As their journey progressed, the astronauts saw their home planet dwindle steadily in size. From lunar orbit, earth appears smaller than a golf ball held at arm's length.

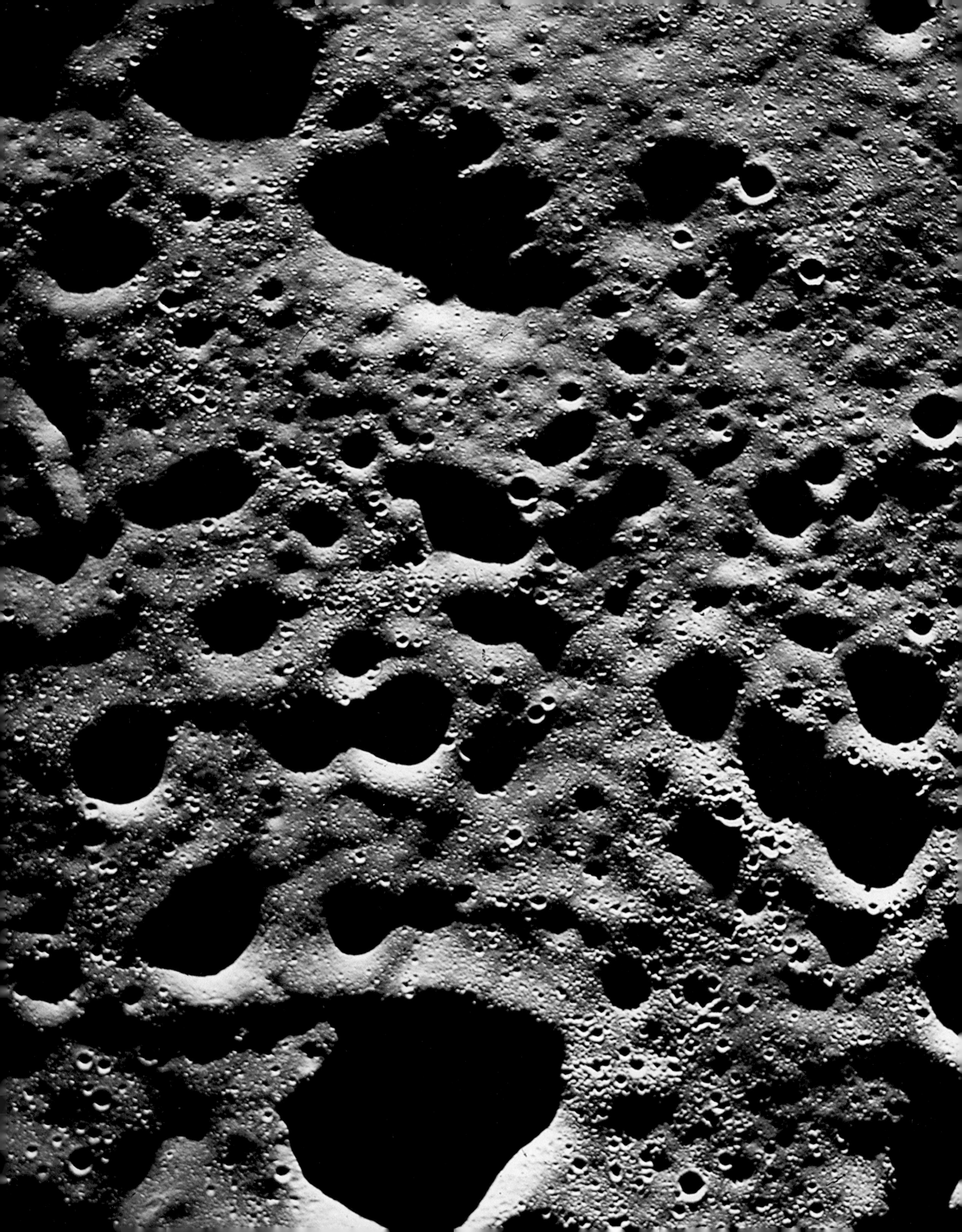

SUNDAY, DECEMBER 22
1 A.M., HOUSTON TIME
18 HOURS MISSION ELAPSED TIME

Borman awoke after about five hours of fitful sleep. He didn't feel well. He told Lovell and Anders he had a headache and took a couple of aspirin, then he just floated in his couch and watched the instrument panel. A few minutes went by, and the next thing Lovell and Anders knew he was retching. Anders handed him a plastic bag, and Borman went down into the lower equipment bay and threw up. Lovell flashed Anders a knowing look: Borman must be motion sick.

The episode was beginning to make Anders feel a bit sick himself, when suddenly he spotted a greenish sphere, about the size of a tennis ball, ascending slowly out of the equipment bay in a flurry of tiny bits and globules. The sight of it made him want to gag. But when it drifted closer he noticed that the blob was shimmering and pulsating in three directions at once in some kind of complex fluid vibration made possible in zero gravity. At that moment the scientist in him took over. He was about to go for a camera when suddenly the blob split in two. As if to affirm Newton's laws of motion, the twin spawns headed away from each other in exactly opposite directions, giving Anders a flash of recognition: *Conservation of momentum!* One scooted away whence it had come and the other headed right for Lovell. The man was cornered. The blob hit him on the chest and then, overcome by the forces of surface tension, spread out on his coveralls as flat as a fried egg.

By now a horrible stench had rolled out of the equipment bay. Anders left Lovell to his predicament and reached for an oxygen bottle on the wall of the cabin, meant to be used in case there was a fire. *To hell with that,* Anders thought; he slapped the mask on his face and turned it on full. Meanwhile, Borman's troubles weren't over; now he was struck with diarrhea. What a mess—Lovell and Anders had to help chase down stray bits of vomit and feces with paper towels. In a strange, detached way, Anders was reminded of hunting butterflies. ☾

Anders floated in his sleeping bag, eyes closed, trying to relax. He was tired. Before now he would have thought that sleeping on a bed of air would have been the best imaginable, but it wasn't working out that way. Like Borman, he found it difficult to take his mind off the flight. He missed the pressure of a pillow against his head and the security of a blanket drawn up around him. The bag was clearly designed for someone as big as Lovell; Anders was bouncing around inside it like a lone pea in a pod. He steadied himself and

IN 1968

In the Heat of the Night **wins the Oscar for best picture; George Kennedy is judged best supporting actor for** ***Cool Hand Luke.***

Student strikes over Columbia University's Pentagon connections close down the school.

Pan American and Aeroflot begin the first direct commercial air service between the U.S. and USSR.

The Berrigan brothers and seven others burn stolen draft records in Catonsville, Maryland.

tried to lay still. Every now and then a residual bit of vomit drifted by and he cowered. And he noticed that his body had not fully adapted to zero g. His heart, accustomed to a lifetime of fighting gravity, was suddenly too strong. As a result he heard a muffled, incessant *boom-boom-boom*—his own blood pulsing in his ears. And each time he was about to fall asleep, he was startled awake by the sensation of falling, just like the feeling he'd had in dreams on earth. His central nervous system seemed to be broadcasting an alarm, telling him what he already knew, that he was in an environment unlike anything he had ever known.

8 A.M., HOUSTON TIME
1 DAY, 1 HOUR MISSION ELAPSED TIME
120,000 MILES OUT

When Anders awoke—he did manage a few hours of fitful sleep—he found Borman much recovered, blaming his illness on a twenty-four-hour virus. Anders suggested he reveal the incident to mission control, but Borman replied, "I'll be damned if I'm going to tell the whole world I had the flu." Anders finally convinced his commander to put a short summary on tape; the message could be sent to earth via a special telemetry channel. That way, no one would hear except the few managers who listened to the tape.

"I'll go ahead and dump this," Anders radioed Houston. He couldn't come out and say what was on the tape, but he had to find some way of getting them to listen to it soon. He suggested, "You might want to listen to it in real time, to evaluate the voice." Then there was nothing to do but wait for a response. Hours went by with no word from earth about the message. Eventually Anders found out why: it was hours before the flight controllers even had a chance to hear it. (So much for putting messages on tape, Anders thought.) Finally, at 1 day, 4 hours Mission Elapsed Time, Mike Collins called up on a special frequency:

"Apollo 8, this is Houston. We're on private loop right now, and we'd like to get some amplifying details on your medical problems. Could you go back to the beginning . . ."

"Mike, this is Frank. I'm feeling a lot better now. I think I had a case of the twenty-four-hour flu. . . . " Borman recapped the whole episode for Collins, and, to Borman's surprise, Chuck Berry came on the line to talk to him directly; that almost never happened. Unbeknownst to Borman, the episode had triggered serious talk of canceling the mission. Berry worried that Borman had a virus, and that it was only a matter of time before his crewmates caught it. But Borman told the earthbound flight surgeon that he

felt much better, and that neither Lovell nor Anders had been affected. "We're all fine," he said. ☾

Minutes later, in consultation with Berry and other managers, Apollo program director Sam Phillips decided to let the flight continue. Even if Phillips had decided otherwise, Apollo 8 was too far away for the SPS to manage a swift about-face maneuver. Borman, Lovell, and Anders were committed now: even if they had to abort their mission, they were going to go around the moon.

2:01 P.M., HOUSTON TIME
1 DAY, 7 HOURS, 10 MINUTES MISSION ELAPSED TIME
140,000 MILES OUT

"Are you receiving television now?"

"Apollo 8, Houston. We just got it."

"You are getting it?"

"Okay, Apollo 8. We have a good picture."

Frank Borman had fought to keep the small television camera off Apollo 8—he wanted neither its added weight nor the demands on his time—but he had lost that battle. And just now, as he conversed with Capcom Ken Mattingly, the big screen at the front of mission control flickered to life, and there was Jim Lovell, apparently upside down, at the navigation station, making star sightings and making lunch. The picture was fuzzy, and it was in stark black and white—but to many who saw the brief telecast on the afternoon of December 22, it seemed a small miracle: a live glimpse inside a moonbound spaceship. For this first telecast from Apollo 8, Anders handled the camera while Borman narrated. ☾

"Jim, what are you doing here? Jim is fixing dessert. He's making up a bag of chocolate pudding. You can see it come floating by." The narrow bag tumbled in the middle of the cabin. To the astronauts it was natural to see such things. But Anders had to wonder—how must it look to those who were watching? There were people who didn't believe Apollo 8 was real to begin with, that it was all a hoax perpetrated by the government. And it crossed Anders's mind that live television of three men floating inside a spaceship was as close to proof as they might get.

The irony was that apart from the weightlessness—and the view—it was pretty hard to convince *himself* that this was really happening. For one thing they didn't seem to be going anywhere. That was the paradox of this flight: faster and farther than anyone had ever gone, with no sense of motion. They might as well have been in the simulator, and yet they were more than

During Apollo 8's first television show, beamed to earth from halfway to the moon, Bill Anders demonstrates the delights of weightlessness by twirling his toothbrush in midair. Lovell ended the show with a happy birthday wish to his 73-year-old mother.

140,000 miles from home. They could *hear* the distance in every conversation with mission control —the pause between question and answer, while radio signals spanned enormous distance.

"This transmission is coming to you approximately halfway between the moon and the earth," Borman continued. "We have about less than forty hours to go to the moon. . . . I certainly wish we could show you the earth. Very, very beautiful." Unfortunately, that attempt failed. When they turned the camera on the brilliant blue and white planet, Houston reported only an unintelligible blob of light. ☾

Now Borman trained the camera on Anders. "You can see that he has his toothbrush here. He's been brushing regularly." Anders twirled the toothbrush and it spun magically in the air in front of him until he snatched it back. "It looks like he plays for the Astros, the way he tries to catch those things." The Astros hadn't been doing well.

"Hey, Frank, how about a couple of words on your health for the wide world." Deke Slayton's voice. Slayton rarely came on the air, but the illness had caused a big flap in Houston and had made it into the news, and now was the chance to show the world that the crisis had passed. Inside Apollo 8, Borman smiled and waved at the camera. "We all feel fine," he said.

Only fourteen minutes after the telecast began, it was time to end it. Borman would now set the spacecraft back on its slow, thermal-control spin, and that meant the high-gain antenna could no longer track the earth. Before they signed off Lovell ducked into view. He wanted his mother to get a good

look at him; today was her seventy-third birthday. Lovell looked at the camera and grinned. "Happy birthday, Mother," he said.

MONDAY, DECEMBER 23
6 A.M., HOUSTON TIME
1 DAY, 23 HOURS MISSION ELAPSED TIME
187,000 MILES OUT

Bill Anders was alone on watch, floating in Borman's seat on the left-hand side of the cabin. The flight was turning out to be somewhat different than he'd imagined. He had never expected to be bored. Make sure it works, Borman had told him, but everything was working just fine without him; he found himself wishing something would go wrong so he would have a chance to fix it. Not that Anders had grown complacent, far from it. He was still having trouble sleeping, in part because he couldn't stop worrying about the systems. Anders didn't want anyone else messing around with the instrument panel, and it seemed to him that Lovell was a little carefree about throwing switches. Once or twice Anders was floating in his sleeping bag, unable to sleep, and heard mission control call, "Apollo 8, we'd like you to switch to the secondary evaporators . . ." Then he saw Lovell's hand reaching for the wrong switch. Anders stopped him—"*Ah, Jim, it's the other one—*"

"I thought you were asleep," said a surprised Lovell.

Not much chance of that, Anders thought. ☾

2 P.M., HOUSTON TIME
2 DAYS, 7 HOURS MISSION ELAPSED TIME
207,000 MILES OUT

Could Jules Verne have imagined the view from Apollo 8? The earth was so far away now that Jim Lovell could hide it behind his outstretched thumb. The feeling this evoked in the pit of his stomach was hard to convey. It was that delicious mix of exhilaration and apprehension that comes from testing yourself in dangerous conditions. (No matter how nonchalant an astronaut might act, Lovell knew, that apprehension was always there; you always wondered whether the engine would work.) And in particular, more so on this flight than either of his previous space missions, it was pure awe. Everything he had ever known was on that blue marble, and it was getting smaller by the moment. None of this came through in Lovell's voice just now, as he became the tour guide for the second telecast from Apollo 8. At last they had succeeded in showing the earth to itself. The black and white image was only a crude facsimile of the real thing, but at least it would convey some of what it meant to be a space traveler. Lovell keyed his mike. ☾

"Houston, what you are seeing is the Western Hemisphere. At the top is the North Pole; just below the center is South America, all the way down to Cape Horn. I can see Baja California and the southwestern part of the United States . . ."

By his own admission, Lovell was addicted to spaceflight. He'd logged eighteen days on his two Gemini missions—more than any other astronaut. Now, in space for the third time, he felt as if he'd come home again. He was glad for that familiarity; it gave him the ability to relax and absorb the experience—especially the view.

And then there were the stars: unmoving, unblinking landmarks along a dark and distant shore.

Mike Collins in mission control asked, "Could you give me some ideas about the colors . . . ?"

Lovell gave it a try. "Okay. For colors, waters are all sort of a royal blue; clouds of course are bright white. The reflection off the earth appears to be much greater than the moon. The land areas are generally a sort of dark brownish to light brown in texture. . . . "

What Lovell had seen in the past two days brought him an entirely new sense of scale. On Gemini, his references were continents and oceans; now he had to think in terms of celestial bodies. The earth was a little ball off in one direction, and the moon in another, and the sun still another. And then there were the stars: unmoving, unblinking landmarks along a dark and distant shore. After two trips into space, Lovell had come to think of them as his friends. Now, on Apollo 8, the stars were his *raison d'être.*

When Apollo was first conceived it was thought that the astronauts would act not only as pilots but as onboard navigators. But the task proved so time consuming and ate up so much space in the memory of the command module's computer that planners decided to let this work be done by computers in mission control. Still there had to be a backup, in case Apollo 8 lost communications with earth. In that contingency, Lovell would use the stars to help himself and his crewmates get home. With the command module's 28-power sextant, he could measure the angle between selected stars and the earth's edge, enter the data into the computer, and let the electronic

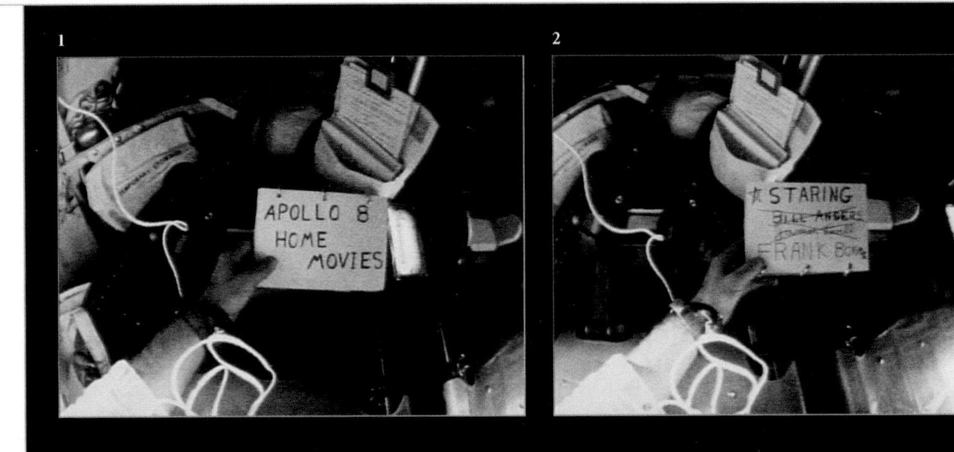

Mission commander Frank Borman stars in the Apollo 8 home movies *(1 and 2)*, which were made onboard with the 16-mm camera; later, Borman occupies the commander's couch *(3)*. Jim Lovell watches a flashlight floating in the command module's lower equipment bay *(4)*, and Bill Anders waves for the camera *(5)*. In a second segment *(overleaf)*, Anders displays the mission emblem *(7)*, and Lovell prepares to launch a roll of tape across the cabin while Borman looks on *(8)*. In the next frame, Lovell and Anders float above their couches *(9)*. And the last frame shows Lovell at the eyepiece of the on-board sextant *(10)*.

brain compute Apollo 8's location relative to the earth and the moon. In principle, it was the same as the shipboard navigation he'd used as a midshipman at Annapolis—but the setting was undeniably different. Already, he had tested his skill with practice star sightings, and the results were within a few thousandths of a degree of perfection. Later, in lunar orbit, he would take navigation sightings on craters and other landmarks over the far side. Those data, crucial to helping Kraft's trajectory people analyze Apollo 8's orbit, could only be obtained by the man onboard.

Lovell was proud of his role on the first circumlunar voyage. Of course, none of them, himself included, was immune to the stress of the mission. The difficulty in sleeping, for example. Lovell had never slept well the first night in space, and this flight was no exception. Eventually you get over it. But he also knew that of the three of them, he probably felt the least amount of pressure. Bill Anders had the rookie's burden of doing his job while adapting to a strange environment. And, like all rookies, he took his work very seriously. And Frank Borman had the heaviest burden of all. There was no doubt in Lovell's mind that command narrows one's focus, because he'd experienced that on Gemini 12. Borman's seeming uptightness—in training as well as now—was probably the commander's syndrome at work. But for Jim Lovell, who had hungered for command, there was an unexpected blessing in not getting it. He was able to enjoy the first flight around the moon more than either of his crewmates, and it was turning out to be every bit the adventure he'd hoped.

●◐◐○○○◑●

Minutes after the second telecast ended, Borman, Lovell, and Anders passed the most significant milestone since leaving earth—and yet, they were completely unaware until Houston mentioned it. Some 38,900 miles from its destination, Apollo 8 reached the top of the gravitational hill and crossed over

3

4

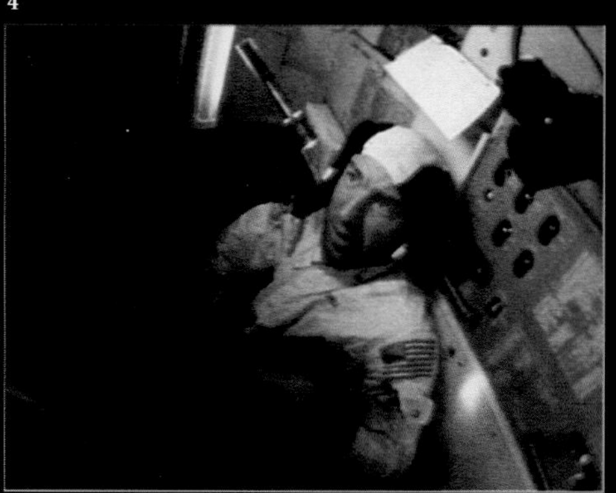

5

into the lunar sphere of influence. At that moment Apollo 8 was traveling only 2,223 miles per hour, but in mission control, Kraft's flight controllers saw the craft begin to speed up. But Borman's crew felt nothing. They saw no change in the visible universe; outside there was the same, dull, starless black. To Bill Anders, the lack of tangible milestones made the voyage seem even longer. At one point, Capcom Jerry Carr asked what they could see and Anders replied, "Nothing. It's like being on the inside of a submarine."

The moon itself was nowhere to be seen. Anders had looked forward to watching it grow ever larger as they closed in until it became a huge, cratered ball in the sky, like a science fiction vision. But he had not seen the moon once on the whole trip out, not even a glimpse. Because of their angle of approach the moon was lost in the sun's glare. It was an act of faith even to convince himself that when they arrived, the moon would really be there.

Anders thought of the moon as his specialty. Before the flight, Borman was so involved with the journey that he had neither the time nor the inclination to worry about the destination. Lovell, who had his tracking tasks, had spent time familiarizing himself with lunar landmarks. But of the three, Anders had the most chance to think about and study the moon. Geologist-astronaut Jack Schmitt had spent many hours with him going over features of interest, and he'd met with other geologists about the observations and photographs he would try to obtain. Anders had his own personal flight plan for the twenty hours he would spend in lunar orbit.

Anders still remembered how, years before, he and the rest of the Fourteen had visited Kitt Peak National Observatory in Arizona, where they saw the moon's forbidding face—that place where he longed to walk—projected onto a large white table. But even then, blurred by the churning desert air, it was not fully revealed. He could only guess what it would be like to see it up close, with only a window in the way. Unfortunately, the windows on Apollo 8 were in pretty bad shape. The largest ones were clouded because of a seal-

6

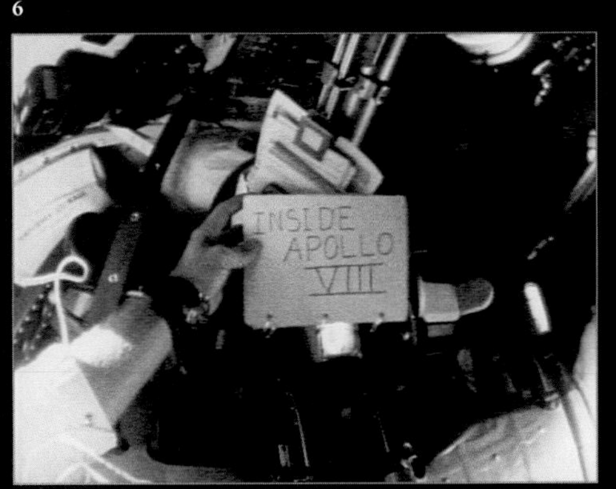

7

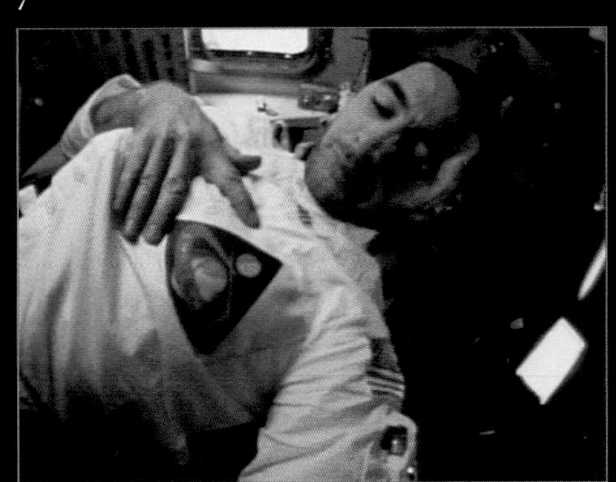

8

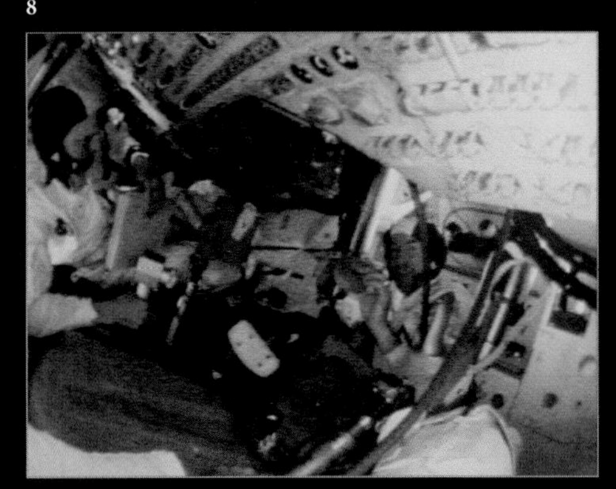

ing compound that had partially decomposed in the vacuum of space. Anders's side window looked as if it had been smeared with an oily rag. Only the two small, forward-looking viewports on either side of the hatch (the so-called rendezvous windows) had stayed relatively clear. Anders wondered how much he'd be able to see of the moon when they got there. He didn't have long to wait.

TUESDAY, DECEMBER 24
2:55 A.M., HOUSTON TIME
2 DAYS, 20 HOURS,
4 MINUTES MISSION
ELAPSED TIME

One of the paradoxes of Apollo 8 was that the three men on their way to the moon were far less able to determine their status than the flight controllers on earth. For all Borman's worries about whether Apollo 8 was staying on the free-return trajectory, there was absolutely no way for him to find out except to ask mission control. Kraft's trajectory specialists were able to detect tiny changes in Apollo 8's path by measuring the Doppler shift in its radio signals. In principle, it was the same as the change in pitch that a stationary listener hears from a passing train. Even a tiny change in the frequency of Apollo 8's signals meant something to Kraft's people. Their data was so good that when they plotted the curve you could see a little wiggle in it, due to the spacecraft's slow thermal-control spin. And they had nothing but good news. The trajectory was nearly perfect. Only two minor midcourse corrections had been necessary so far, and it looked as though Apollo 8 would get to the moon without making any more. The perfect marksman's shot that everyone had hoped for was about to happen.

Meanwhile, a constant stream of telemetry beamed from Apollo 8 to earth was picked up by the giant radio dishes of the Manned Space Flight Network, then transmitted across land lines and via satellites to Houston, where an army of flight controllers kept watch on hundreds of different com-

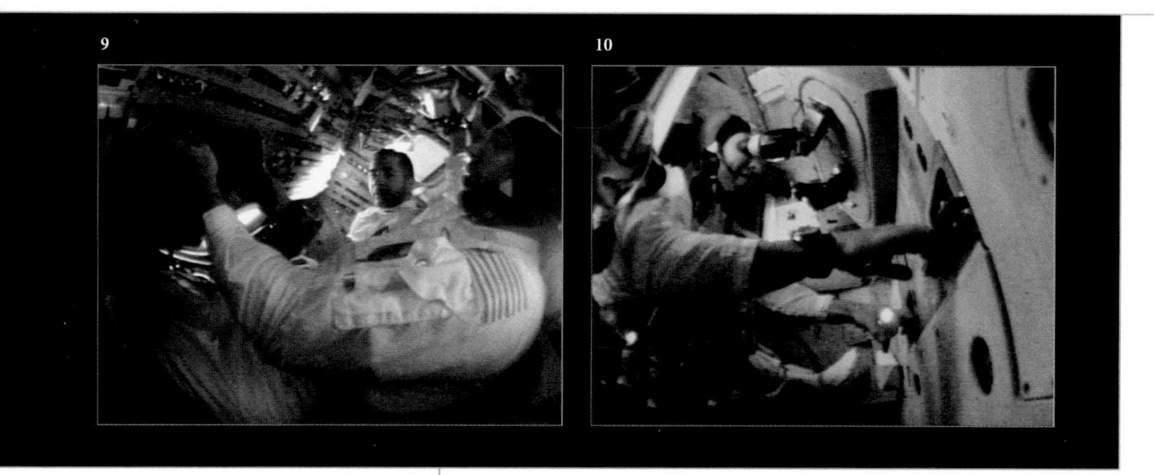

ponents. Borman was amazed at how much they could tell about his spacecraft, more than 200,000 miles away. He had no idea, for example, whether the fuel lines in the SPS engine were as warm as they should be or frozen solid, but the systems people did, and their reports were terrific. The fuel cells were functioning even better than expected; the computer was running like clockwork; there was plenty of maneuvering fuel left. Borman had wanted a perfect spacecraft before he'd commit to the Lunar Orbit Insertion burn, and now he had it. Jerry Carr radioed the word to Apollo 8: "You're Go for LOI. You're riding the best bird we can find."

With characteristic caution, Borman had already turned the spacecraft to the precise orientation for the burn, in fact he had done it two hours ahead of time. The moment of truth, the crucial Lunar Orbit Insertion burn, would come when Borman's crew was out of radio contact, with only themselves and their machine to rely on. If the firing went as planned, this radio blackout would last 45 minutes. But if there was a malfunction and Borman decided to abort the mission, Apollo 8 would come around a good bit sooner than that. In mission control, Kraft's trajectory people would know how things had gone simply from the moment they picked up Apollo 8's telemetry.

Strapped in their couches Borman and his crew waited out the last minutes of a three-day journey. Each of the three astronauts knew they were cutting it very close to aim nearly a quarter of a million miles across space to a world 2,160 miles across, zip just ahead of its leading edge, and go into orbit just 69 miles from its surface. (The joke around the simulator was, wait till you see the 70-mile-high mountain on the far side of the moon.) Sixty-nine out of 234,000 left very little room for error. It was understandable that Borman's crew wanted something more than numbers to assess the accuracy of their path. Before the flight, the trajectory people had told them that they would not be able to see the moon as they came in. Deprived of the one

seat-of-the-pants method a pilot has—eyeballing the target—they asked for something else. There was one answer, and it was Loss of Signal. "LOS," as it was called, was the moment when Apollo 8 would slip behind the moon and lose radio contact with earth. Once the craft was on its way to the moon the controllers would be able to predict the time of LOS down to the second. If it happened precisely as mission control predicted, Borman's crew would know that all the calculations were right after all.

"One minute to LOS," advised Carr. As he spoke, the mission clock read 68 hours, 57 minutes, 4 seconds.

"Ten seconds to LOS," Carr radioed. "You're Go all the way."

"Thanks a lot, troops," Anders said.

"We'll see you on the other side," added Lovell.

Borman watched the mission clock intently. At precisely 68:58:04 he and Lovell and Anders heard static in their headsets. He could hardly believe it—right to the second. He said aloud, "That was great, wasn't it? I wonder if they turned it off." Anders laughed; he could just imagine Kraft saying, "No matter what happens, turn it off."

The men were running through the checklist for the burn when suddenly the spacecraft was enveloped by darkness. Anders realized they were deep in the shadow of the moon. As his eyes adapted, he saw that the sky was full of stars, so many he could not recognize constellations. He craned toward the flat glass to look back over his shoulder, where they were headed, and he noticed a distinct arc beyond which there were no stars at all, only blackness. All at once he was hit with the eerie realization that this hole in the stars was the moon. The hair on the back of his neck stood up. *Come on, Anders,* he told himself; *you're not supposed to feel this way.*

III: "IN THE BEGINNING..."

TUESDAY, DECEMBER 24
3:53 A.M., HOUSTON TIME
2 DAYS, 21 HOURS,
2 MINUTES MISSION
ELAPSED TIME

Falling in darkness, Apollo 8 was pulled toward its rendezvous with the moon at more than 5,000 miles per hour. The spacecraft was turned so that its big SPS engine pointed forward, into the direction of flight. Borman, Lovell, and Anders would need every bit of its power, because Apollo 8 would have to slow down in a hurry, or else speed right past its goal. It would take just 4 minutes to slow Apollo 8 to about 3,700 miles an hour, slow enough to go into orbit. Inside the command module, Borman's crew set to work, running through the checklist to bring the SPS engine to life. With 10 minutes to go, rapid-fire conversation, in the jargon-rich language of spaceflight, buzzed

in the command module cabin as Anders called off to Borman each item on the checklist: ☾

"TRANSLATION CONTROL POWER, ON."

"On."

"ROTATIONAL HAND CONTROLLER NUMBER 2, ARMED."

"Armed."

"Okay. Stand by for the primary TVC check. . . . "

About 3 minutes before the scheduled ignition, Apollo 8 suddenly flew into sunlight once more. Lovell glanced through the hatch window and said, "Hey, I got the moon."

Borman asked, "Do you?"

"Right below us."

Anders looked up from his checklist at his smeared window. It looked as if streams of oil were descending slowly across the glass. *Dammit,* he thought, *whatever that stuff is, now it's running down the window!* But then his eyes refocused and he realized he was looking at *mountains.* They moved slowly past, lit by the slanting rays of the sun, trailing long black shadows. The mountains of the moon. He said quietly, "Oh, my God."

"What's wrong?" Borman said anxiously.

"*Look* at that."

"Alright, alright, come on," Borman said, "you're going to look at that for a long time." ☾

With seconds to go, the computer gave its flashing "99" message and Lovell pushed the PROCEED button in response. Four seconds later the engine lit, pressing the men into their couches. They heard a clattering noise as a stray piece of gear fell to the cabin floor. Slowly acceleration mounted. There was no noise at all, just a gentle vibration and a smooth, steady push. Even though they were held in their couches with just a fraction of normal gravity, after three days in weightlessness it felt like 3 g's. Anders scanned the gauges: tank pressures, valve positions, fuel quantities. "Pressures are coming up nicely," he told Borman. "Everything is great."

Time seemed to slow down. Each man knew the engine must fire for the prescribed duration—no more, no less. If the engine shut down prematurely, or if it didn't deliver the proper amount of thrust, they could end up in a weird, errant orbit. If it fired even a few seconds too long, Apollo 8 would lose so much energy that it would crash into the moon. By the 2-minute mark the burn had begun to seem very long. Borman said aloud, "Jesus, four minutes?"

"Longest four minutes I ever spent," Lovell said as the engine roared silently in the vacuum.

IN 1968

Soviet armed forces invade Czechoslovakia.

France tests its first hydrogen bomb.

Demonstrators at the Democratic convention in Chicago turn riotous; police brutalize demonstrators, reporters, and bystanders.

Presidential candidate Robert Kennedy dies after being shot by Sirhan Sirhan at a Los Angeles hotel.

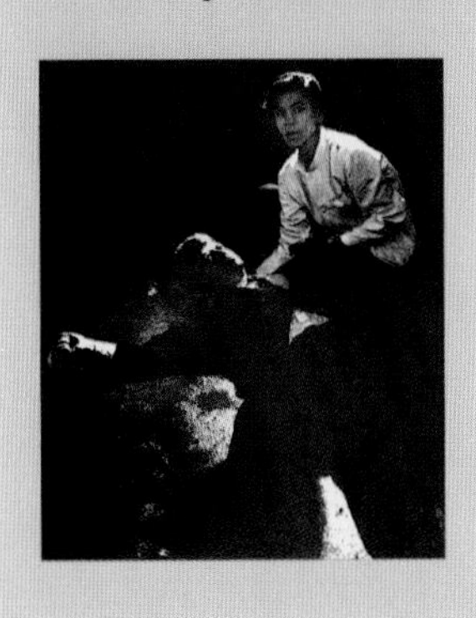

Nearly 70 miles above the far side of the moon, the astronauts looked down on a desolate, pockmarked world. This view shows an area roughly 16 miles wide by 20 miles long.

Two more minutes passed without mishap, and then Anders counted down the last few seconds. Borman knew the computer was programmed to shut down the engine automatically but he wasn't taking any chances. At zero he pushed the shut-off button just in case.

"Shutdown," Borman said. Suddenly they were weightless once more. "Okay," Borman sighed, "go ahead."

They ran through their deactivation checklist like clockwork. Lovell queried the computer for the dimensions of their orbit. They were circling the moon in an ellipse that ranged from 69 miles, at a point above the far side, to 194 miles at the opposite point above the near side. A little over 4 hours from now, at the start of the third orbit, Borman's crew would fire one more 11-second blast to change their path into a 69-mile circle. For now, the SPS had done its work beautifully. Within a few tenths of a mile, the orbit was perfect.

"That's it," Anders said. "Dig out the flight plan."

Apollo 8 drifted above the far side of the moon while three visitors looked down at a scene of total desolation. It appeared devoid of color, apart from various shades of gray. With no atmosphere to soften the view, it was a scene of unreal clarity. If they hadn't known, the men would not have been able to tell whether the moon was sixty-nine miles away or six. Everywhere there were craters: smooth round bowls, misshapen gouges, gentle hollows, tiny BB-shot holes in the gray moon, shoulder to shoulder, one on top of another. Large craters bore on their ancient walls the scars of smaller craters. Every so often a lonely mountain rose from the bleakness, its slopes rounded and pockmarked. Everything else—every rise and fall of the landscape—was formed from the rim or the shoulder or the floor of a crater. The place looked like the deserted battlefield of the final war.

"*Whew.* Well, we answered it," said Borman with a laugh. "They're meteorites, aren't they?"

"It looks like a big beach down there," Anders said. That's what it reminded him of: beach sand darkened by the cold embers of bonfires, churned up by a big game of volleyball, but now deserted. ☾

Lovell, meanwhile, got out the map and tried to figure out where they were. That wasn't easy. They were over a part of the moon where the sun was almost directly overhead, and the moonscape was without definition, like a bleached, rocky ocean. Finally Lovell spotted a huge crater with a dark floor, like a mountain lake. That was Tsiolkovsky crater, named for the Russian scientist who had dreamt of space flight over a century before. Lovell knew it as soon as he saw it.

Minutes later, Apollo 8 crossed over onto the lunar near side. It took several tries for Anders to make contact with earth, but once the high-gain antenna locked on to the signal it was amazing how clearly Jerry Carr's voice came through, as if he were somewhere nearby. And after Lovell passed down the essential data on the burn, Carr spoke for a curious world: "What does the ol' moon look like from sixty miles?" ☾

"Okay, Houston," Lovell radioed. "The moon is essentially gray. No color. Looks like plaster of paris—"

"Or a beach," Anders prompted.

"—or sort of a grayish beach sand. We can see quite a bit of detail. . . ."

Frank Borman did not join his crewmates in their excited descriptions. He was more concerned with the health of his engine. In Houston the engineers were poring over strip charts of data from the burn; Borman wanted to know as soon as possible what they found out. And he wanted mission control to give him a go-ahead for each new orbit, otherwise he would prepare to leave. *Let Lovell and Anders rhapsodize about the moon,* Borman thought; *I've got to think about getting us back.*

●●◐○○○○◑●

Seen from earth on December 24, the moon was a ripening crescent. Most of the near side was in darkness, but that meant that almost all of the moon's hidden face, never before seen by human eyes, was in sunlight. Inside Apollo 8, Bill Anders manned a pair of Hasselblad still cameras and a 16 mm movie camera, his goal to record on film as many lunar mysteries as possible. His photography plan was packed with objectives, and Anders had gotten to work minutes after Apollo 8 reached orbit. When Borman, like any tourist, asked to take a picture, Anders became the rigid one—he didn't want to take any pictures that weren't in the photo plan. Now, armed with his map and his checklist, Anders scanned the parade of craters searching for his assigned targets, and whatever else might look intriguing. The command module had never been designed as an observation platform, but it was turning out to be much worse than he'd anticipated. Only the two small rendezvous windows were reasonably clear, but the view through them was disappointingly restricted. It was sight-seeing in a Sherman tank. To make matters worse, the best maps of the far side, drawn from unmanned probe photos, weren't that accurate. Even when he *thought* he knew where he was, he couldn't find anything he recognized among the swells and hollows. And when he managed to get his bearings, it was all too easy to lose track in the scramble of setting up camera gear, changing film magazines

IN 1968

Arthur Ashe Jr., a lieutenant in the U.S. Army, becomes the first African American to win the men's U.S. tennis championship.

At the 10th Summer Olympics in Mexico City, long jumper Bob Beamon breaks the world record by almost two feet.

Olympic medal winners Tommie Smith and John Carlos are suspended from games for giving the Black Power salute during U.S. national anthem.

and switching lenses. At first he hesitated to take pictures, but he decided if he was going to come home with anything he'd better just aim the camera and fire away. By the end of the third orbit, six hours into the twenty-hour lunar visit, he'd already taken many of the targets on his list, but there was still a lot left to accomplish.

It was a place of such unrelenting sameness–crater upon crater, hill upon battered hill–that to see it with his own eyes was almost an anticlimax.

The irony was that the far side of the moon was turning out to be very different from the place he had envisioned. Like most of the astronauts, he went to see the film *2001: A Space Odyssey* when it opened in the fall of 1968, and somehow, through the weeks of training, poring over the unmanned probe pictures, it was still Arthur C. Clarke's moon that stayed in his mind: a place of drama, with towering, sharp edged mountains, cliffs, and cracks. Instead he'd come nearly a quarter of a million miles to see dirty beach sand. It was a place of such unrelenting sameness—crater upon crater, hill upon battered hill—that to see it with his own eyes was almost an anticlimax. Anders realized, with some disappointment, that the moon was a less interesting world than he had imagined. ☾

10:37 A.M.

Perhaps it is true that our most electrifying experiences are the ones that take us by surprise. Even on the first flight around the moon, in which everything was figured to the second, rehearsed in painstaking detail, an event that no one anticipated became the most moving of all. Apollo 8 was drifting over the far side for the fourth time. Borman prepared to turn the spacecraft so that Lovell would be able to sight the moon through the command module's sextant.

"Alright," Borman announced, "we're going to roll." He nudged the hand controller and the craft turned slowly until it was right side up. When the maneuver was finished, Anders glanced out the window.

"*Oh, my God.* Look at that picture over there."

"What is it?" Borman asked.

"The earth coming up. *Wow,* is that pretty." Slowly, beyond the bleached horizon, a radiant half-circle of blue and white emerged, ascending into the black sky.

"Hey, don't take that, it's not scheduled," Borman said, seizing the chance to give Anders some grief about a picture that wasn't in the photo plan.

Anders wasn't listening. He called urgently to Lovell, "Hand me that roll of color, quick, would you?" But Lovell was already joining them at the windows.

"Oh, man, that's great!"

"Hurry. Quick," Anders said. At last he slapped on the color magazine and aimed the camera with its telephoto lens.

Lovell was impatient: "You got it? Take several of them! Here, give it to me." After telling Lovell to calm down, Anders snapped the picture. "Are you *sure* we got it now?" asked Lovell urgently.

"Yeah. It'll come up again, I think," Anders said dryly.

The first witnesses to an earthrise returned to their work, each carrying the impact of the sight. For his part, Anders had been so focused on photographing, observing, and describing the moon since they arrived that it had not occurred to him to look at the earth. When it suddenly appeared, his overwhelming impression was how beautiful it was, even more so beside the barren face of the moon, and how very small.

12:30 P.M.

Apollo 8 had been in lunar orbit for more than eight hours, and Borman was in need of sleep. He left Anders in charge of the systems while Lovell attended to his landmark tracking. Here too, the computer did amazing things. Once Lovell had determined Apollo 8's position and entered it into the computer, he had only to give it the coordinates of the next target and the sextant automatically swung to the right place. It even moved to track the landmark, compensating for the spacecraft's swift motion. The results were breathtaking. He felt as if he were flying only a few miles above the surface. Peering down into craters, he spotted landslides, even a few boulders. To his surprise, he found he could see detail even in the shadows. If Anders found the moon less intriguing than he'd hoped, Jim Lovell did not share his disappointment.

Lovell's most important target lay on the near side, in the eastern region of the Sea of Tranquillity. There, mission planners had picked out a possible site for the first lunar landing. One of the main objectives of the mission was to reconnoiter East 1, as it was called, from orbit. Lovell had studied the approach that a lunar module would make before landing, and had picked out landmarks along the way. Some of them—the ones that had been discovered on the unmanned probe photos—had no names, and Lovell, following the explorer's prerogative, had named them. Now, as Apollo 8 flew over the Sea of Tranquillity, Lovell was pleased to find the familiar craters and mountains so easy to recognize. Over East 1, Lovell searched for boulders and other potential obstacles to a descending lunar module and found none. The lighting conditions were even better than he expected. It seemed a fine place for a team of astronauts to try to land. With only 69 miles between himself and the moon, Lovell wished he were making the journey. ☾

CHRISTMAS EVE
TIMBER COVE, TEXAS

As night fell outside Houston, Marilyn Lovell left her home, got into her car, alone, and headed for St. John's Episcopal Church. Knowing she would be too busy to attend the scheduled Christmas Eve mass, she had arranged with Father Raish for a private service. She needed this, especially after the ordeal of yesterday afternoon.

Even before her husband left earth, Marilyn Lovell's life had become an emotional roller coaster. She was the only one of the three Apollo 8 wives who decided to witness the launch personally, and she had packed up her four children, from teenage Barbara to little Jeffrey, not quite three years old, and headed to Cape Kennedy, where a friend had arranged accommodations on the beach. Two nights before the launch, Jim had found time to stop by for a visit, and the two of them had driven out to see that magnificent rocket, ablaze in floodlights, that would propel him to the moon. He'd explained to her what the launch would look like, that the Saturn would veer off to one side as it lifted off, to avoid hitting the launch tower. She thanked God that she had known what to expect; otherwise it would have scared her to death. He'd seemed so confident that night, so excited, so ready to do what he had been trained to do. She wished she could have faced the prospect of his moon trip as well as he did.

From the time she and the kids arrived back in Houston, it seemed the house was full of people. Marilyn was glad of that, glad for all the activity. It

Photographed through the command-module window, earth rises above the bleak and battered moonscape. Speaking on television from lunar orbit, Jim Lovell called his home world "a grand oasis in the big vastness of space."

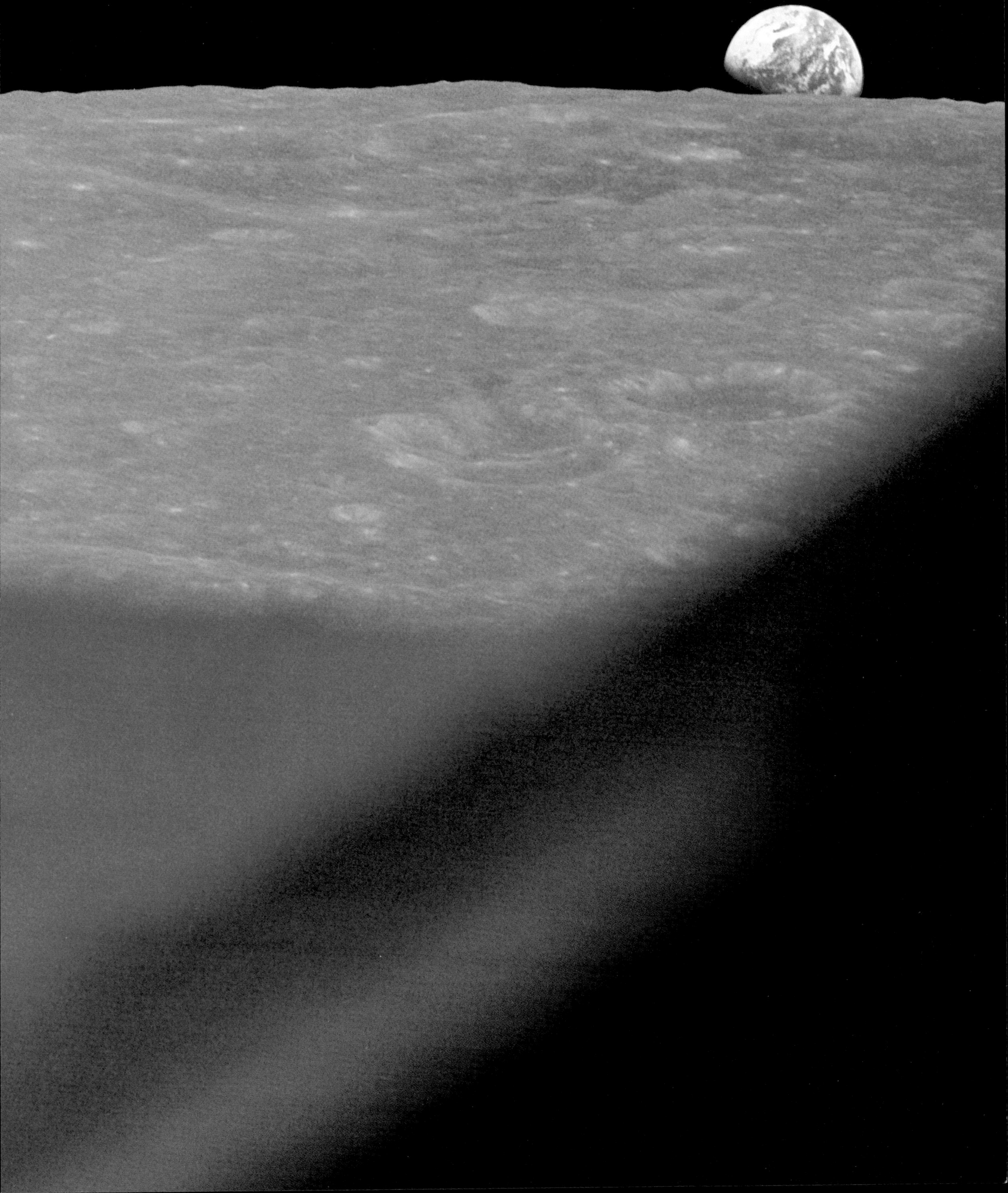

was one of the best customs of the space community; in times of greatest stress, the astronaut wives really came together and helped each other. There was always someone coming by with food, or offering to watch the kids or run an errand.

The squawk boxes—small speakers NASA had installed to pipe in the conversations between Apollo 8 and mission control—were a constant presence. She had kept one ear open, listening for her husband's voice, thrilled to hear him describe what he was seeing. Best of all were the television transmissions. It had been wonderful to see Jim on Sunday, smiling, wishing his mother a happy birthday. But her children had their own ways of absorbing the experience—or not absorbing it. Yesterday afternoon, when she got the children together for the second telecast, thirteen-year-old Jay was outside; she was barely able to get him into the house once the transmission began. When he did sit down, he was decidedly moody. He complained that he couldn't recognize his father's voice. He asked, "How fast are they going now?" The answer—that Apollo 8 had slowed to only a few thousand miles per hour—disappointed him. ☾

Whatever her children might think, Marilyn was spellbound to see pictures of the cloudswept earth, and to hear Jim talking about it. On the screen, the TV networks superimposed Apollo 8's distance from home. Already it was 200,000 miles, hard enough for her to fathom, and it kept increasing as she watched.

No sooner had the TV show ended than her children dispersed, and for the first time since the mission began Marilyn was alone, with nothing to distract her from her darkest fears. On the squawk box, she heard Jerry Carr saying that Apollo 8 had crossed into the moon's gravitational influence. Marilyn wasn't particularly attuned to the technical details of her husband's work, but she understood one crucial thing: it was only a matter of hours until the men would fire their rocket engine and go into orbit, and that engine had to work perfectly, along with the entire, complex machine her husband was flying. Twenty hours later that engine would have to be perfect one more time, or she would never see Jim again. She had known this all along, of course, but she had pushed the thought out of her consciousness. She said nothing about it to Jim. And she certainly wasn't going to discuss it with any of the other wives. There were certain unwritten rules here, carryovers from the test pilot business, and one of them was that you just don't talk about things like that.

And so Marilyn had succeeded in banishing her terror—until yesterday afternoon. Suddenly overwhelmed, with no one there to see, she broke down. Several minutes later the doorbell rang; the daughter of a friend had come by

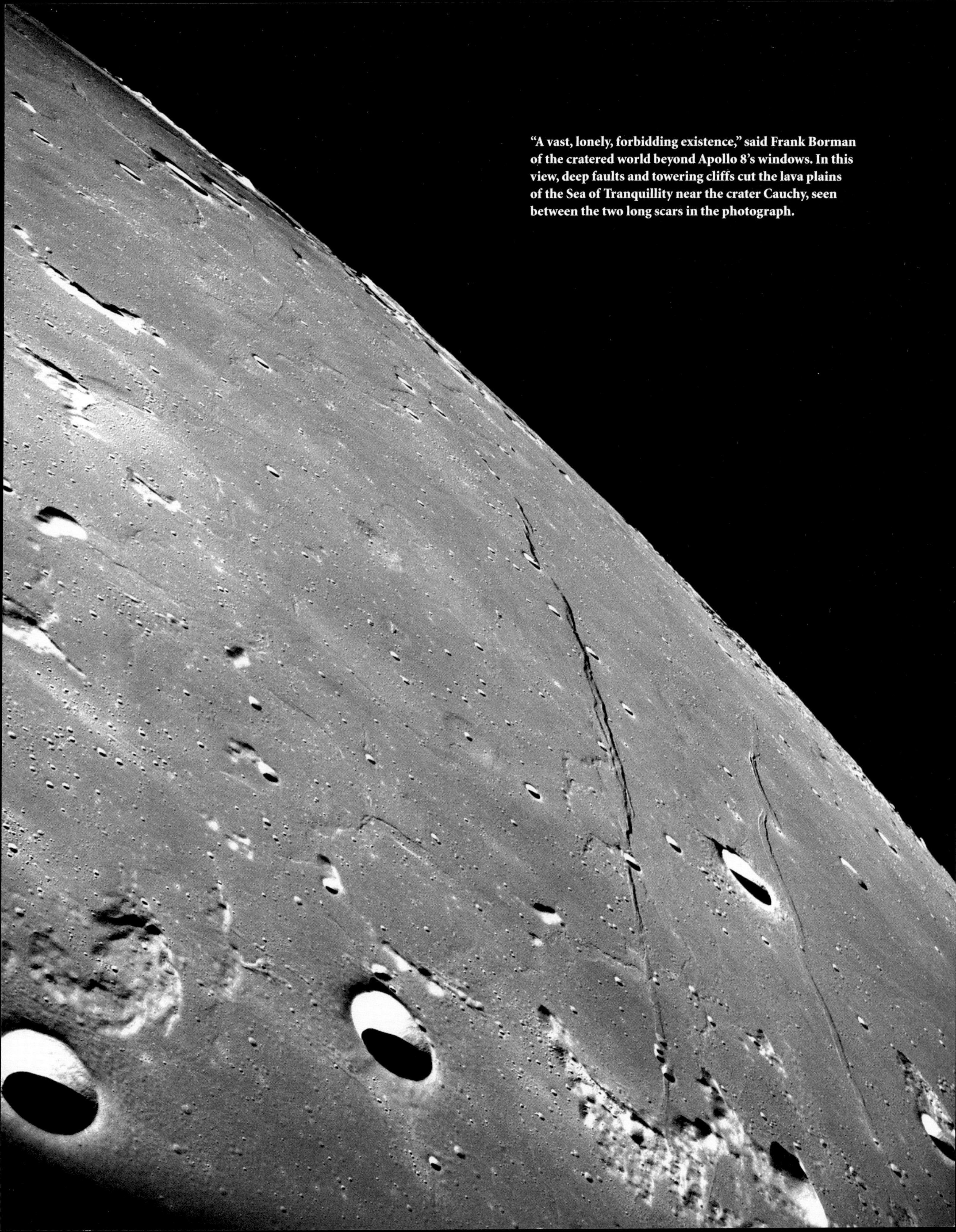

"A vast, lonely, forbidding existence," said Frank Borman of the cratered world beyond Apollo 8's windows. In this view, deep faults and towering cliffs cut the lava plains of the Sea of Tranquillity near the crater Cauchy, seen between the two long scars in the photograph.

with some food. Seeing that Marilyn had been crying, the girl told her mother about it, and by evening Marilyn's house was full of people once more. Her neighbors had come to spend the night with her, to share her anxiety, and her hope, for what was about to happen.

Marilyn barely slept. During the night she came out of her bedroom to find the living room floor strewn with bodies, the sleeping forms of her dear friends. At three-thirty in the morning, everyone waited by the squawk box in silence as Apollo 8 headed for lunar orbit. She heard her husband say, "See you on the other side," and then, after an agony of waiting, his voice came through the static once more to announce success.

All day today, Marilyn had tried to keep up with what was happening. She knew Jim was making landmark sightings, and she had a special interest in them. One night at the Cape before the launch, when she and Jim were alone, he had given her a present: a large black-and-white closeup photo of the moon from one of the unmanned probes.

"What's this?" she asked.

She heard her husband say, "See you on the other side," and then, after an agony of waiting, his voice came through the static once more to announce success.

"I just wanted you to see where I'm going to name a mountain for you." Near the upper-right-hand corner of the photo, Jim pointed to a triangular-shaped mountain sticking up from the dark plains of the Sea of Tranquillity. Jim said it was one of the most important landmarks leading up to the landing site. He was going to call it Mount Marilyn. The name wouldn't be official, not unless it was approved by the International Astronomical Union, but that didn't bother her. As far as she was concerned, nobody could take away such a splendid gift. But on this Christmas Eve, after the trauma of the past three days, Marilyn needed something more.

When Marilyn arrived at St. John's, the organist was practicing, and the church was filled with the sounds of Christmas hymns. Candles glowed everywhere; Father Raish had arranged to have them lit for her. At the altar she celebrated a private communion; she left feeling renewed. And on the way home, she looked up and to her amazement, there was the crescent moon. She could barely comprehend it: Jim was *there.*

3 DAYS, 8 HOURS, 55 MINUTES MISSION ELAPSED TIME

"Well," Lovell yawned, "did you guys ever think that one Christmas Eve you'd be orbiting the moon?"

"Just hope we're not doing it on New Year's," said Anders with fighter-pilot gallows humor.

If Lovell got the joke, he didn't show it. "Hey, hey, don't talk like that, Bill," he said quietly. "Think positive."

Borman was awake now. It was getting to be a long day. They had been in orbit for nearly fourteen hours and they still had six hours to go before they would leave the moon. He could tell that his crew was tired. Lovell was hard at work on his landmark tracking. There was weariness in his voice whenever he spoke. And he was making mistakes. Several times he punched the wrong commands into the computer, triggering warning tones and startling Borman and Anders. Anders was probably tired too, racing around to keep up with his photo plan. Borman knew how tired *he* had felt a few hours ago, before he got some rest. *The flight plan is just too full,* Borman thought. They still had a TV show to do during the ninth revolution. Then the Transearth Injection burn. Compared to that burn, Lovell's navigation was secondary, and so were Anders's photographs. The most important thing was getting home. Borman knew what he had to do.

On the radio, they heard Mike Collins in mission control asking about some of Lovell's landmark sightings.

"Apollo 8, Houston. We'd like to clarify whether you intend to scrub control points one, two, and three. . . ."

Borman keyed his mike. "We're scrubbing everything. I'll stay up and keep the spacecraft vertical, and take some automatic pictures, but I want Jim and Bill to get some rest."

Anders couldn't believe what he was hearing. The last thing he wanted to do was waste time sleeping in lunar orbit. He still had stereo pictures to take, dim-light photography, and filter work, and there were the targets north-of-track he had to finish up. He didn't feel tired. Was Borman serious?

Borman looked at the overcrowded flight plan. "Unbelievable, the details those guys put in here," he said to Anders. "A very good try, but completely unrealistic. I should have warned you."

Grounded by a bone spur in his spine, Mike Collins lost his seat on Apollo 8 to Jim Lovell. By the time of the flight, Collins had recovered from corrective surgery and was serving as a capsule communicator in mission control.

"I'm willing to try it," Anders said gamely.

"No," Borman said. "You try it, and then we'll make another mistake."

Lovell started to speak but Borman cut him off—"I want you to get your ass in bed! Right now!"

"I can do another rev," Lovell said.

"No, get to bed. Hurry up. I'm not kidding you, go to bed."

Anders thought of all the unexposed film. He hid his frustration and asked his commander, "What do you want me to do?"

"Go to bed. We'll get that thing going when we get to daylight," Borman said, indicating the camera. "Then you guys go sack out for two hours."

So that was it. Like it or not, Borman had the authority to send them to bed, in the interest of keeping his crew alert in a dangerous situation. Just now, Mike Collins radioed, "We agree with all your flight plan changes. And have a beautiful back side; we'll see you next time around." No one in mission control was going to argue with Borman's decision.

The radio fell silent once again as Apollo 8 coasted out of contact with earth and into total darkness. Lovell had gone to his sleep station, but Anders remained in his couch, tending the cameras, hoping he might hold Borman off long enough to take some more pictures when they came into sunlight again.

"We're doing fine," Borman said quietly. "Why don't you go to bed?" Anders was about as close to arguing as he'd ever been. But in a spacecraft almost a quarter of a million miles from home, an argument with his commander would have been tantamount to mutiny. Still, he tried to hang on.

"This is a closed issue," Borman said. Anders asked about the movie camera. "I'll just click it on when the time comes," Borman said. "You should see your eyes. Get to bed. Don't worry about the exposure business, goddammit, Anders, get to bed. *Right now.*" Anders had no choice.

"You want me to take some pictures?" Borman offered. "Okay, I'll take care of it all." As Anders headed for his sleeping bag Borman was saying, "A quick snooze, and you guys will feel a hell of a lot better."

An explosion of light came again over the far side, and the sun cast long shadows behind gray mountains and inside countless holes. In his sleeping bag, Anders craned his head to look past his couch through the tiny rendezvous window. Now he did feel tired; he could hardly keep his eyes open. Ironically, this was the best view of the moon he'd had on the whole flight. And something on the stark ground caught his eye, a feature that stood out from the pulverized sameness. He was all but certain he was looking at a region of old lava flows. This was what he'd been looking for, some sign of vol-

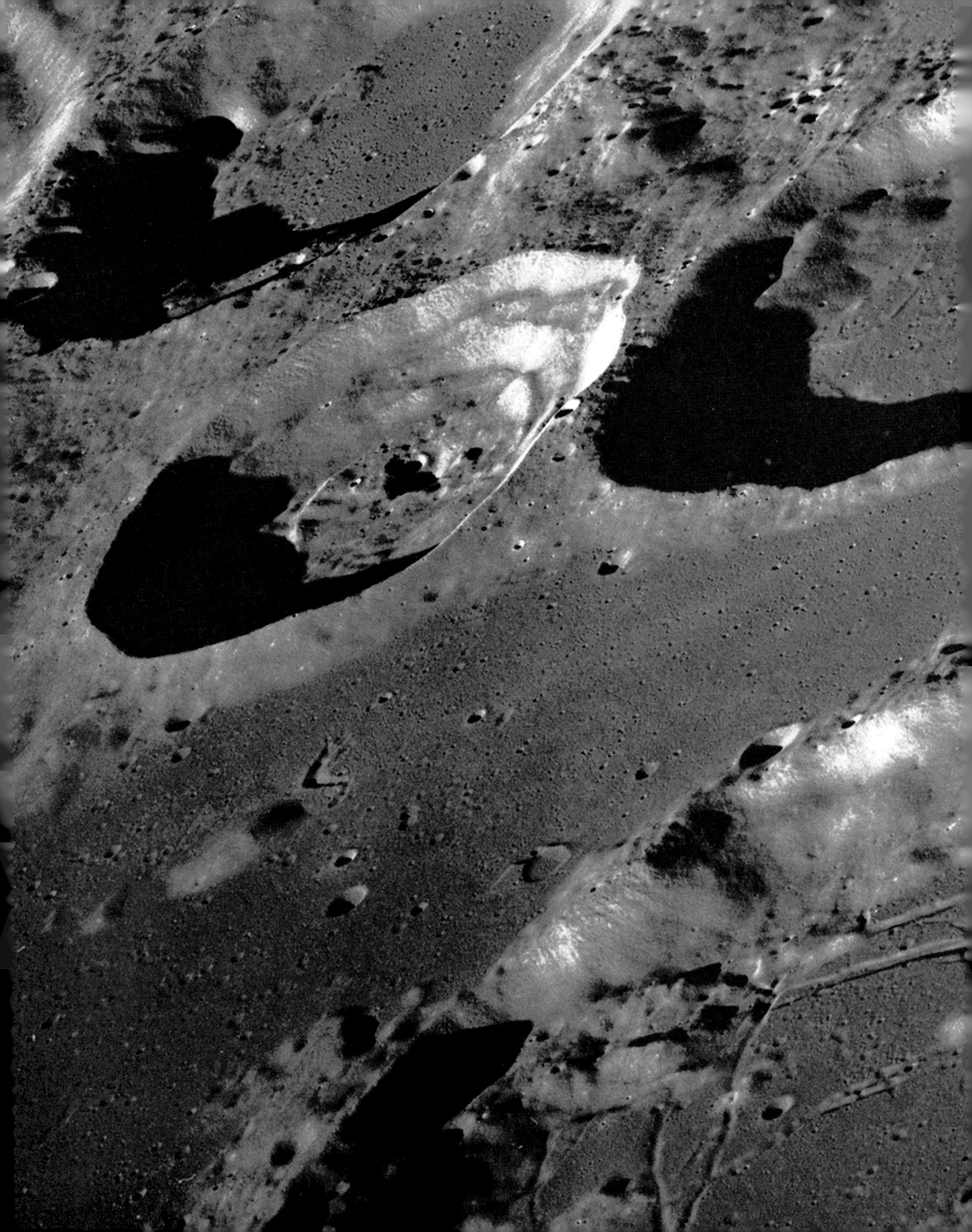

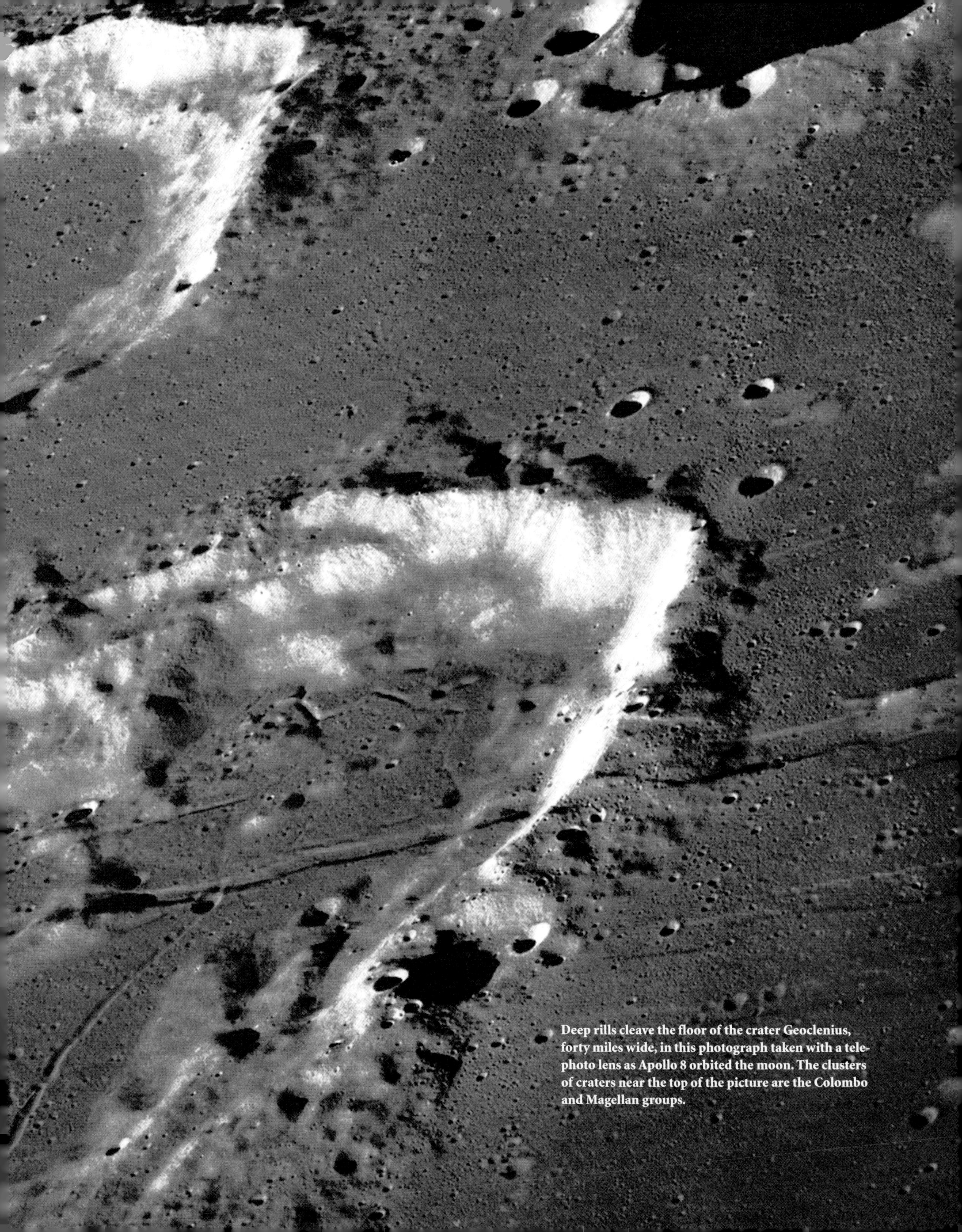

Deep rills cleave the floor of the crater Geoclenius, forty miles wide, in this photograph taken with a telephoto lens as Apollo 8 orbited the moon. The clusters of craters near the top of the picture are the Colombo and Magellan groups.

canic activity in the highlands. He could hear the Hasselblad clicking away on automatic in Borman's window; he hoped they were getting this. Even if they were, Anders was aware that he was bringing home something more important. From lunar orbit, the earth looked no bigger than the end of his thumb, and yet, on a cosmic distance scale a quarter of a million miles was nothing at all. He knew that if he were to go a hundred times farther out—so far into the lonely dark that the earth would shrink to a point of light—he would barely have left home. He couldn't help but think that the cosmos would continue to turn as it always had if suddenly there were no earth. But how little that mattered when it appeared, blue and radiant, rising beyond the lifeless

We came all this way to explore the moon,
and the most important thing is
that we discovered the earth.

moon. In that moment he saw a thing of inexplicable fragility; later he would liken it to a precious Christmas tree ornament. And if the earth was only a mote of dust in the galaxy, that blue planet was everything to him and the creatures living on it. On his way into a fitful sleep, Anders began to realize: *We came all this way to explore the moon, and the most important thing is that we discovered the earth.*

●◐◐○○○○◑◑

While Anders and Lovell slept, Borman floated in the commander's left-hand seat. Spaceflight was quiet. There was none of the constant vibration of an airplane, the steady whine of jet engines. Whenever he fired a maneuvering thruster, there was a thump of solenoids opening the valves, but otherwise, as long as they kept the cabin fans turned off, there was only quiet.

Borman knew Anders was upset about the unplanned sleep period. Borman liked Anders, but he understood that Anders just didn't have the experience to always see the big picture; Borman had been flying ten years longer. The real role of the commander on these missions wasn't to fly the spacecraft; there was precious little of that. It was to make the crucial decisions. And if ever there were a crucial need, it was to keep his crew sharp for the Transearth Injection burn.

But there was something else, before the burn, that Borman had come to realize was extremely important. The flight plan called for two TV transmissions from lunar orbit; the second was set for the ninth rev, a couple of hours from now. The Public Affairs people had told him, "There will be more people watching those shows than have ever listened to a single human being in all of history. Say something appropriate." And with the help of a friend in Washington, Borman found something. He had it reproduced on fireproof paper and placed in the back of the flight plan; after that he didn't give it a minute's thought. ☾

Before the flight, he'd barely thought about the spiritual impact of going to the moon. But now that he was here, he couldn't deny it. To see the moon so desolate, looking like the earth must have looked before life—or how it would look after nuclear war—was more sobering than he could have anticipated. But what moved him most was his own planet: the only color in the universe. To see the earth rising beyond the moon on Christmas Eve was all the confirmation of a Creator that Borman needed. Now that he was here, he was glad to have that TV camera: he wanted to share his new perspective with humanity. What the three of them were about to do was perfect. It was time to get Lovell and Anders up for the telecast.

8:11 P.M., HOUSTON TIME, CHRISTMAS EVE
3 DAYS, 13 HOURS, 40 MINUTES MISSION ELAPSED TIME

"Here it comes!" Borman nearly shouted when he saw his world ascend once more from behind the desolate moonscape. On that earth, in town squares and living rooms, pubs and offices, half a billion people were tuning in for a broadcast from the three men circling the moon.

"This is Apollo 8, coming to you live from the moon," Borman began. "Bill Anders, Jim Lovell, and myself have spent the day before Christmas up here, doing experiments, taking pictures, and firing our spacecraft engines to maneuver around. What we'll do now is follow the trail that we've been following all day . . ."

He'd gone over the plan with Lovell and Anders before they started transmitting. First, each man would say what impressed him most; then, just before Apollo 8 flew into darkness, they would give a joint reading of the message.

"The moon is a different thing to each one of us," Borman told his huge audience. ". . . I know that my own impression is that it's a vast, lonely, forbidding type of existence or expanse of nothing. . . ." His words came out with an awed, sober cadence. "And it certainly would not appear to be a very inviting place to live or work."

In a moment of Christmas Eve magic, the astronauts broadcast from lunar orbit a TV show, viewed here by three children in Virginia, which included a Bible reading from Genesis.

When Lovell's turn came, he spoke eloquently of two worlds, the lonely one he was orbiting and the "grand oasis" he had left behind. Anders described the moon's spectacular appearance near lunar sunrise and sunset, where long shadows made the landscape look jagged and forbidding. For about twenty minutes, the three men took their viewers on a tour of the landmarks passing below them. Over the Sea of Crises they flew, and then the Sea of Fertility, and then the Marsh of Sleep, and then, at last, the Sea of Tranquillity. In the distance, they could see the place of long shadows; soon they would be crossing into night. Anders gave the introduction. ☾

"We are now approaching lunar sunrise. And for all the people back on earth, the crew of Apollo 8 has a message we would like to send to you." Anders held the flight plan in front of him and began to read:

In the beginning, God created the heaven and the earth; and the earth was without form and void, and darkness was upon the face of the deep; and the spirit of God moved upon the face of the waters.

And God said, "Let there be light," and there was light.
And God saw the light, that it was good.
And God divided the light from the darkness.

Now it was Lovell's turn:

And God called the light Day, and the darkness He called Night.
And the evening and the morning were the first day. . . .

As Lovell read, Anders thought, "We're trying to say something fundamental. This isn't just another space mission; it's a new beginning, for all of us."

. . . And God called the firmament Heaven.
And the evening and the morning were the second day.

The shadows lengthened on the Sea of Tranquillity as Borman closed the reading:

. . . And God called the dry land Earth, and the gathering together of the waters called He Seas.
And God saw that it was good.

"And from the crew of Apollo 8, we close with, Good night, Good luck, a Merry Christmas, and God bless all of you, all of you on the *good earth*."

EL LAGO, TEXAS

The night was crisp and clear in Houston. When the verses from Genesis came down a crescent moon shone high overhead, and after the telecast ended more than one witness went outside to look at it. In El Lago, Susan Borman had a house full of friends and relatives. Susan was a model of composure—in other words, she seemed no different now that her husband was circling the moon than she was at any other time.

The other wives could not look at Susan Borman without feeling some amazement. Always, she was an impeccable military wife: white gloves for formal occasions, hair always perfect, always neat and well dressed. Her clothing alone was an accomplishment on an astronaut's salary. Valerie Anders, raising four children, had to scrimp on her wardrobe, and when her mother saw her on television she would ask Valerie later, "Why are you always wearing the same dress?" The answer was that it was the only nice one she had.

This Christmas Eve, Susan had carried herself with characteristic poise. She'd been up all the previous night, listening to the conversations between Houston and the moon on the squawk boxes. In the morning she'd gone to church, pausing outside to speak with reporters, and then back home by midday.

After the evening telecast, Valerie Anders left her house, also full of people, to be with Susan during the critical Transearth Injection burn. Now the two women and several visitors waited by the squawk box in the Borman kitchen. They heard the voice of Ken Mattingly, the young, serious astronaut serving as Capcom, read up a long list of numbers and technical shorthand. It was almost midnight when Mattingly advised, "Three minutes to LOS"—Loss of Signal—as Apollo 8 flew behind the moon for what everyone hoped would be the last time.

"All systems are Go, Apollo 8," Mattingly said. Susan heard her husband say, simply, "Thank you." Then the squawk box fell silent. And now came the worst moment of her long, private agony.

IN 1968

Jacqueline Kennedy marries Aristotle Onassis.

President Nixon wins the U.S. presidential election.

Stanley Kubrick's *2001: A Space Odyssey* coincides with America's plans to go to the moon.

Rowan and Martin's zany TV show, *Laugh-In*, becomes a weekly must-see for millions of viewers.

Her ordeal had begun the previous August, as soon as she learned of Frank's moon mission; from that moment she'd been sure he wasn't going to come back. She was furious with the NASA managers for sending him. There was nothing new about this terror; it had been with her off and on ever since Frank started flying. But it didn't take long for her to learn what was required of her: Keep smiling; keep your fears to yourself. In the squadron, she and Frank had seen some of the men wash out because of wives who couldn't keep their worries to themselves; after a while it affected a man's concentration. She knew it was wrong to complain, but when they were at Edwards, and Frank was assigned to fly zoom tests in that silver beast they called the F-104, in a pressure suit she knew was outdated, she'd pleaded with him not to go. He'd said to her, in exasperation, "*Look*—There is something you've got to get through your head. There's more to this life than just living." And she had only begun to understand it then, that nothing mattered to him more than carrying out his mission.

Over the years, she embraced her role, and even picked up some of her husband's fighter-pilot bravado. When someone in his squadron got killed, she told herself, "that would never have happened to Frank." She did the same thing when they came to NASA—until the Fire. When Ed White died—this magnificent physical specimen—Susan realized that no one, not even Frank, could have gotten out of that burning command module alive. And she knew that as long as he stayed an astronaut he was waiting his turn to die. That was when Susan began to escape into alcohol. No one knew; she drank in private, and was careful never to appear intoxicated around her family. It wasn't hard to hide her drinking from Frank; he spent most of 1967 at North American. She told herself her own problems were minor compared to the stress he was under, and when he came home for an overnight visit she made sure there were no troubles to greet him. She blamed her unhappiness on herself. And when Frank accepted the circumlunar mission, she acted as if nothing were wrong.

But it was one thing to pull off the charade for her husband; it was another to manage it under the media microscope. On launch day, Susan's facade finally cracked. Not with the reporters who were camped on the lawn; she did fine with them. It happened when the cameras invaded her house. Producer David Wolper, who was making a documentary on Apollo 8, had asked NASA for permission to film Susan and the children as they watched the launch and the Translunar Injection burn. Dreading this, Susan had voiced her reluctance, but Frank told her, "I'm sorry, but NASA wants us to do it. It's for the good of the program, and that's the way it's going to be." Wolper arrived at dawn with a back lot's worth of camera equipment. They put microphones in the kitchen cabinets to catch bits of candid conversations. And they left with footage of a very anxious Susan Borman.

However worried she might have looked on camera, that was nothing compared to the anguish inside her, and when Chris Kraft paid a visit a couple of nights later, she made no attempt to cover up. "If you think the Fire was bad, wait until these guys get stranded in lunar orbit!" It would take days for the condemned men to die, circling until their oxygen ran out. She could just imagine what the press would do with that story. She could imagine the NASA man who would come to tell her that Frank was dead, and she could picture the big memorial service they would hold. With a strange kind of logic, she decided that no government official would write her husband's eulogy; that was her job. And she told Kraft that too.

But that was her only outcry. Her husband was commanding the first circumlunar voyage, and if she was expected to play the smiling, confident leader of his support crew, then that is what she would do. Twenty hours ago, just before Apollo 8 slipped behind the moon for the first time, she had sent

Crossing the far side of the moon, Apollo 8 fires its engine to break out of lunar orbit and into a trajectory toward earth. The time: just past midnight, Christmas day, 1968.

him a special message, via Jerry Carr in mission control: "The custard is in the oven at 350."

"No comprendo," she heard Frank say, and then a moment later, "Roger." He understood; it was an old line from their Edwards days. Frank used to say to her, "You worry about the custard and I'll worry about the flying." She wanted him to know she was playing along. But now, with only silence from the squawk box and nothing to do but wait, it was all she could do to keep her composure as the time neared when Apollo 8 was to reestablish radio contact. She could not have known that in mission control Kraft was enduring his own agony of waiting, and that even the engineers who knew the SPS engine like the back of their hands were sweating out this silence like nothing before.

IV: "IT'S ALL OVER BUT THE SHOUTING"

WEDNESDAY, DECEMBER 25
MANNED SPACECRAFT CENTER, HOUSTON

The clocks in mission control crept toward a Mission Elapsed Time of 89 hours, 28 minutes, and 39 seconds, and the tension was palpable. If the Transearth Injection burn went as planned, Apollo 8 would reemerge at that time, 19 minutes past midnight on Christmas Day. If the engine didn't fire, contact would come as much as 8 minutes later.

The mission clock read 89:28:39. Seconds passed in silence. Suddenly a cheer went up from the flight controllers: Telemetry from Apollo 8 began to register on their screens. It took a few more minutes for earthbound antennas to lock onto the signal, and finally, they heard Jim Lovell's voice:

"Houston, Apollo 8. Over."

"Hello, Apollo 8," Mattingly replied. "Loud and clear."

"Please be informed there is a Santa Claus." ☾

"That's affirmative," Mattingly responded gratefully. "You are the best ones to know."

●◐◐○○○◑◑

Once the mission is done, go home. Not only was that Frank Borman's attitude, it was most astronauts'. Last August, in Chris Kraft's office, Borman had asked the trajectory people to get him home in two days, but the speed required would have trimmed the margins of accuracy on the trajectory dangerously close, as well as subjecting the command module to undesirable stresses during its reentry into the earth's atmosphere. Borman settled for 2½ days.

The voyage began with a spectacular view of the entire moon as Apollo 8 climbed away from it like a jet on afterburner. Then there was nothing to do but tend the systems, listen to news reports from mission control—headlined by messages of congratulations that were pouring in from around the world—and catch up on sleep. Around midafternoon, the men gave a televised tour of their home away from home. While Anders demonstrated how to prepare a freeze-dried meal in space, Borman told Mike Collins in Houston, "I hope you all had better Christmas dinners today than us." But Borman spoke too soon. When the TV show was over, they discovered a surprise waiting for them in Apollo 8's food locker, wrapped in foil and tied with red and green ribbons: real turkey with stuffing and cranberry sauce. This was a so-called wetpack meal developed by the military, one of the innovations

Borman had fought to keep off the flight. It was also by far the best meal of the voyage. And there was another surprise, courtesy of Father Slayton: three tiny bottles of brandy. Borman was annoyed. "Put it back," he told his crew. He wasn't about to risk someone in the public raising a ruckus; if they made a single mistake on the rest of the flight the brandy would get the blame. The bottles went unopened (Lovell would say later that neither he nor Anders had any intention of opening them). ☾

But there were other packages, and these were meant to be opened: Christmas gifts to the three men from their wives. Susan Borman had sent cuff links made from a pair of St. Christopher's medals that had gone through World War I with the late husband of a dear friend. From Marilyn Lovell there were cuff links and a man-in-the-moon tie tack, and from Valerie Anders, a gold "8" tie tack, replete with moonstone.

Valerie Anders *(left)* and Susan Borman rejoice as their husbands emerge from behind the moon, heading for home. They would splash down off Hawaii 58 hours later.

For Bill Anders, the trip home was a long, quiet, and boring fall. At one point, Mike Collins mentioned that his son Michael had asked who was driving up there. Anders replied, "I think Isaac Newton is doing most of the driving right now." This should have been a welcome chance to catch up on all the sleep he'd missed on the way out, but Anders, at least, wasn't having much luck there. Somehow it worked out that when he was trying to sleep Borman and Lovell were awake, and they got into small talk. Because of Borman's bad ear, they yelled a lot—"YOU THINK THE OILERS HAVE A CHANCE?" Anders was only thankful that in zero g he could survive on so little sleep, because he wasn't getting much.

At the other end of this fall lay the high-speed reentry into the atmosphere. When Anders was four years old the circus came to the small California town of Vallejo, where his family was visiting, and his grandfather took him to see it. There was an enormous tent, and a man climbed up a ladder—to his young eyes it looked about eight stories high—and dove off into what seemed a tiny tub of water. The boy talked about it for weeks. Anders hadn't thought of that circus dive in many years, but it came back to him when he looked across the lunar distance at the earth: *That's what that guy did in Vallejo!* He couldn't help but think, "I sure hope we hit that thing."

Flying the reentry was a task reserved for the command module computer. If it worked, Borman would just sit there and monitor. If it broke down, he would have to take over and fly it, and he'd worked hard to help create the techniques to do that. They'd probably be off target for splashdown—nobody could fly it as well as the computer—but they'd be alive. The tough part was going to be flying through the periods of high g's. It was hard enough in the training runs in the centrifuge, when you could barely lift your arm, but after six days of weightlessness it would be even tougher. Flying the reentry was one piloting job Borman would just as soon not have.

Even a perfect reentry would subject the command module to extreme stress. In Gemini, the ride down from earth orbit was long and slow, but Apollo 8 would be coming in at 25,000 miles per hour, and the forces of heat and deceleration would be far greater. Temperatures around the command module would soar to 5,000 degrees Centigrade, and their lives would depend on the heat shield on the craft's blunt end. It was made from a substance called phenolic epoxy resin whose protection came not from resisting the intense heat—no one had found an alloy that could do that and still be light enough to use on a spacecraft—but from giving in to it. Just as the boiling water in a kettle absorbs the heat of the stove and keeps the pot from overheating, the heat shield would become white hot, then char and melt away, taking with it the awesome heat of reentry. And when the fiery plunge through the atmosphere was finished, there would still be one more critical event, the blossoming of three 80-foot parachutes to lower Apollo 8 to the waters of the Pacific.

Flying the reentry was one piloting job Borman would just as soon not have.

For Frank Borman, that would mark the end of his astronaut career. He had decided beforehand that Apollo 8 would be his last mission. Slayton had all but offered him the first landing, but Borman turned it down. He appreciated Slayton's confidence, he told him, but he doubted he could get his crew ready for a landing mission in time. He knew they would be disappointed—Anders in particular. But Borman had decided it was time to move on. If Apollo was a war, then a crucial battle was almost won; let someone else have the final victory.

In Borman's mind, the truth of it—which would have come as a great surprise to Anders and Lovell—was that Apollo 8 had been less difficult than he'd expected, far less stressful than Gemini 7. Apollo 8 was turning out to be a wonderful finale to his test flight career. There was only one thing Borman wanted now. After the ordeal of Gemini 7, nothing compared with the high he felt standing weak-legged on the carrier deck, his mission accomplished. And now, as Apollo 8 sped homeward, that was what Borman was looking forward to most of all.

FRIDAY, DECEMBER 27
9:31 A.M., HOUSTON TIME
6 DAYS, 2 HOURS,
40 MINUTES MISSION
ELAPSED TIME

Borman: Look who's coming there, would you?
Anders: Yeah.
Borman: Just like they promised.
Lovell: What?
Borman: The moon.

With just six minutes to go until reentry, the brief appearance of an old friend, rising beyond the dark curve of the earth's night side—at exactly the moment the trajectory specialists had predicted—was welcome reassurance that Apollo 8 was aimed right for the middle of the corridor. Just minutes earlier Borman had flipped a switch to cast off the now unneeded service module. Unprotected, it would meet its end as a shower of meteors over the darkened Pacific. Inside the command module, the conversation sounded like a movie script:

"Well, men, we're getting close!" Borman said.

"There's no turning back now," Anders said.

"Old mother earth has us," Lovell said. He was right; though they could not sense it yet, the men were returning to the earth as they had left, at fantastic speed.

"It's getting hazy out there," Anders said. "Does that mean anything? Every time you fire a thruster."

Now Borman gave the spacecraft to the computer. From here on, the autopilot would fly them in. He glanced out the window. "God, it *is* hazy out there, isn't it?"

"That's sunrise," Anders said.

"Yeah, that might be sunrise," Lovell added quickly. But suddenly they all knew it wasn't sunrise at all, but something far more strange: the glow of ionized gas. The command module was slamming into the outermost fringes of atmosphere so fast that atoms were being stripped of their electrons, creating a glowing plasma. Borman and Lovell had seen a similar glow on their Gemini reentries, but never this bright. And it was only beginning. "God damn," Borman said, "this is going to be a real ride. Hang on."

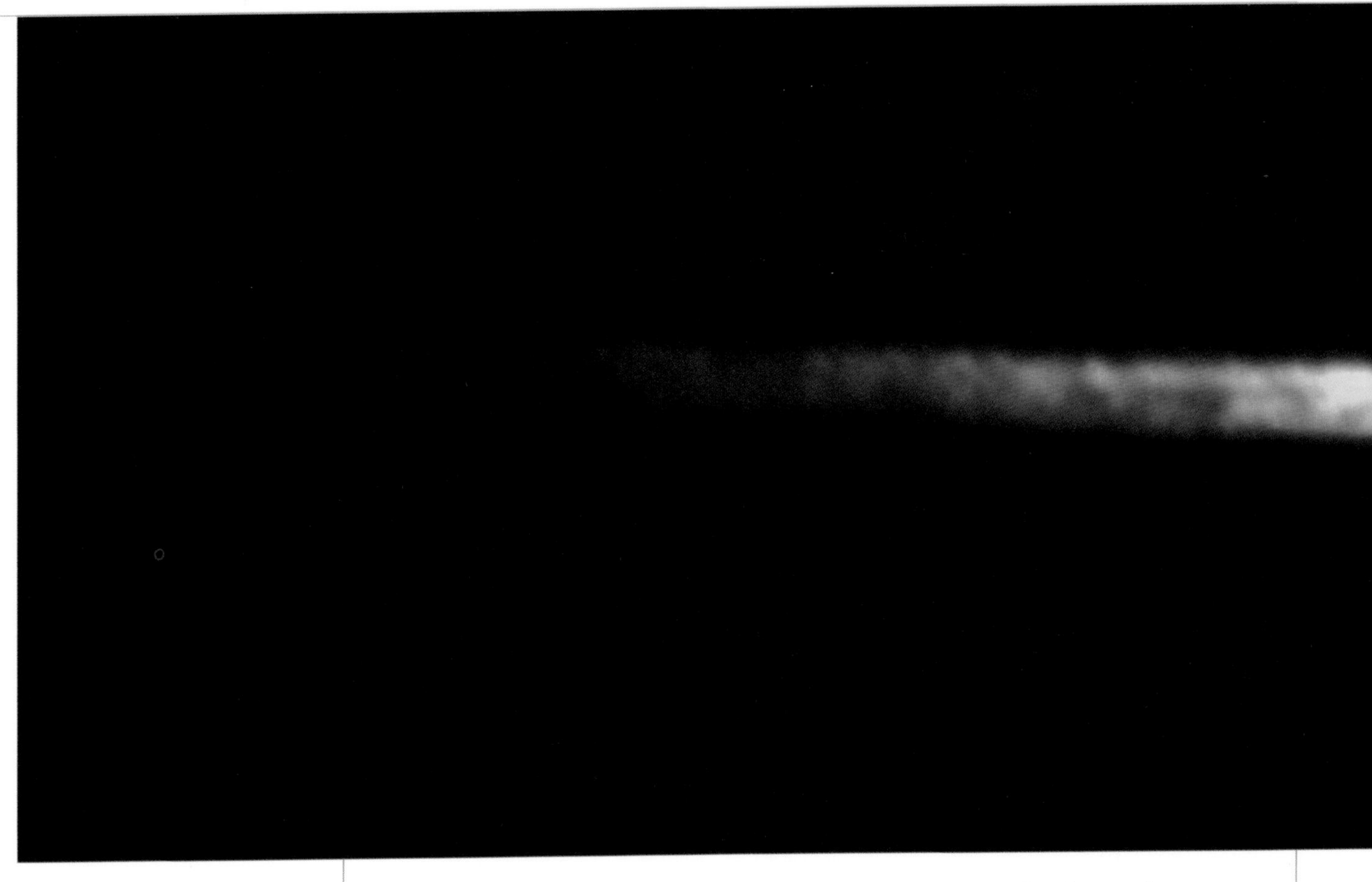

Lacking a protective heat shield, Apollo 8's cast-off service module becomes a glowing meteor as it reenters the atmosphere. This photograph was taken by an airborne tracking camera.

Still they were weightless in their couches, but soon, they knew, the command module would begin to decelerate. A light on the instrument panel would come on when the g-forces measured 0.05; at that moment, the real reentry would begin. From then on things would happen fast and furious.

"Got it! O-five *g*," said Borman. "Hang on!"

"And they're building up!" called Lovell. G-forces mounted as the command module slammed into the denser layers of air.

"Call out the g's," Anders reminded Lovell.

"We're one g." After six days in weightlessness 1 g felt like 3. Seconds later the men were pressed into their couches with tremendous force. Lovell groaned with the sudden deceleration.

"Five!" Lovell's voice was thick with the strain of five times his normal weight.

"*Six!*" An elephant was sitting on their chests. But the worst was over. Within moments the g-forces began to slack off. The command module had slowed to orbital speed now. Borman, Lovell, and Anders were captives of the earth once more.

A cold white light flooded the windows, as bright as daylight. It was like flying inside a neon tube. Borman had never seen anything so weird. He

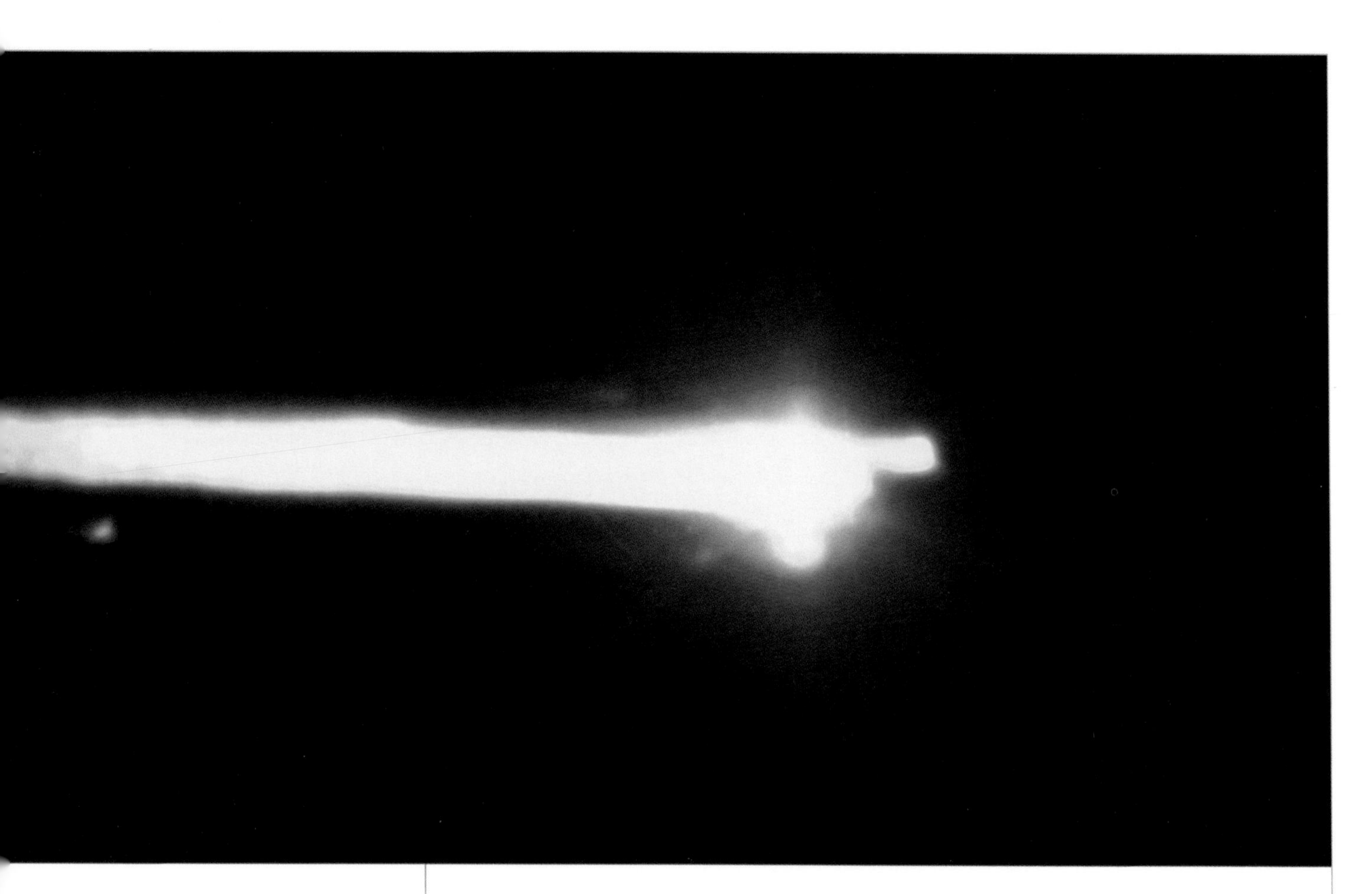

glanced over at Lovell and Anders, who were bathed in this unearthly glow.

Outside, Anders could see flaming objects, undoubtedly tiny pieces of the burning heat shield, flying past his window. Every so often what looked to be a fist-size chunk shot by and Anders thought, "Jesus, we can't take too many of those . . ." He kept waiting to feel heat building up at his back; it never came. While the heat shield charred and melted, inside the cabin the temperature hardly rose a degree. Far below, to those who were fortunate enough to see it, the command module was a glowing meteor in the night sky. ☾

The computer, meanwhile, was doing a miraculous job. Under its control the command module turned and soared along a precisely crafted roller-coaster ride. Apollo 8 dipped down into the denser air, then ascended briefly for a respite from the heat and the g-loads, then dipped and climbed once more. At last, it headed down for the final descent, falling through the pre-dawn darkness like a stone. Borman said, "It's almost all over but the shouting."

Now they were back in radio contact with Houston. At about 100,000 feet the altimeter sprang to life. The command module plunged toward the Pacific at a thousand feet a second. All that was left was the parachutes.

The command module cleared 30,000 feet and there was a loud crack as the parachute cover flew off. Then another crack.

"There go the drogues," Borman said, announcing the release of three small, stabilizing parachutes. Now there was a loud whoosh of air as a vent opened to let cabin pressure equalize with the outside.

"Should be approaching ten K," Anders called. "Stand by for the mains in one second."

They heard a crack.

"You see it?" asked Lovell.

"Can't see a thing," Borman said, peering up into the night. But the altimeter had slowed its readings; the chutes had to be okay. Now a loud mechanical groan filled the cabin as the excess fuel was dumped overboard and the command module thrusters spat flame; in their pale light Borman and Anders glimpsed three great canopies of red and white—beautiful, perfect parachutes.

Now their headsets filled with the chatter of recovery helicopters. "*Apollo 8, Air Boss 1, you have been reported on radar as southwest of the ship at twenty-five miles . . . Welcome home, gentlemen, we'll have you aboard in no time.*"

As Borman, Lovell, and Anders went through the checklist for splashdown—

"CABIN PRESSURE RELIEF VALVES, CLOSED!"

"Got it!"

"DIRECT 02, OPEN!"

"Open!"

—they were all but drowned out by the radio traffic.

"*This is Recovery 2, I see the chutes, I see the light, level with me at precisely four thousand feet . . .*"

Anders was yelling at the top of his lungs— "Floodlights to postlanding!" —for Borman and Lovell to hear him.

"*This is* Yorktown. *Affirmative, we do have capsule in sight—*"

"Turn him down," Anders said. "Christ, we can't get anything done."

"Alright," Borman asked, "anything else we missed?"

"Negative," said Anders. "Stand by to release the mains." That was a bit of teamwork that Borman and Anders had worked out in advance: when they hit the water, Anders would push the right circuit breaker, and Borman would flip a switch to cast off the chutes. Otherwise the wind might drag them through the water until they tipped over.

"Brace yourselves," Anders called.

"Well, wait," Borman said, "we've got two thousand feet yet."

"I don't know if we have or not," Lovell said. "They were reporting us lower."

"Oh, they were?"

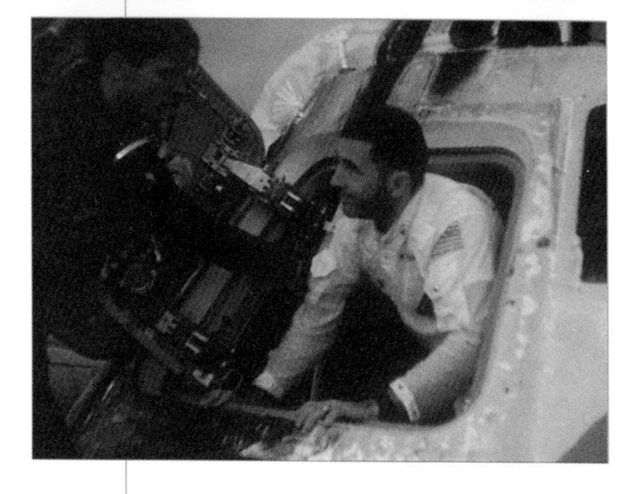

One at a time, the three astronauts are plucked from the sea. Borman *(top)* is hoisted to the recovery helicopter, as Lovell *(center)* waits in a life raft. Anders *(bottom)* emerges from the command module, aided by a navy swimmer.

Moments later the command module hit the water with a tremendous jolt. Water poured in through an open vent, so much that for an instant Anders thought the spacecraft's hull had cracked with the impact. The water poured all over Borman, who was poised to release the chutes, and before he could do anything the wind dragged the command module upside down.

Now it was up to three balloons in the spacecraft's nose to inflate and set them upright. Meanwhile Borman, Lovell, and Anders hung in their straps, upside down, in a dark, warm, suddenly quiet spacecraft, tossed about by a rough sea. Ten-foot swells set them pitching and rolling while helicopters circled overhead, waiting for the first light of dawn before dispatching swimmers.

Anders wryly took stock of the situation: Here they were, in the middle of the Pacific, hanging upside down in the dark, like bats, with bits of trash from the command module floor raining down past them. Not quite what came to mind for returning space conquerors.

Borman had never had a strong stomach for sea voyages, and it wasn't long before he was sick. This time Lovell and Anders showed him no mercy, saying, "What do you expect from a West Point ground-pounder?"

At last they were upright again, and the swimmers had arrived to secure a flotation collar around the command module.

From one of the circling helicopters came a voice: "Apollo 8, is the moon made of Limburger cheese?"

"No," radioed Anders, "it's made of *American* cheese."

The command module, scorched by its high-speed reentry, is lifted onto a wheeled cradle aboard the recovery ship U.S.S. *Yorktown.* Post-flight inspection revealed no unexpected damage to the spacecraft.

Belowdecks aboard the U.S.S. *Yorktown,* the returned moon voyagers receive a congratulatory phone call from President Lyndon Johnson. Garbed in government-issue pajamas for the event, the astronauts soon changed into NASA-blue flight coveralls.

The words of the returned moon voyagers belied the condition inside Apollo 8. Borman, Lovell, and Anders were positively grungy. For one thing they needed a shave, Borman less so than his crew, but he'd taken so much grief about his scrawny beard after two weeks on Gemini 7 that this time he'd arranged for an electric razor to be waiting for him in the recovery helicopter. After six days of living in a flying toilet they didn't even notice how bad it smelled inside Apollo 8. But when that first swimmer opened the hatch, he reeled backwards as if he'd been kicked in the head. Inside, the three men noticed a strange smell too: fresh sea air.

Life rafts waited outside the hatch, and minutes later the chopper lowered a Billy Pugh net and lofted each man, one at a time. As Borman ascended he glanced down at spacecraft 103, his ship, bobbing in the Pacific, and felt gratitude. And a few minutes later he stood smiling with the crew on the deck of the U.S.S. *Yorktown* while hundreds of sailors cheered, waving American flags in the morning light.

It was the middle of the night when Borman, Lovell, and Anders stepped off the plane at Ellington Air Force Base, clean, rested, dressed in caps and blue coveralls. Susan was there with her sons to greet Frank and take him home. They arrived in El Lago to a celebration of candles and welcome home signs. Her husband had flown around the moon—would anything ever be the same? Well, yes; her husband would. As they went into the house Frank spotted the dog's dinner sitting in a corner, uneaten. He turned to his sons and said, "Why is that still sitting there? You know the rules." Not that she was surprised. But she had to laugh.

●◐◐○○○◑◑●

In the last days of 1968, there was a single image—pure, awesome, even holy—to counter a year's worth of violence. It was a photograph of the earth, rising beyond the battered and lifeless face of the moon. Apollo 8 was more than a successful space mission; it was a bright moment for a nation experiencing its first pangs of self-doubt. Even as Vietnam threatened to become a war America could not win, here was an American triumph. Not long after Borman, Lovell, and Anders were back in Houston, Borman got a telegram from someone he had never met. It said, "You saved 1968."

And for NASA, 1969 held great promise. The way was clear now for them to go the rest of the way, to meet Kennedy's challenge, to land on the moon. In his Christmas cards, Alan Bean, backup lunar module pilot for Apollo 9, wrote, "It's going to be a wonderful year for all of us."

TRIUMPH AND TRAGEDY

At the close of 1968, earth seemed a tranquil, inviting place from the vantage point of lunar orbit. But as recalled in the photographs on the following pages, a closer look reveals a year of strife and turbulence.

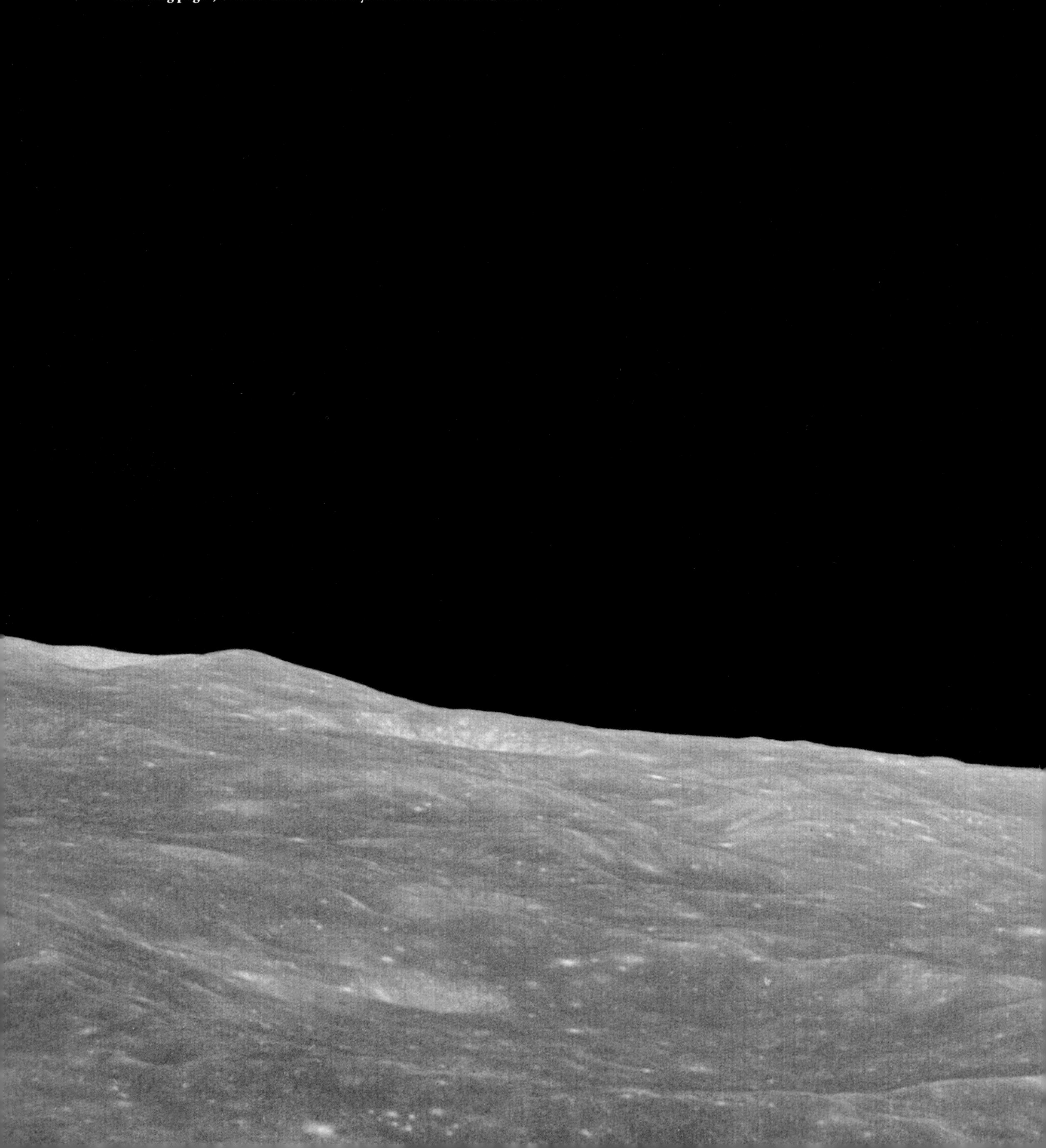

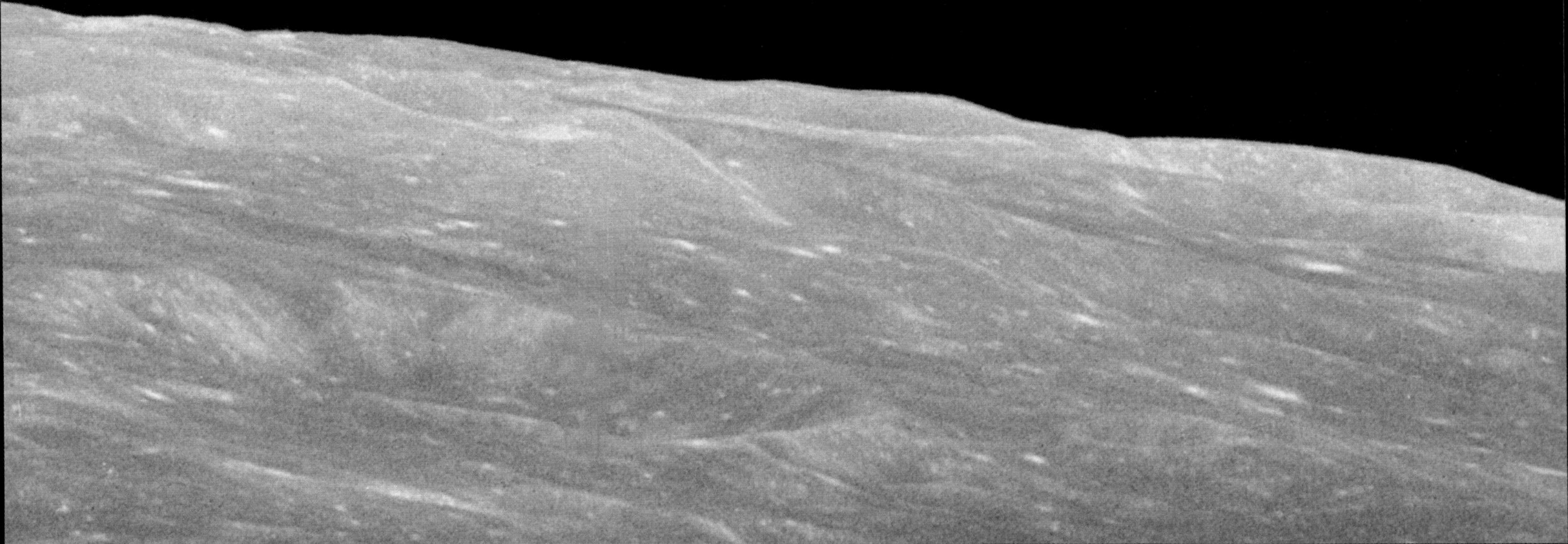

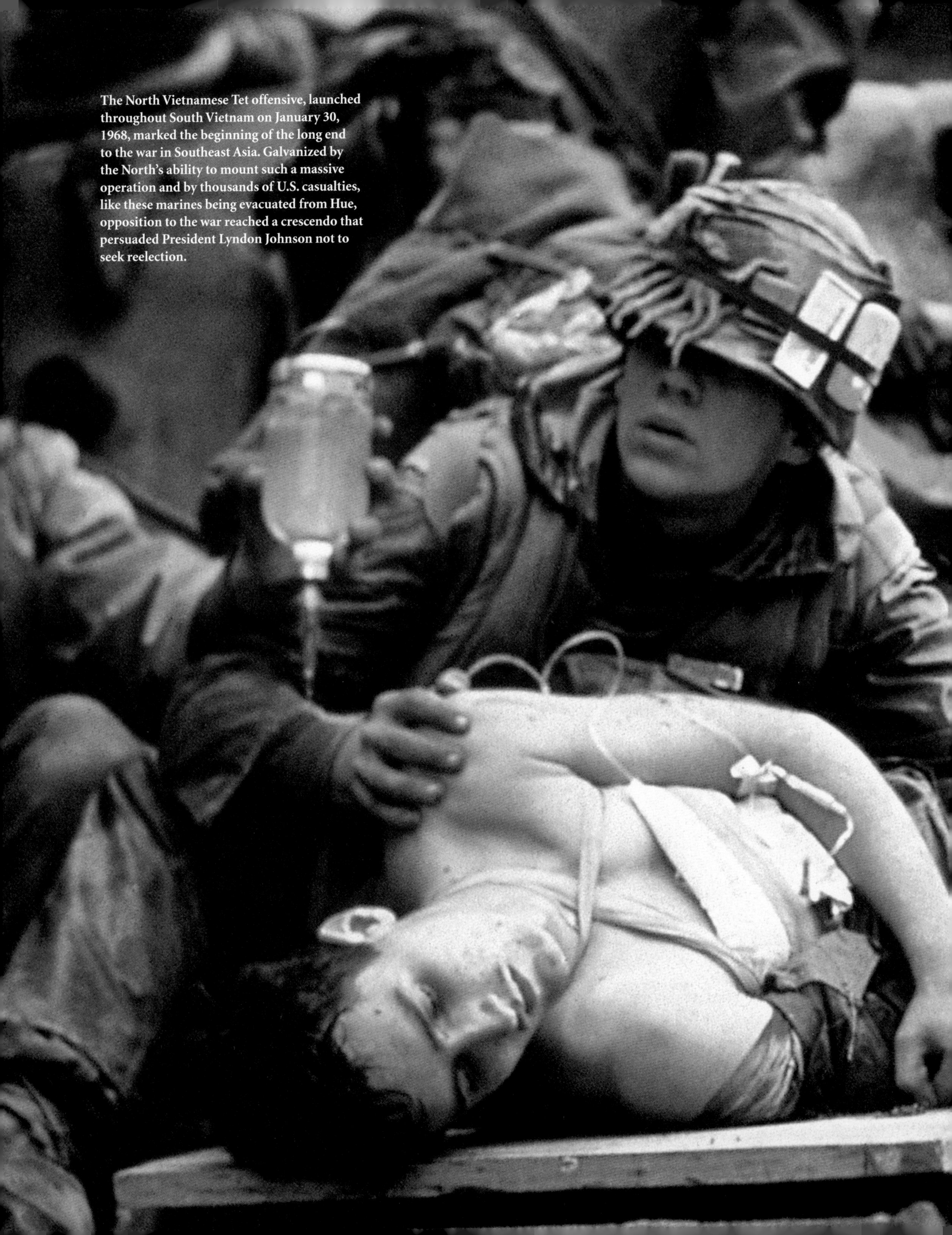

The North Vietnamese Tet offensive, launched throughout South Vietnam on January 30, 1968, marked the beginning of the long end to the war in Southeast Asia. Galvanized by the North's ability to mount such a massive operation and by thousands of U.S. casualties, like these marines being evacuated from Hue, opposition to the war reached a crescendo that persuaded President Lyndon Johnson not to seek reelection.

I AM
A MAN
I AM
A MAN
I AM
A MAN
I AM
A MAN

The struggle for equality in America sometimes became a battle. Outraged at being paid far less than whites for the same work and at being subjected to other indignities, these African American sanitation workers in Memphis walked off the job in March 1968. When the picketing turned violent, police responded with tear gas and clubs; one black youth was killed and 60 people were injured.

On a hotel balcony in Memphis, civil-rights leader Martin Luther King Jr. lies mortally wounded on April 4, 1968, as aides point in the direction of the shot that felled him. King's assassination—and that of presidential candidate Robert Kennedy in a Los Angeles hotel just a couple of months later—weighed heavily on the American psyche.

"BEFORE THIS DECADE IS OUT"

I. THE PARLAY

The first moon voyagers were back on earth, and the impact of their voyage was considerable. In one masterstroke Apollo 8 had given the United States a clear lead in the space race. Publicly, at least, the Soviets were now talking about missions in earth orbit without even mentioning manned lunar flights, as if that had been their plan all along. At NASA Headquarters, no one was ready to write off the competition; intelligence reports said the Soviets were still working on a "super booster" for their own moon landing program. But in Houston, everyone felt the surging confidence that took hold of the Manned Spacecraft Center. For the first time, John Kennedy's end-of-the-decade lunar landing deadline seemed within reach. Indeed, if everything went according to plan, NASA would make it with months to spare. The architects of Apollo were talking about a landing for Apollo 11, slated for a July liftoff—earlier than anyone, including the oddsmakers in the Astronaut Office, would have dared guess. ☾

Working underwater in a giant tank to simulate weightlessness, an astronaut practices an emergency space walk from the lunar module to the command module. The maneuver would be necessary to circumvent any docking problem the lunar module might have upon returning from the moon's surface.

To land on the moon—after only four manned Apollo missions? To many the plan seemed wildly optimistic, considering the tremendous challenges that lay ahead. The lunar module, which had passed an unmanned

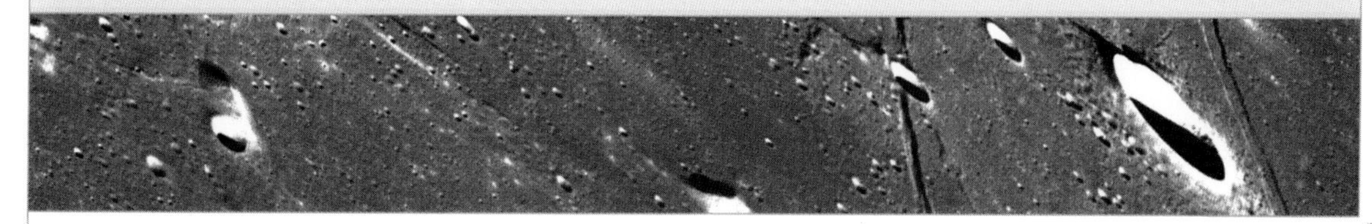

Apollo 9 commander Jim McDivitt *(center)* and his crew, lunar module pilot Rusty Schweickart *(left)* and command module pilot Dave Scott *(right)*, take a break from training, outside the command module simulator behind them.

test in earth orbit, had never been flown with a man aboard. The space suit designed for the first moonwalks had never been tested in the vacuum of space. No Apollo crew had attempted the intricate and crucial rendezvous between the lander and the command module, which would have to be practiced not only in earth orbit but around the moon. Remarkably, NASA was planning to soar over those hurdles with just two missions: Apollo 9, an earth-orbit flight crammed to the hilt with tests, and Apollo 10, a full-up "dress rehearsal" of the landing mission in lunar orbit.

The lunar module's manned debut was the primary goal for the earth-orbit Apollo 9 mission, commanded by Gemini veteran Jim McDivitt and set for a late-winter launch. McDivitt's team—veteran Dave Scott and a red-headed, irreverent rookie named Rusty Schweickart—would practice the critical rendezvous maneuvers and give the entire Apollo spacecraft a thorough workout. Schweickart would perform Apollo's first space walk, to test the space suit and life-support backpack. By any measure, Apollo 9 would be an extraordinarily ambitious mission. In May, if nothing went wrong on Apollo 9—and that was a big if—Tom Stafford's all-veteran Apollo 10 crew would head for lunar orbit for the dress rehearsal. While John Young circled alone in the command module, Stafford and his lunar module pilot, Gene Cernan, would steer their LM within 50,000 feet of the moon and in so doing demonstrate every maneuver for the landing mission, save the final descent to the surface. Then they would rejoin Young and head home.

Both of these missions were staggering in their complexity. By comparison, Apollo 8 had almost been simple. Both would have to be nearly perfect for Apollo 11 to attempt a landing in July. Attempt was really the right word; there was no assurance that the first mission assigned to land would succeed. Nevertheless, as 1969 opened there was more than a little speculation, both inside and outside the Astronaut Office, on who would get the chance to try. Of course, anyone who went through the arithmetic of Slayton's crew rotation had a good idea of who it would be. The crew that backed up Apollo 8 would skip 9 and 10 and fly the next mission, Apollo 11. Pete Conrad had figured that out some time ago, and he didn't like the answer.

Since early 1967, Conrad and his crew had been the backups for Jim McDivitt's team, training for the first manned LM flight. Conrad and his lunar module pilot, Al Bean, had racked up hundreds of hours in the Lunar Module Simulator, helping the engineers wring out the bugs in its computer software. Conrad hoped that with that kind of experience under his belt, if the timing worked out just right, he might one day find himself in command of the first lunar landing. Conrad would have readily admitted it: he wanted the first landing as much as anyone in the Astronaut Office. And the numbers were on his side; he was backing up Apollo 8, and that meant he'd fly Apollo 11, a feasible, if optimistic bet for the landing. But Conrad's fortunes changed when NASA decided to send Apollo 8 around the moon. In August 1968, when Frank Borman's crew swapped places with Jim McDivitt's, so did their backup teams, and Conrad's place in line went to Neil Armstrong.

Now it looked as though Armstrong's team, a crew with hardly any lunar module experience, was about to be assigned to the landing. That's what bothered Conrad most. After all, he and Bean knew more about flying the LM than any other astronauts. One day Conrad took his misgivings to Slayton, who said what he always said: Any crew can fly any mission. And besides, Conrad knew, Slayton wasn't about to pull him off backup duty for Apollo 9, where he was sorely needed. Of course, there was still a chance that his own mission, Apollo 12, might turn out to be the first landing, and around the space center in January 1969 more than a few bets were on Conrad. What

Conrad did not know, because Slayton told next to no one, was that Slayton had almost thrown his any-crew-can-fly-any-mission credo out the window.

For years the media had tried to break the code of Slayton's crew selections, theorizing that he must be picking the best astronaut for each mission, carefully matching crewmen for some combination of skills and temperaments. Slayton also knew that some NASA managers wanted him to hand-pick crews. And he had always resisted. He knew anything other than an orderly crew rotation would make a shambles of Astronaut Office morale. But what about the first lunar landing—was that important enough to put the very best astronaut in command? NASA's upper echelon had always considered Jim McDivitt and Frank Borman as prime candidates for that mission, and in truth, both men were highly regarded by Slayton. Late in 1968, Slayton reasoned that as the only veteran lunar crew, Borman's team would have an edge that might make the difference between success and failure. And on the chance that Borman did not succeed, Slayton was ready to put McDivitt's crew right onto Apollo 12 instead of Pete Conrad's. Slayton knew he'd face a firestorm of protest from the pilots waiting in line—no astronaut had ever been ushered directly from one prime slot to another—but he was willing to put up with that. But Slayton's plan fell by the wayside in the fall of 1968 when Borman turned down the offer.

Many years later, the myth would endure that somehow NASA had hand-picked Neil Armstrong to command the first lunar landing mission. In one version it was because he was the best test pilot among the astronauts; in another, because he was a civilian and NASA was eager to avoid any connotations of militarism on the first landing. Both theories were wrong. In truth, the crew for the ultimate test flight was chosen not by design, but by chance, just as Slayton had always said it would be.

On Monday, January 6, 1969, Slayton summoned Armstrong to his office and told him that he was planning to assign his crew to Apollo 11. Buzz Aldrin would be his lunar module pilot, and Mike Collins, who had fully recovered from his surgery, would be his command module pilot. It wouldn't be official for a few days, but assuming it was approved, they would be in line for the first lunar landing.

Neil Armstrong had gone to work that Monday morning with only a suspicion, not an assurance, of what might come his way. Looking back, he would say that he was very pleased, but not wildly elated, as he left Slayton's office. The sobering reality was that a host of untried procedures would have to be defined, evaluated, and finally boiled down into neatly scripted check-

Buzz Aldrin was a young fighter pilot when he married Joan Archer, an aspiring actress, on a rainy day in 1954. Five years later the future astronaut traded the flight line for the classroom as a graduate student at MIT, where he theorized about orbital rendezvous.

lists before he and his crew could even begin their most intensive training. There were just seven months for him, Collins, and Aldrin to master the complexities of the most demanding space mission ever attempted. Even more significant in Armstrong's mind on this January day was the lunar module, and the tremendous challenges facing Apollos 9 and 10. The odds that both missions would be flawless, as NASA's plan required, were small indeed. Armstrong was not at all sure that when July came, landing on the moon would really be his mission.

"It was a day, the first of many, I'll bet, of walking on eggs, or normalcy tinged with hysteria." Joan Aldrin didn't usually write entries like that in her diary, but that is how she described life on the day after she learned her husband was on the crew of Apollo 11. In 1954, on one of their first dates, Buzz had informed her that people would land on the moon sometime in this century. And she had thought, How ridiculous.

When she met Buzz—the nickname came from his baby sister, whose efforts to pronounce "brother" came out "buzzer"—Joan was working on an acting career. She did a little television work—a walk-on part on "Playhouse 90," a few lines on "Climax." Years later, in Houston, she would help open the Clear Creek Country Theater, but late in 1954 she put all that on hold to marry this young blond fighter pilot, and to go with him to Germany. And aside from wild statements about moon trips, Joan thought she had married a fighter pilot like the others in his air force squadron. She learned otherwise when he traded the flight line for the classroom in 1959, as a graduate student at MIT. At first Buzz had intended to stay only long enough to earn a master's, then apply to the test pilot school at Edwards. But the academic life appealed to them both, and soon he was working on his doctorate. He was also talking about the astronaut program. Buzz knew NASA required test pilot credentials, but he was betting that would change. In the meantime, he was hard at work on his thesis on piloting techniques for space rendezvous. He chose it knowing that rendezvous would become a critical part of NASA's moon program, when astronauts returning from the lunar surface would have to link up with their command ship.

There were days, during training, when McDivitt would go home and tell his wife it couldn't be done, it would never all come together in time.

At MIT, Joan discovered a side of Buzz she hadn't seen before: his extraordinary intelligence. He would pace the floor and talk excitedly about his thesis work. She wanted to support him in any way she could, and that meant she listened, but she didn't comprehend very much of what she heard. There were many people who didn't understand Buzz. His mind was on things that were far removed from day-to-day existence. Like many very smart, intense people, he wasn't too comfortable socially. He understood the intricacies of orbital rendezvous, but he didn't know how to make simple, light conversation. But he was always steady, always calm, always logical. He was her anchor, and she was, quite simply, in awe of him.

Not until years later, when Buzz finally got his chance to fly in space, did he show a crack in his armor, and even then, she didn't immediately recognize it. Buzz came home from Gemini 12 elated at the flight's success, but that was followed by a brief and minor depression. She had never seen him like that before. When she thought about it, she realized that he was going through the same letdown she experienced whenever a show closed; she didn't know when she would get to do another one.

Now Buzz was going to be in another show, and it was the big one. When he told her, she put up a calm and happy front, but whenever she thought about the lunar landing she felt sick. This was something new for Joan. She hadn't been afraid when Buzz was getting ready for Gemini 12, despite the fact that he was going to spend a total of five hours walking in space, more than any previous astronaut. Space walks had been done before, she told herself, and Buzz would just do his better. But this time the old confidence wasn't working. No one had ever landed on the moon.

MARCH 7, 1969
MANNED SPACECRAFT CENTER

In the quiet of mission control, Buzz Aldrin stood behind the Capcom's console and listened to the voices from space. He felt at home here; in some ways he preferred the company of the flight controllers to the other astronauts'. He had never shared the fighter-jock bravado of his colleagues, but among Kraft's men he felt the bond of common academic interests. With them, Aldrin had helped to work out the techniques for the events now unfolding 155 miles up. The climactic moment of Apollo 9 was under way.

"Okay, Dave, I can see you. Boy, are you bright . . . Okay, I'm at nine hundred fifty feet, ten feet per second."

To the rest of the world, the ten-day earth-orbit mission of Apollo 9 must have seemed decidedly mundane, coming as it did on the heels of the first flight around the moon. But few astronauts would have been surprised to learn that Jim McDivitt had fumed down the circumlunar mission to stick with this one. They wouldn't have said so publicly, but many saw Apollo 8 as little more than a ride—no real flying involved. But Apollo 9 was a test pilot's feast. In truth it was far more difficult, more ambitious, and in some ways more dangerous than Apollo 8. For the engineers, the demands of getting two manned spacecraft ready for flight were headache enough. The simulators kept breaking down. There were days, during training, when McDivitt would go home and tell his wife it couldn't be done, it would never all come together in time. Miraculously, it did, and McDivitt's crew was launched on March 3. They figured that if they accomplished half of what was in the flight plan, they'd call the mission a success.

By March 7, the mission's fifth day, they'd already come close to doing it all. Besides a complete shakedown of their command module *Gumdrop*, they'd activated the docked lunar module *Spider* and test-fired its descent engine. Schweickart had suffered a debilitating bout with motion sickness that almost canceled his space walk, but he recovered in time to go outside, wearing the lunar space suit and backpack, for thirty-eight minutes. And today came the climax. While Scott remained in *Gumdrop*, McDivitt and Schweickart climbed into *Spider* and undocked. For six hours the two men flew a craft with no heat shield, venturing up to 111 miles from the safety of their

IN 1969

Publishers of the *Saturday Evening Post* announce its suspension.

Soviet cosmonauts transfer from one orbiting spacecraft to another for the first time.

President Nixon is sworn in as president.

Former president Dwight Eisenhower dies at age 78.

CAUTION
FINGER TORQUE ONLY

Emerging from the Apollo 9 command module *Gumdrop,* 155 miles above the earth, Dave Scott prepares to photograph space-walking Rusty Schweickart, who snapped this view while floating outside the lunar module *Spider.*

command ship. Then they fired their engine and headed back to Scott, beginning the long and intricate dance called space rendezvous. As Buzz Aldrin knew only too well, an astronaut flying a rendezvous couldn't use the instincts he'd honed in airplanes. This was a completely different way of flying.

Rendezvous was a domain ruled by the arcane statutes of orbital mechanics. A spacecraft in orbit is like a ball bearing whizzing around inside a deep, curved funnel. A ball thrown into the funnel will "orbit" at a height and speed that depends on the amount of energy it has. A ball with a lot of energy will circle at the upper end of the funnel; one with less energy will circle lower down. If the ball is down toward the neck of the funnel, it will circle faster than one up near the mouth, even though it has less energy. If the balls are spacecraft around the earth, the funnel becomes the invisible "well" of gravity which all orbiting objects must fight in order to stay in orbit. The closer the spacecraft is to earth, the stronger the force of gravity, and the faster it must go to balance that pull. The farther away it is, the weaker gravity's pull, and the slower the spacecraft travels.

From a pilot's standpoint, the analogy makes clear the most important rule of orbital mechanics: Height and speed are inextricably linked. To slow down, the spacecraft must be kicked into a higher orbit (by adding energy with a burst from a rocket). Conversely, speeding up requires dropping into a lower orbit (by using the rocket as a brake). A pair of astronauts who start out behind their target must lower their orbit until they catch up, but not for too long, or they will overtake it. If they start out ahead, they must raise their orbit, go slow for a while, and then descend in time to meet their target. Every burst of speed, every bit of braking changes their height and therefore their speed. Catching the moving target—and staying there once they've arrived—becomes a feat of great complexity.

At MIT, Aldrin had foreseen the importance of rendezvous to NASA's moon program. He knew that special onboard computers were being developed for Gemini and Apollo to help astronauts fly a rendezvous. But what if the computer broke down? With the help of specialists in MIT's Instrumentation Laboratory, Aldrin worked out the techniques the pilot could use to take over and fly the final stages of a rendezvous by hand. At NASA, Aldrin had been instrumental in designing the rendezvous schemes for Gemini and Apollo. He would look back on those months, working with mission planners, as one of the most demanding and rewarding experiences of his life.

Now, listening to the words from Apollo 9, he knew that McDivitt and Schweickart were on their final approach to the command module. Inside *Gumdrop,* Dave Scott could see the returning lunar module:

Schweickart, anchored to *Spider's* front porch,tests the space suit and backpack designed to be used on the surface of the moon. His gold-plated sun visor reflects the lunar and command modules, joined together and glistening against the earth.

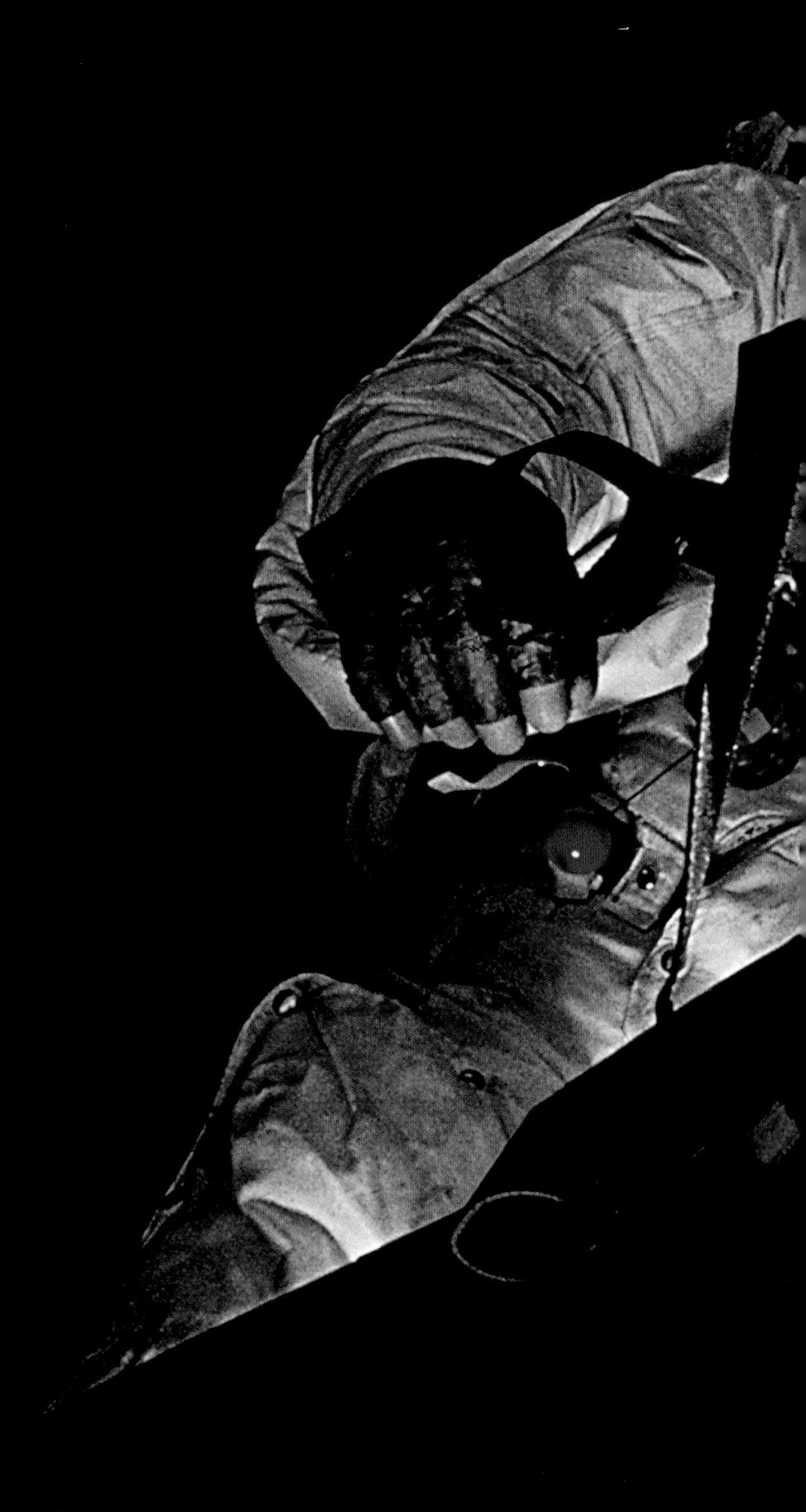

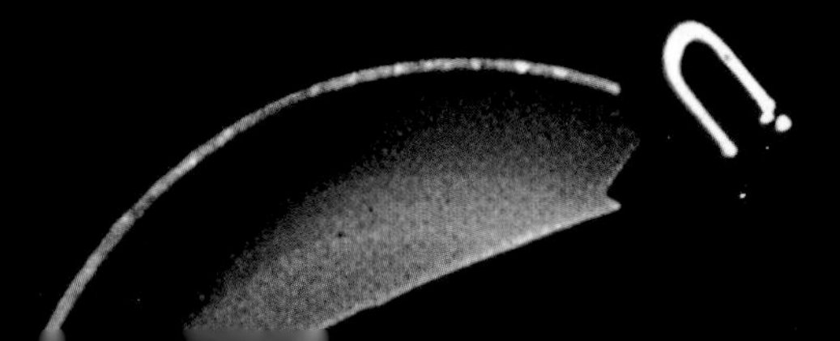

In this multiple-exposure photograph, an astronaut in a command module simulator practices docking maneuvers. During the linkup, the probe on the nose of the command module *(white)* will fit into a conical recess on the roof of the lunar module, seen at right in the picture.

"You're upside down again!"

"I was just thinking, one of us isn't right side up," McDivitt answered.

"Boy, you've got contraptions hanging out all over."

"That's show biz. Okay, I have us at about three hundred seventy feet. . . ."

Around the space center, Buzz Aldrin—the first astronaut with a Ph.D. after his name—was known as an intense, brilliant theoretician whose interest in rendezvous was nothing short of fanatical. The other astronauts called him "Dr. Rendezvous," and Aldrin lived up to the name. For anyone who would listen, he offered a discourse on his ideas brimming with orbital-mechanics jargon, whether his listener could keep up with him or not. Once, when Aldrin's wife was away, Walt Cunningham invited him to dinner, then was called out of town unexpectedly. Somehow Aldrin never got word of the change. At the appointed hour he showed up for dinner, where he enthusiastically lectured Cunningham's wife, Lo, on out-of-plane errors and line-of-sight velocities until the small hours of the morning.

It was hard to think of anything besides rendezvous that had been the subject of more meetings and discussions within and outside the Astronaut Office. Knowing the ins and outs of rendezvous was almost synonymous with being an accomplished astronaut. No astronaut knew more about it than Aldrin; and yet, to his regret, he had always felt like something of an outsider in the office. Aldrin was sure it was because of his résumé—not only was he not a test pilot, but he wasn't a member of the navy clique that he felt dominated the astronaut corps.

Other astronauts would later say it had more to do with his personal style. To them his self-confidence had a way of seeming like arrogance. He could be direct to the point of bluntness. Aldrin saw no reason to conceal his pride at his academic achievements, but the Old Heads rolled their eyes when he showed up wearing a tie clasp fashioned from a pair of air force wings, with his Phi Beta Kappa key dangling from it. And if the other pilots knew less about the theory of space rendezvous than he did, then they also understood that an astronaut didn't have to be a theoretician in order to fly.

Knowing the ins and outs of rendezvous was almost synonymous with being an accomplished astronaut.

If Aldrin drove the Old Heads up the wall, he was also right more often than they would have cared to admit. But he sometimes fell prey to that well-known syndrome of technical organizations known as Not Invented Here. When the astronauts were concerned about lighting conditions during the rendezvous's final stages, Aldrin suggested that the lunar module fly upside down as it approached the command module. In zero g, of course, it would make no difference, and it would keep the sun's glare out of the lander's windows. The Old Heads scoffed at the idea—*Fly upside down?* But eventually when it was time to make a decision, they went ahead and quietly adopted the plan, as if that had clearly been the way to go all along.

When Aldrin finally did get a chance to fly in Gemini 12, he got a chance to put his rendezvous expertise to good use. He and Jim Lovell had just reached orbit and were chasing their Agena target rocket when their rendezvous radar malfunctioned, which in turn prevented the men from obtaining crucial data from their computer. While Lovell did the flying, Aldrin got out a set of rendezvous charts—which his own efforts at MIT and NASA had helped to create—and computed the maneuvers himself, and the pair completed their rendezvous successfully. And after a string of aborted space walks on earlier flights, Aldrin made the most successful walks of the Gemini

program. Chris Kraft himself would later say Aldrin had turned in a superb performance.

"Dave, I just can't see it. Let me get in a little closer."

As *Spider* maneuvered into position for docking, Scott radioed instructions to guide McDivitt, who was blinded by the sun's glare: "Just keep coming easy like that. . . . You ought to go forward and to the right a little. . . . There you go. . . ."

At last the two ships met. A buzzer sounded as a set of docking latches on the two craft snapped shut. A relieved McDivitt said, "I haven't heard a song like that in a long time!" Apollo 9 had fulfilled all its major objectives. At that moment, Aldrin knew that Apollo 10 would also succeed, and that he and Armstrong would attempt to land on the moon. On March 24, NASA made it official. For Aldrin and his crewmates, training now went into high gear.

FRIDAY, APRIL 18, 1969
BUILDING 9, MANNED SPACECRAFT CENTER, HOUSTON

Inside the cavernous expanse of Building 9, a lone figure clad in a bulky white space suit and backpack stood at the center of a small crowd of technicians. He stooped slightly under the burden of the equipment necessary to keep him alive in the vacuum of space. After the technicians had attended to the necessary adjustments, he strode stiffly to a silvery mockup of a lunar module, and stood in one of its bowl-shaped footpads, ready to begin the strange spectacle of a practice moonwalk.

"All set, Neil, if you read me."

"Yeah, I read you," said Armstrong, answering the technician who was acting as Capcom.

"Okay, proceed."

Armstrong held on to the ladder that was attached to the mockup's front landing leg, and lifted his left foot over the footpad and onto a bed of sand. As he had practiced many times over the past few weeks, he tested his weight on the sand as if it were the moon, and described what he saw.

"Okay. Checking the bearing strength, and we're leaving a one-quarter-inch footprint. And we have a poorly sorted sand-and-gravel aggregate which does not stick to the boot. Range of the groundmass is from one centimeter down to below the resolution of the eye." Now he stepped off the footpad and took a few steps. "My balance seems to be very much like earth simulations." It was an optimistic thing for Armstrong to say, weighed down

Flying free of the command module *Gumdrop,* McDivitt and Schweickart prepare to give *Spider* a workout in earth orbit. One feature they couldn't test were the probes extending from the lunar module's footpads, which were designed to signal contact with the lunar surface.

Before an audience of NASA engineers, a suited Neil Armstrong rehearses collecting a contingency sample. The first task of his upcoming moonwalk, gathering a few scoops of moon dust, would ensure that the astronauts would not return from the moon empty-handed in case of a hasty departure.

by 200 pounds of gear, but he had reason to expect that the real thing would be easier. In the past two months he had glimpsed what it might be like to walk in the moon's one-sixth gravity, inside the KC-135, a converted air force cargo plane with a long, padded cabin. By flying a carefully planned parabolic trajectory, the pilot of the KC-135 could create about half a minute's worth of simulated lunar gravity, giving Armstrong time to practice scooping up rocks, or handling gear, or simply get accustomed to the surreal lightness that came with every step. With a few dozen parabolas per flight, a session on the KC-135 was enough to let Armstrong taste the freedom that the moon would offer him once he was there. But the bulk of his work was here, in Building 9, where a moonwalk was as hard as a day on a construction site.

Like a polar bear lumbering about in his pen, Armstrong went about his work while a small crowd of technicians and training specialists looked on. As he moved, he struggled not only against the suit's weight but its stiffness. Pressurized at 3.5 pounds per square inch, the suit was a rigid balloon in which every movement required effort. The gloves were clumsy, and it was especially difficult for Armstrong to manipulate a camera or grasp a geologic hammer. Simply opening or closing his hand was like squeezing a tennis ball. As he reached into a pocket on his space suit thigh and pulled out a collapsible long-handled scoop, technicians with headphones could hear his controlled but labored breathing. "We're beginning the contingency sample," Armstrong said. "I have the collector . . ."

Standing at the sidelines, a fully suited Buzz Aldrin watched his commander at work and waited for the proper moment to join him. It had been decided in the summer of 1968 that the first landing mission would feature a single moonwalk; by February 1969 its duration had been decided at about 2 hours and 40 minutes. For months now, the people in the space center's Crew Systems division had been working with him and Armstrong to make every minute on the lunar surface as productive as possible. The two men would collect rock and dust samples, lay out a few simple scientific instruments, and take pictures. All of it had to be rehearsed to the letter, until the tasks became second nature. Never in the history of exploration had 2 hours and 40 minutes been so carefully planned.

"Okay," Armstrong said, "we're ready for the second man to come down now. . . ." Aldrin lumbered onto the sandy training area and began his work. Some of the NASA doctors, Aldrin knew, were predicting that he and Armstrong would have trouble adjusting to lunar gravity, that they would wear themselves out working in pressurized suits. Aldrin was sure they were wrong. With nearly five and a half hours of space walking under his belt, he'd proved that an astronaut could do useful work outside a spacecraft, in the three-dimensional ice rink called zero g. Aldrin was certain that walking on the moon, with the luxury of a gravity field, would be even easier. It was the part that would come afterward that worried him.

Media attention had always been the single biggest adjustment for the astronauts. Most of the Original 7 had been woefully unprepared for the barrage of attention that greeted them the day they were introduced to the press. The spotlight had dimmed somewhat since then, but not much, and most of the astronauts had learned to deal with the constant demands of their

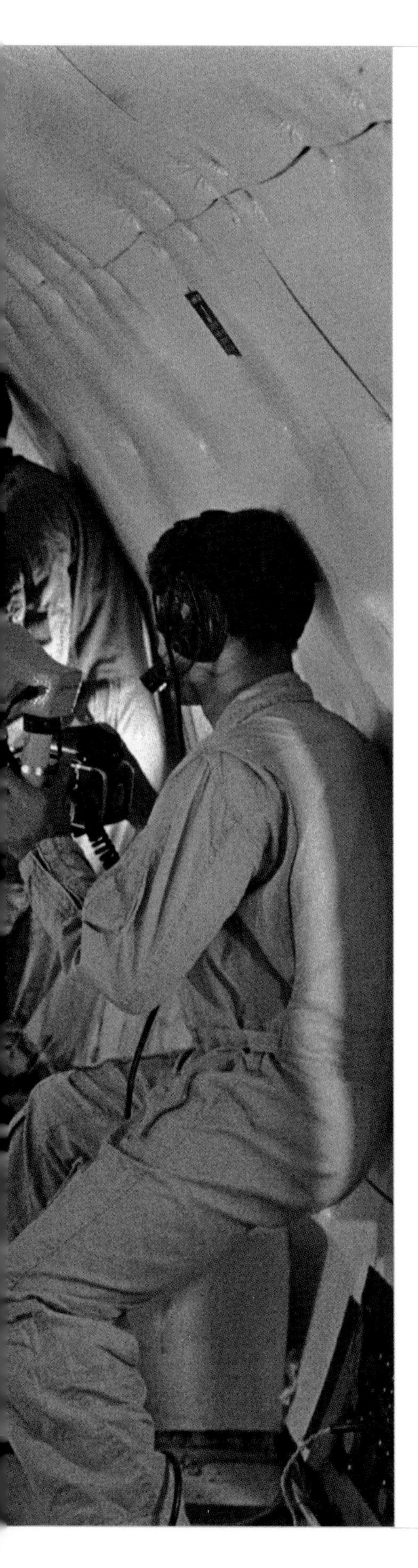

Aboard an air force KC-135 jet, Buzz Aldrin experiences a re-creation of moon gravity, one-sixth the strength of earth's. The cable attached to Aldrin's space suit allows him to communicate with a nearby technician.

celebrity. There were even some, like Gene Cernan, who seemed to enjoy talking to the press about their experiences. But it wasn't so easy for Aldrin. Publicity was a double-edged sword. Whatever enjoyment he got from the recognition, it was far outweighed by his discomfort. He hated the question that awaited him on his return from Gemini 12: What did it feel like in space? He could go on at length about the technical aspects of the experience—his mobility during the space walk, or the computations he made during the rendezvous—but don't ask him to plumb his own feelings. It was the worst kind of bind. Sitting under the TV lights, he couldn't just shrug off the question with a pat response; he was too much the perfectionist for that. And so he was caught, struggling to find an honest answer, when the only one he could think of was, "I don't know."

Aldrin's unease was so great that given a choice, he would have preferred to be on the second lunar landing, or the third, instead of the first. For one thing, he would be able to put his scientific talents to greater use on a later mission, when the emphasis would be more on exploration. Even more important, there would be less attention. No one, save Aldrin's wife, knew that when he learned he would be on the lunar landing crew, he had briefly considered bowing out. He'd rejected the idea just as quickly. No astronaut had ever resigned from a mission. If he asked to leave this landing crew, he might never get onto another one.

Years later, Aldrin would say his mixed feelings were thrown into sharper focus by one issue in particular. In fact, it was the first question asked at the first press conference after the crew announcement. "Which one of you gentlemen will be the first man to step onto the lunar surface?" Deke Slayton had been sitting up on the stage with Armstrong, Aldrin, and Collins, and he told the press that the question hadn't been decided. Armstrong added, "It's not based on individual desire," and said it would be worked out as the training progressed, according to how best to accomplish the mission objectives. Of course, they all knew very well that there were very few astronauts who *didn't* want to be first to walk on the moon.

At one time, there had been a preliminary version of the checklist for the moonwalk, and on it the words "LMP EGRESS" came before "CDR EGRESS"—in other words, the lunar module pilot would go outside ahead of his commander. Aldrin could see sound reasons for having it that way. Armstrong had his hands full with the landing, and with the mission commander's responsibility for the flight's overall success. Instead of adding more to that workload, didn't it make sense for Aldrin to take the lead when it came to the moonwalk? It had been the same way in Gemini: The copilot always made the space walk while the commander stayed inside and flew the spacecraft. And there was the matter of physical conditioning. The moonwalk was going to be physically demanding, and Armstrong was anything but an exercise fanatic. More than one astronaut remembers that Aldrin paid a visit, checklist in hand, explaining the operational merits of having the lunar module pilot get out first. ☾

There was nothing extraordinary about an astronaut campaigning for a pet cause, but this was no ordinary issue, and the Apollo crew commanders who heard about Aldrin's idea had strong reactions. For them, it was simple: The commander gets out first. Of Aldrin, they wondered, "Who is he kidding?" The Gemini precedent didn't apply, because a lunar module sitting on the moon wouldn't be in flight—it would be *in port*. And as any naval officer knows, the protocol on such matters is clear: When the ship comes to port the skipper is always first down the gangplank.

But in the end, there was something more decisive, outside the whims of human emotions. It was the lunar module's front hatch. The lunar module cabin had about as much space as a large broom closet. For one astronaut to go outside, the other man would have to hold the hatch open and stand back in his corner. The departing astronaut would then get down on his stomach and wriggle through an opening only 32 inches square. Getting out of the LM wasn't like going down a gangplank; it was more like being born. Years earlier, with the first landing still a long way off, Grumman engineers had designed the hatch so that it opened from left to right—that is, toward the lunar module pilot's side of the cabin. The only way Aldrin could get out ahead of Armstrong would be if the men first changed places. That would have been possible for two men in street clothes, but not encased in pressurized suits and massive backpacks. One day Armstrong and Aldrin tried it, fully suited, in a lunar module mockup, and they damaged the cabin. Deke Slayton was there; he saw the damage, and he told himself the situation would have to change.

Mission planners had quietly come to the same conclusion in February, but well into the spring, Armstrong and Aldrin had not learned of any decision. But Aldrin heard rumors that Armstrong would be named to go out first, because he was a civilian and because NASA wanted to keep the first

IN 1969

Yasir Arafat becomes head of the Palestine Liberation Organization.

Golda Meir is elected as the first woman prime minister of Israel.

Glen Campbell wins best album Grammy for "By the Time I Get to Phoenix."

The Concorde supersonic airliner makes its first test flight.

steps on another world free of any militarism. The implied slight against the air force, and the thought that the matter would be decided on political grounds, instead of on operational considerations, angered Aldrin. But he knew there wasn't any point in maneuvering. It was best to be direct about it; he would take the issue to Armstrong. Their time on the Apollo 8 backup crew had left them with a sort of friendship—not exactly drinking buddies, but hardly strangers. Still, as he entered Armstrong's office he wasn't sure how his commander would take what he was about to say. ☾

"Neil, we've got to come to some kind of decision on this," Aldrin said, somewhat self-consciously. "I'm sure you know how I may feel about it . . . " The words were out. There was an awkward silence; Armstrong looked uncomfortable—was he embarrassed at Aldrin's directness? At last he spoke:

"I'm aware of the historical significance of the decision, Buzz, and I'm not about to rule myself out."

Years later, however, Armstrong would have no memory of their conversation. He had been aware that Aldrin had a certain focus on the issue, he would recall, and he felt sorry, because he didn't care one way or the other. In fact, as Aldrin himself had suspected, Armstrong had nothing to do with the decision, and he was never asked for his opinion until after it was made. Slayton came by to say the plan was for him to go out first, and did he agree? In Armstrong's mind the new plan was not only more practical but far safer. He told Slayton, "Yes, that's the way to do it."

In Slayton's mind, the decision was clear, hatch or no hatch. He recommended, and the managers agreed, that Armstrong, as Apollo 11's commander, and as a senior astronaut, should be the first man to set foot on the moon—assuming, of course, that he and Aldrin were able to get there. By mid-May, that rested with the success or failure of Apollo 10, which was about to leave the earth.

II. "WE IS DOWN AMONG 'EM!"

MONDAY, MAY 19, 1969
1:49 P.M., HOUSTON TIME
ABOARD APOLLO 10,
128,000 MILES FROM EARTH

"If people want to know what kind of men go to the moon, there's a good look at one right there." Having prepared his earthbound audience, Gene Cernan pointed the television camera at his commander, and a gum-chewing Tom Stafford, his balding head hidden under his communications hat, smiled back. Cernan said, "Could you believe it?"

"Some people still don't," drawled Stafford.

On earth, where Stafford's grinning countenance filled television screens

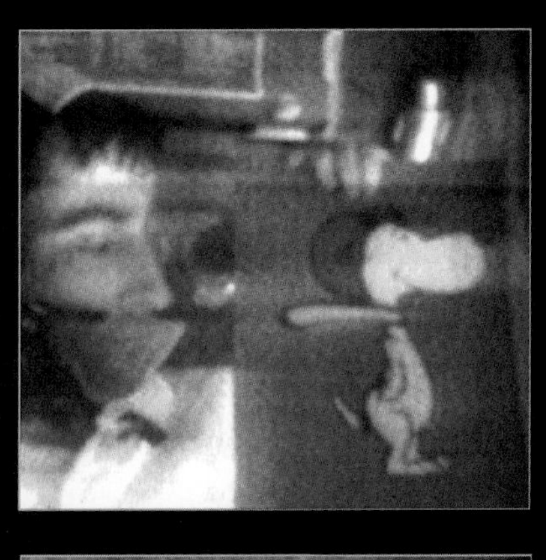

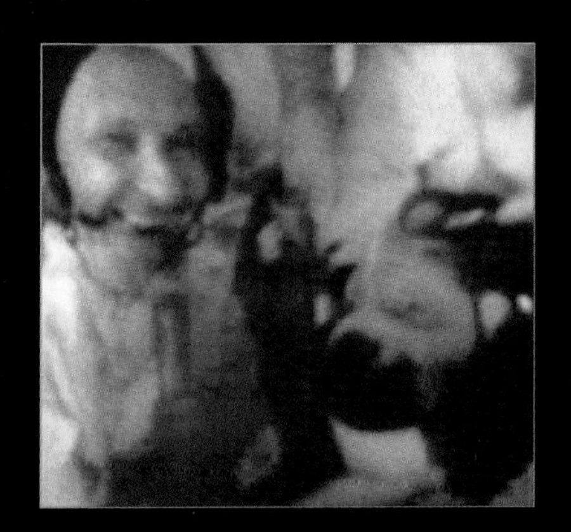

Scenes from Apollo 10's onboard color television camera: John Young displays the comic strip characters Snoopy *(top)* and Charlie Brown, namesakes of the spacecraft. At bottom, Tom Stafford clowns with Young in weightlessness.

across the world, Capcom Bruce McCandless played straight man: "I'm surprised you all have not set this to music."

Cernan answered, "Oh, you want music? Well, we'll give you some music. . . ." On the monitors, the image blurred and shifted, then came clear again, and there was the earth; this time, thanks to Apollo 10's new TV camera, it was in full color. "Here it comes," Cernan said, as one of his crewmates activated a small portable tape recorder. "This is just so that you guys don't get too excited about the TV and forget what your job is down there." And then, the sounds of Frank Sinatra and the Nelson Riddle Orchestra came floating down from halfway across the translunar gulf:

Fly me to the moon, let me play among the stars
Let me see what spring is like on Jupiter and Mars
In other words, hold my hand
In other words, darling kiss me. . . .

The second voyage to the moon was under way, complete with background music.

There was a time, in the first part of 1968, when Tom Stafford thought he might have a chance of making the first landing. And more recently, there had been others at NASA, most notably George Mueller, head of the agency's Office of Manned Space Flight, who had pushed for a landing on Apollo 10. From the beginning, Mueller tried to quicken the pace of the moon program. It had been Mueller who insisted on "all-up" testing for the Saturn V, that is, launching a completely assembled booster instead of testing one stage at a time, as the more conservative engineers had wanted. Mueller's impatience had probably saved months in the race with the end of the decade. On the eve of Apollo 9's spectacular success, Mueller, like others at NASA, was asking, why in the world send the entire Apollo spacecraft to the moon—with all the risks involved—and not try to land? ☾

One reason was that Stafford's lunar module, built before Grumman enacted a super weight-saving program, was too heavy to land. There was some talk of letting Stafford use Armstrong's LM, the first one built that was light enough for a landing, and postponing Apollo 10 a month to allow the switch. But some, particularly Chris Kraft, raised strong objections. There were too many unknowns, he said. His trajectory people didn't understand the moon's lumpy gravitational field well enough yet to predict what mascons would do to the paths of the orbiting command and lunar modules. Would they pull the lander off course for its descent to the surface? And while the astronauts explored the moon, would mascons pull the command module off course for the rendezvous ahead? NASA needed more navigation data from Apollo 10 before it could commit the next crew to a landing. Furthermore, Kraft's flight controllers needed experience in communicating with two separate spacecraft at lunar distance. Sam Phillips, the tough, exacting air force general who served as Apollo program director, listened to all sides of the argument and decided that the dress rehearsal was not only desirable, but crucial. ☾

Tom Stafford agreed. He'd wanted the first landing as much as anyone else, but he wasn't about to campaign for a mission he knew was beyond accomplishing. Now was the time to find the hidden unknowns and solve them, so that Apollo 11 would be able to concentrate on the landing itself. And Apollo 10 wasn't simply a repeat of Apollo 9 in a different place; new procedures were required for a rendezvous in lunar orbit. Stafford and Cernan would take their lander and descend to 50,000 feet above the lunar surface, where they would make a critical test of the landing radar. Then, from this close vantage, they would scout Apollo 11's proposed landing site in the Sea of Tranquillity before rejoining Young in the command module. By any measure, the dress rehearsal was a grueling mission; it seemed to Stafford's crew that they had more to do on Apollo 10 than all the others combined.

With five Gemini flights among them, Stafford and his crew were the most experienced team yet sent into space. They had a camaraderie that came not only from having trained together for more than two years, but from a long personal history with one another. Stafford, a veteran of the first space rendezvous in 1965, went on to command Gemini 9 with Cernan as his copilot. He had come to know Cernan as a quick study with a sharp mind. Cernan also liked to party, and he liked the attention that came with being an astronaut; once or twice Stafford had to pry him away from some reporter and back to work. It was Cernan, as lunar module pilot, who had the major responsibility for the lander's systems. Stafford and John Young went a bit further back; they had been shipmates aboard the battleship *Missouri* twenty years earlier. Besides being a good pilot and a hell of an engineer, Young had a wonderful dry wit—a real down-home, country-boy type of delivery. And he

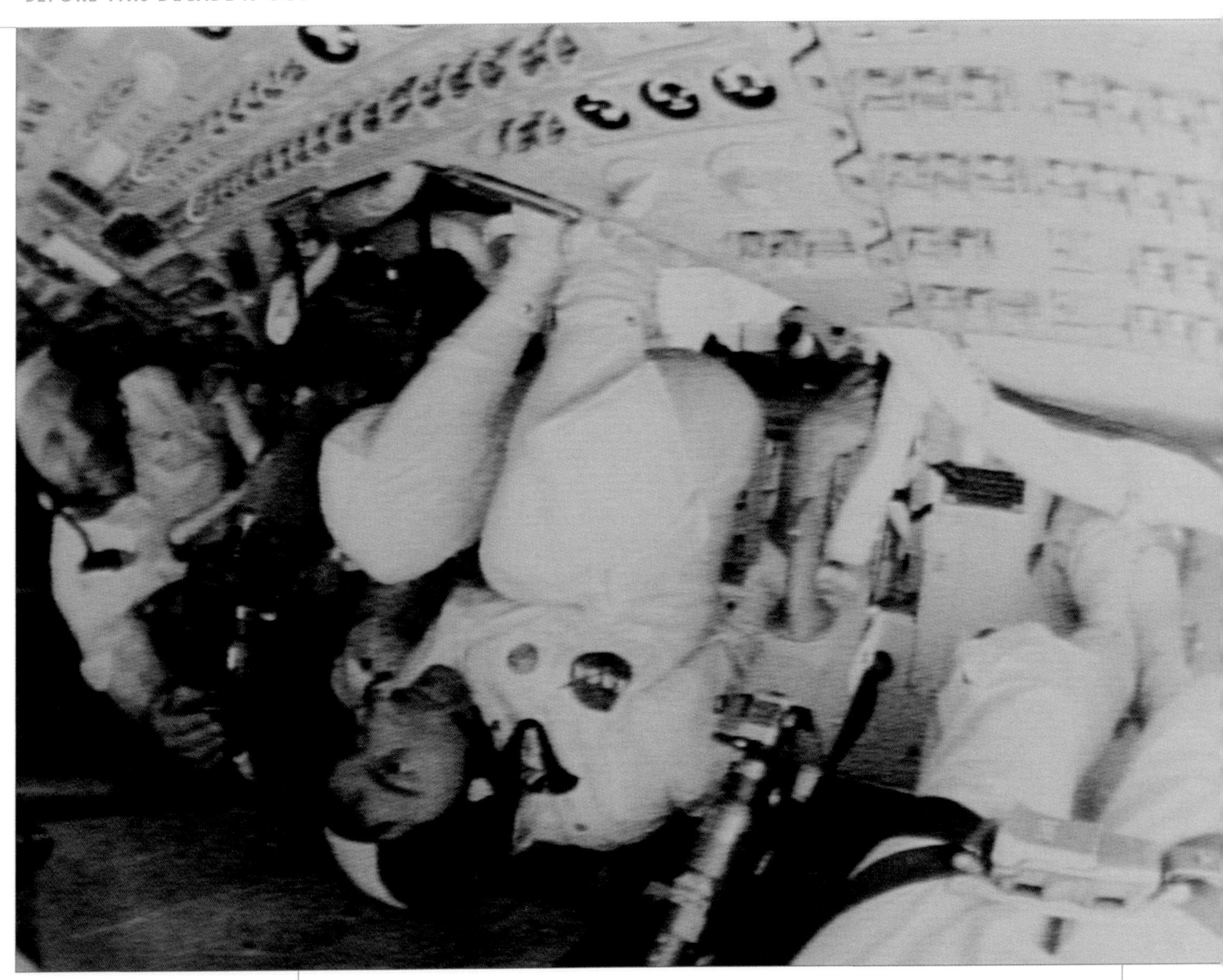

Indulging in zero-g acrobatics inside *Charlie Brown,* Young curls into a somersault above his couch while Stafford looks on. Weightlessness made the command module's tight quarters seem roomy.

was an expert on the command module systems; Stafford felt lucky to have him on the crew.

Now, a day into their journey to the moon, Stafford, Young, and Cernan were following the trail blazed by Apollo 8. None of them believed for a moment that the second lunar voyage might somehow be routine, and their suspicion was confirmed as soon as their Saturn V thundered off from Pad 39-B. Having heard the reports from Borman's and McDivitt's crews, Stafford and his crewmates thought they knew what to expect, but the ride was even more violent than they'd imagined. And the launch was only a prelude to the real heart-stopper. Three minutes into the Translunar Injection burn, the cabin began to vibrate with a strange buzzing sensation that the men both heard and felt. Something was wrong with the booster.

"Okay," Stafford radioed to Houston, "we're getting a little bit of high-frequency vibrations in the cabin. Nothing to worry about." Those words were for the benefit of mission control. In reality, Stafford was sure the mission was over, and so were his crewmates. Cernan, in the right-hand seat, was already thinking ahead to the abort maneuver they'd have to perform to get back to earth. The vibrations worsened until Stafford could barely read the instruments. His heartbeat quickened as he tried to puzzle out what might be happening. The only thing he could think of was a form of aerodynamic shaking that pilots call flutter. If it's bad enough, flutter can rattle an airplane to pieces. But how could flutter occur in a vacuum? Stafford held the abort handle in his left hand; with a twist he could shut down the booster and end the mission. But he told himself, "No way. We've come this far—if she blows, then she blows." The three men held their breath as the nearly six-minute burn continued. Stafford anxiously eyed the computer's velocity readout, which was climbing toward the 35,000-plus feet per second needed to send Apollo 10 to the moon. He said silently, "C'mon, baby . . ." At last the booster shut down, on time, and Apollo 10 was right on course. Later, engineers would blame the vibrations—which were well within the spacecraft's design tolerance—on some of the booster's pressure-relief valves, and would remedy the situation for future missions. Apollo 10's first heart-stopper turned out to be false alarm, but Stafford and his crew understood: they could take nothing for granted. ☾

WEDNESDAY, MAY 21
9:49 P.M., HOUSTON TIME
3 DAYS, 11 HOURS
MISSION ELAPSED TIME

It had ridden into space nestled within the third stage of the Saturn V, hidden from view by a set of protective panels, waiting to be extracted from its berth. That had happened shortly after Translunar Injection, when John Young separated the command module *Charlie Brown* from the booster and pulled away. At the same time, the protective panels departed from third stage, exposing the lunar module *Snoopy,* its metallic skin gleaming in the sunlight. Young then activated a special docking probe in *Charlie Brown*'s nose, and slowly steered the 32-ton command ship back toward the lander. The two ships met in a scene Freud would have loved, as Young guided *Charlie Brown*'s docking probe into a conical port, called the drogue, in *Snoopy*'s roof. Tiny "capture latches" on the tip of the probe fit inside a hole at the drogue's center. Young retracted the probe, drawing the two ships together, until twelve spring-loaded "docking latches" snapped shut, firmly mating the two craft.

During the three-day voyage, *Snoopy* had waited, dormant, docked to *Charlie Brown*'s nose. Now, on Wednesday evening, Young opened the command module's forward hatch and removed the docking mechanism, opening up a short tunnel between the two craft and clearing the way for Gene Cernan to enter the lander. When he did, pushing off the command module's floor and floating through the tunnel, it was like entering another world. Opening *Snoopy*'s hatch, he found himself staring at the floor of the tiny cabin, as if he were hanging by his toes. He tucked his body into a spin until he was upright in the small, gray space. The lander was a strange machine, even

Opening Snoopy's hatch, he found himself staring at the floor of the tiny cabin, as if he were hanging by his toes.

from the inside, but Cernan had come to know it well. In front of him was a square instrument panel, packed with switches, gauges, and displays. At waist height were two sets of hand controllers, one for each man. There were two small triangular windows, one on either side of the main panel, and a smaller, rectangular rendezvous window in the ceiling on the commander's side. More panels, covered with circuit breakers, lined the side walls. Bundles of wiring and all kinds of plumbing were visible along the floor, exposed because covering them would have added too much weight. They gave the cabin the look of a boiler room. Behind him, atop a small ledge, the can-shaped cover for the lander's ascent engine protruded into the cabin. There were no seats; none were needed. The lunar module was the first true spacecraft: it flew only in the void of space.

On the outside, the lunar module was as alien as the world it would land on. With no need for aerodynamic sleekness, it was an unlikely amalgam of angular shapes and boxes, rocket engines and antennae, doors and landing gear. No spacecraft before it had so completely merged form and function, and the result was a definite shift in appearance toward the organic. In the clean room at the Cape, where Cernan and Stafford had last seen it, the lander was a metal chrysalis, landing gear folded against its sides in preparation for the launch shroud. Now, in space, it was a huge robotic insect, with two triangular windows for eyes and a square hatchway for a mouth, antennae jutting at all angles, and four foil-clad landing legs. Even its name seemed to say it was alive; the astronauts called it "the lem." ☾

The reason the LM looked so strange was simple: it had to be as lightweight as possible. The Grumman designers achieved the biggest weight savings from splitting the craft into two pieces. A boxy descent stage held the rocket used for the landing, as well as the landing gear and supplies needed for the moonwalk; it would be left behind on the moon. Perched atop the box was the angular ascent stage, which contained everything else—the crew cabin with its controls, supplies, breathing oxygen, and electronics, as well as a separate rocket for the ascent from the lunar surface back to orbit. Before long, however, the LM was hopelessly overweight, and the Grumman people were obsessed with making it lighter. In short order, the seats called for in early designs were scrapped, since the astronauts would not need them in weightlessness or the moon's one-sixth g. Then all traces of aerodynamic curves were shaved off. By 1965, with 95 percent of the LM's design finalized, the weight-trimmers were still looking for ounces.

In the end, the lander became a strange mix of strength and fragility. The skin of the descent stage was only a Mylar wrapping stretched over a frame. In the ascent stage, the walls of the crew cabin were thinned down until they were nothing more than a taut aluminum balloon, in some places only five-thousandths of an inch thick. Once, a workman accidentally dropped a screwdriver inside the cabin and it went through the floor. Now, in space, it seemed deceptively flimsy. When the cabin was pressurized the front hatch bulged outward. *That* had scared John Young, who was in the command module wearing nothing but a pair of long johns; he was up there muttering, "I didn't know I was volunteering to go on this damn thing in my underwear . . ." No wonder Jim McDivitt called his LM "the tissue-paper spacecraft." But it was sturdy enough—even more than enough—for the moon. ☾

For about two hours, Cernan was to check *Snoopy*'s systems in preparation for the next day's dress rehearsal. But when he arrived, he found himself floating in a snowstorm of white fiberglass. It had somehow blown out of an insulation blanket on the tunnel wall, and had found its way into the LM via a pressurization valve. Before he could do anything more, Cernan had to vacuum the fiberglass bits out of the air. Stafford arrived to lend a hand and burst out laughing. There was Cernan with bits of white stuck to his hair, his eyelashes, his mouth. All Stafford could think of was chicken-plucking time at the poultry house.

There followed a couple of hours of concentrated work. Cernan was tunnel-visioned into his checklists, and when there was a lull, he stopped and

EVERY POUND A CHALLENGE

Three models of the lunar module show an evolution greatly influenced by the need to trim its weight. In 1962 *(left)* the lander had five legs, four windows, and two round docking hatches. By 1965 *(below)* the craft had lost a leg and two windows, the front hatch had become merely a doorway, and inside the astronauts lost their couches—they would stand during ride in the lander. In the final version *(right)* rounded surfaces became angular, and the lander took on a distinctly buglike appearance.

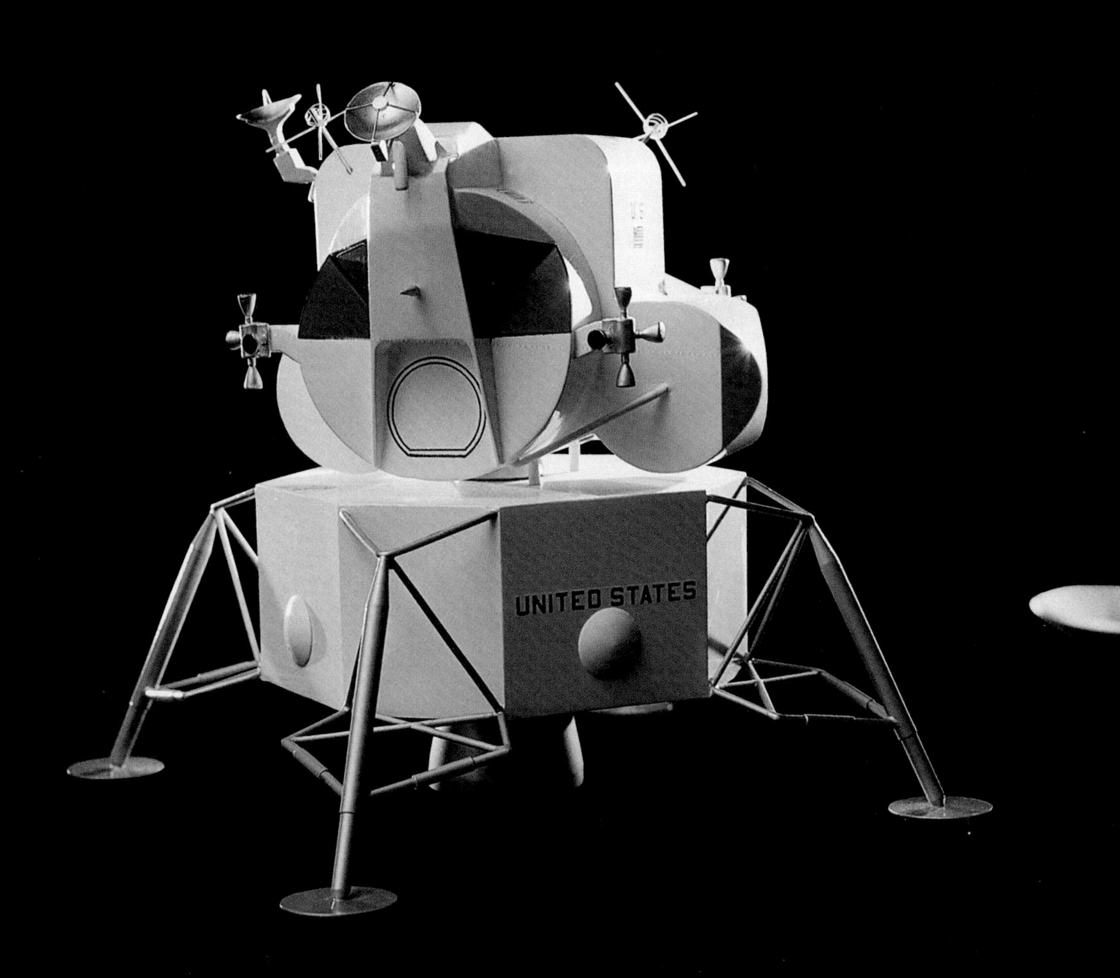

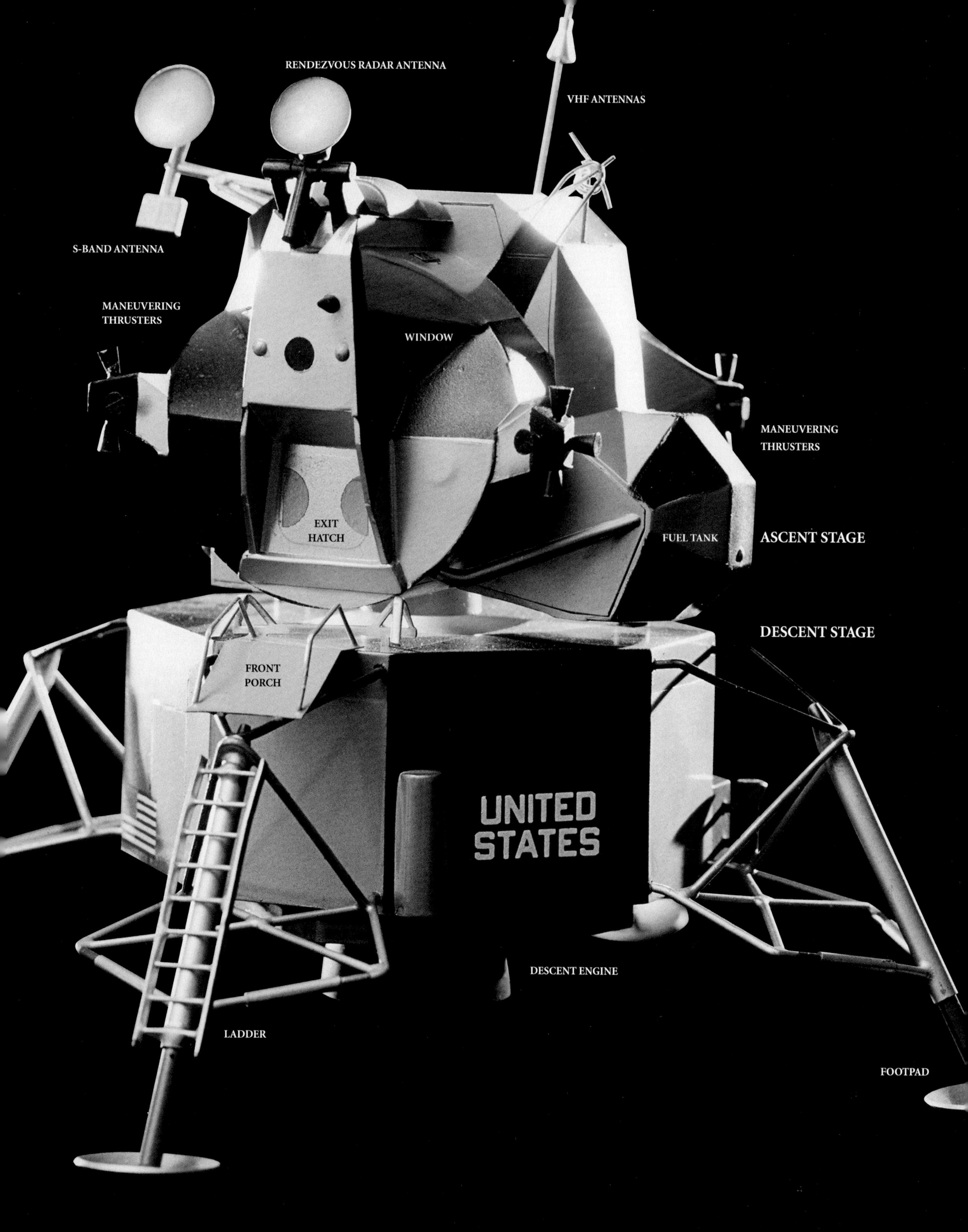

RENDEZVOUS RADAR ANTENNA

VHF ANTENNAS

S-BAND ANTENNA

MANEUVERING
THRUSTERS

WINDOW

MANEUVERING
THRUSTERS

EXIT
HATCH

FUEL TANK

ASCENT STAGE

DESCENT STAGE

FRONT
PORCH

UNITED
STATES

DESCENT ENGINE

LADDER

FOOTPAD

The Apollo 9 lunar module looks deceptively fragile as it undergoes pre-launch preparations in a clean room at Cape Kennedy in September 1968. Exposed sections will eventually be covered by sheets of the metal-coated plastics Kapton and Mylar, and thin sheets of aluminum.

consciously tried to take in what was happening. On his first spaceflight, the Gemini 9 mission three years earlier, Cernan's moment of realization had come during a two-hour space walk. He was floating next to the spacecraft, with no sense of speed as he whizzed around the earth at 17,500 miles per hour, about to make an unsuccessful attempt to test a rocket-powered backpack, and he took a moment to look around at the planet passing beneath him, at the craft that was his sanctuary in the void, at the blinding sun, and the dark of space.

The same thing happened now, inside *Snoopy,* as Cernan looked up from his checklist. The two triangular windows were still covered by translucent yellow shades, but he could see the moon's surface through them, drifting slowly and silently past, eerie because of the yellow tint. He wasn't quite sure it was real. And he said to himself, "Do you realize where you really are?"

THURSDAY, MAY 22
1:35 P.M., HOUSTON TIME
4 DAYS, 2 HOURS, 46 MINUTES
MISSION ELAPSED TIME

Charlie Brown and *Snoopy* flew in formation through the unfiltered lunar sunlight. For about half an hour, the two craft had drifted above the craters while the three men prepared for what lay ahead. An hour from now, Stafford and Cernan would fire *Snoopy*'s descent engine and enter a lopsided ellipse called the descent orbit, with a low point of only 50,000 feet. On a real landing, at that 9-mile altitude, the lander's engine would reignite for the final descent to the surface. But on this mission, Stafford and Cernan would simply coast along, then soar up to 215 miles, then swoop down once more. On the low passes, the pilots would reconnoiter Apollo Landing Site 2, a small patch of moonscape near the southwestern edge of the Sea of Tranquillity. Then Stafford and Cernan would cast off their descent stage and fire the ascent rocket to rejoin Young.

Now, with a burst from *Charlie Brown*'s maneuvering thrusters, Young pulled away. Inside *Snoopy,* Stafford and Cernan saw the command module shrink into the distance.

"Have a good time while we're gone, babe," Cernan called.

"Yeah," Stafford said. "Don't get lonesome out there, John."

Cernan added wryly, "And don't accept any TEI updates."

"Don't you worry," said Young.

It was the kind of banter that often comes to fighter pilots when their necks are out. Stafford and Cernan knew only too well that they were flying a craft that could not take them back to earth, that if something went wrong

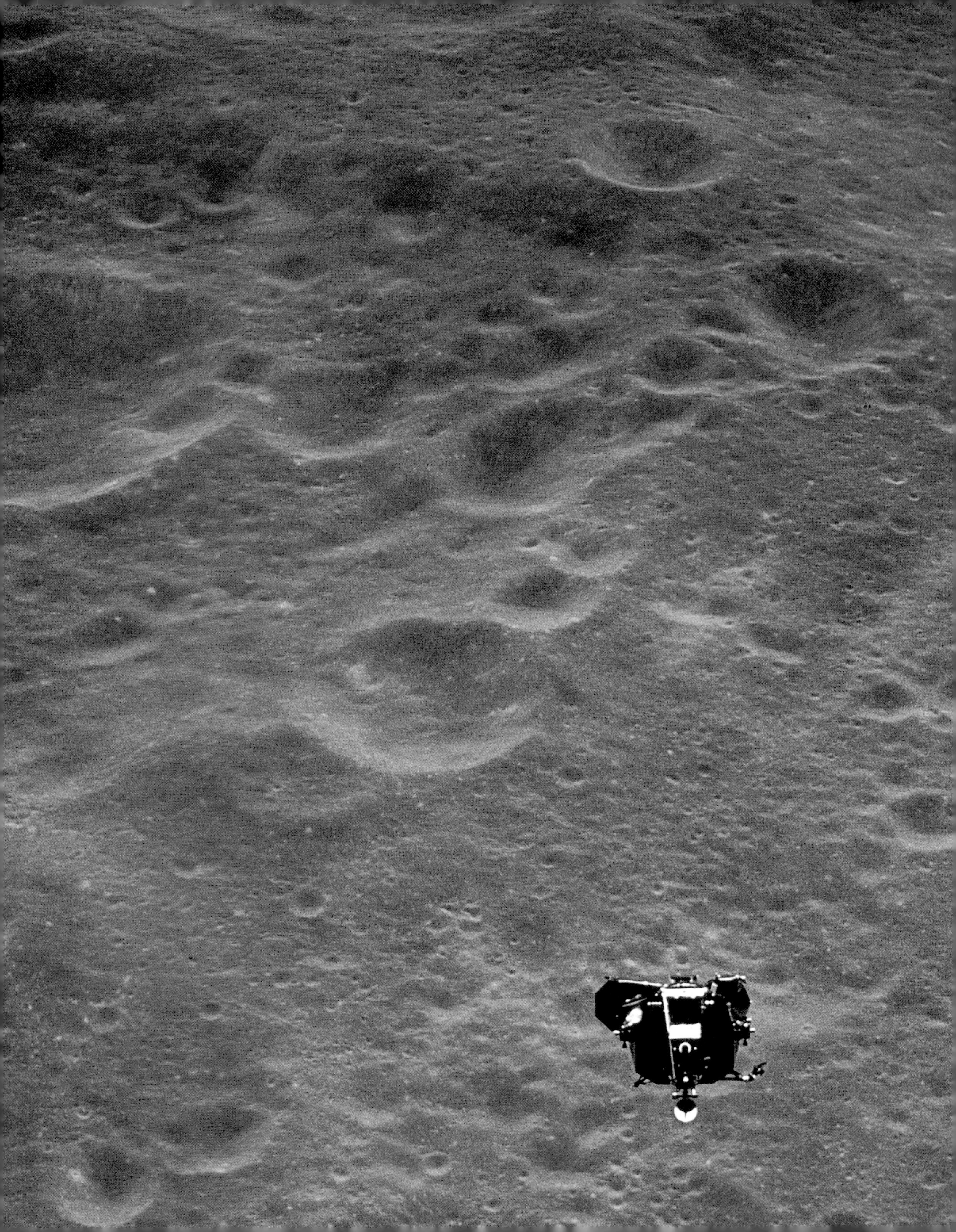

down there at 50,000 feet, Young would have to come get them. Ultimately the lunar landing was a walk across a long, high wire, and they were about to take the penultimate step.

2:35 P.M.

Now came the moment of truth. It was time for Stafford and Cernan to fire *Snoopy*'s descent engine to enter the descent orbit. It would take about half a minute to shift the low point of *Snoopy*'s orbit from 69 miles to roughly 50,000 feet. The timing was critical, for if the burn went even two seconds too long, the lunar module would crash into the front side of the moon. Meanwhile, in *Charlie Brown,* John Young listened in and tracked *Snoopy* with the command module's sextant, ready to come to the rescue if Stafford and Cernan had a malfunction that prevented them from getting into the proper orbit.

Rising steadily from the far side of the moon *(opposite),* Tom Stafford and Gene Cernan return from their lunar barnstorming run. Above, the crater Schmidt typifies the rugged terrain seen by the astronauts.

At the proper moment *Snoopy*'s computer flashed "99" and Cernan pushed the PROCEED button. The engine fired soundlessly in the vacuum.

"We're burning, John," called Cernan. Inside *Charlie Brown,* Young could see the glow from *Snoopy*'s engine. Half a minute later, the burn was over, and *Snoopy*'s computer showed they were right on target.

Almost immediately Stafford and Cernan sensed their diminishing altitude. Within 20 minutes they were dramatically lower. The moon's curved edge flattened out. As *Snoopy* flew onward, mountains appeared on the horizon; as Stafford and Cernan advanced they could see that the mountains were actually the rims of huge craters. Enormous boulders dotted the land—some looked as big as five-story buildings; after the flight Stafford would find out they were ten times that size. There were cliffs that must have been four or five thousand feet high. The mountains of the lunar highlands looked close enough to touch.

Suddenly the men were stunned by the sight of their first earthrise. Now

Neil Armstrong pilots a glider through updrafts south of Houston. For Armstrong, one of the most private men in the astronaut corps, flying could serve as a refuge from the glare of publicity he attracted as Apollo 11's commander.

they were in contact with Capcom Charlie Duke, and Cernan's voice soared. "Houston, this is *Snoopy*! We is Go and we is down among 'em, Charlie!" Down they flew, heading for Landing Site 2. Shadows lengthened as they coasted into the realm of lunar morning. Now they were over Tranquillity's plains, smooth, like wet clay; to Stafford, they seemed almost like the desert near Edwards. Later, he would report that Site 2 looked smoother than he expected, at least in the center of the area. But he would warn that if Armstrong and Aldrin were off-target, and especially a few miles downrange, the terrain would be rougher, and they had better have enough fuel to maneuver around and look for a safe spot.

Stafford and Cernan had flown routinely at 50,000 feet, but this was unlike any airplane ride. No one had ever flown this fast at such low altitude. Following the laws of orbital motion, *Snoopy* had traded its altitude for speed, until it was now nearing 3,700 miles per hour, more than five times the speed of sound. Stafford and Cernan were barnstorming the moon.

5:33 P.M.

Flying upside down and backward, Stafford and Cernan prepared to rejoin the command module by firing the lander's ascent engine. First, they would cast off *Snoopy*'s descent stage by firing a set of pyrotechnic bolts. "That

mother may give us a kick," Cernan warned. "You ready?" Suddenly, Cernan saw the moon's horizon spin wildly; the lander was tumbling out of control. Still on hot-mike, his words heard by the listening earth, Cernan blurted, "Son of a bitch . . . What the hell happened?"

Stafford quickly punched the button to get rid of the heavy descent stage, to give him more authority with *Snoopy*'s small maneuvering thrusters. For eight long seconds he struggled to regain control, and then *Snoopy* was still. After the flight, the incident would be traced to a combination of minor and easily correctable failures. Ten minutes later, in darkness, the ascent engine started Stafford and Cernan on their journey back to John Young. Thirty-one hours after that, early in the morning of May 24, the second team of moon voyagers headed back to earth.

In Houston, Neil Armstrong and his crew were into their final months of preparation. When Tom Stafford's crew came back from the moon, only the landing remained. Everything else was now part of the collective spaceflight experience—because when one team of astronauts and flight controllers learned something on a mission, it was as if all of them had done it. The achievements of one mission were there for the next to build on, and the mistakes would not be repeated—or so everyone hoped. The way was now clear for Apollo 11 to attempt the ultimate test flight.

III. DOWN TO THE WIRE

He was alone, high above the flat Texas landscape, with only the sound of the air rushing past his canopy. From the ground, you might have seen him, wafting on the thermal currents, circling in wide arcs. Whenever he got the chance, Neil Armstrong took time out from the pressure of his life as an astronaut to go soaring. He loved this kind of flying—unpowered, pure, mentally demanding—as much as he did any other; in short, there wasn't anything about flying that didn't interest him. But it was here, apart from the world, harnessing the wind, that Armstrong found his greatest relaxation. Almost from the moment his name was announced as the commander of Apollo 11, the press tried to know him, and for the most part he gave them shy smiles and brief, almost cryptic answers to their questions about his thoughts and feelings. By his own admission, he much preferred to talk about ideas than people. What emerged from their typewriters became a kind of legend: the small-town Ohio boyhood, the love affair with airplanes that began earlier than he could remember, the way he devoured books as a

schoolboy, how he built airplane models and tested them in a homemade wind tunnel, how he got his pilot's license before he learned to drive. They wrote of his achievements in the Korean War, where he flew seventy-eight combat missions and was awarded a couple of air medals. They chronicled his career as a test pilot and an astronaut. But like his sailplane, Armstrong was essentially beyond their reach. There had always been that quality about him, even in the summer of 1955, when he arrived at Edwards Air Force Base, where a small group of fliers was exploring the unknowns of high-speed flight for the National Advisory Committee on Aeronautics. Edwards was a place of blast-furnace heat, howling winds, and utter desolation, but it was heaven on earth for pilots. Dawn came still and clear, spilling over distant mountain ranges onto a smooth, hard expanse of clay that seemed as vast as the cloudless blue canopy above. Armstrong loved the dawn at Edwards and reveled in its short-lived serenity—short-lived, because on any given morning the stillness was shattered by the sonic booms of pilots unleashing the most exotic flying machines in existence. Somehow, the primitive beauty of Edwards was the perfect backdrop to the unfolding of the future. And it was the perfect place for Neil Armstrong.

In the meantime, there was Sputnik, which set off a chain of events that moved spaceflight toward reality faster than Armstrong could have anticipated.

NACA's High Speed Flight Station was perched on the edge of Rogers Dry Lake. The NACA fliers epitomized a new breed of test pilot, engineers as much as aviators, that had emerged since the Second World War. Unlike the air force pilots at Edwards, who seemed to treat a new airplane as if it were a romantic conquest, losing interest as soon as they'd used it to set a record, the NACA pilots delved into lengthy, often tedious analyses that were the heart and soul of test flight. The combination of meticulous research and all kinds of flying opportunities—the mundane as well as the exotic—was what made Edwards a place to be cherished. Armstrong would look back on his years at Edwards as the most fascinating period of his life. ☾

There were two things about this quiet young man, whose clear, boyish face made him look even younger than his twenty-five years, that the other NACA pilots found remarkable. Not his flying; he was a skilled aviator, with

an impressive grasp of aerodynamics, but there were better stick-and-rudder men at Edwards. No, the first thing they noticed was his intellect; everything he did, even casual speech, seemed to be the result of a great deal of thought. The second was his remoteness. Armstrong often kept people at arm's length. He rarely engaged in idle conversation and steadfastly guarded his privacy. He and his wife, Jan, hadn't been there very long before they left the community that clung to the base and moved into a cabin in the Juniper Hills, among the Joshua trees, without electricity or running water. Was such a man knowable? One friend would say years later that at his core, Armstrong was Scottish, with the moral code of a Highlander: rock-solid integrity and a relentless memory of any injustice committed against him or his kin. In time, the NACA pilots realized that Armstrong wasn't aloof; he was shy. Once they got past his great reserve, they found warmth. Once he became a friend, he was a good friend. If he could be reticent, then he could also become so involved in conversation while driving that his passengers nervously eyed the road. Under the serious layers lurked a tart and understated wit. If he was a consummate loner, then it was also true that a post-flight party in full swing usually saw Armstrong at the piano, pounding out a bit of ragtime; he might be the last to leave. But even his friends could only guess at what Armstrong was thinking, what really drove him. As fellow NACA veteran Milt Thompson said, years later, "I knew him, but I didn't know him." ☾

At Edwards, Armstrong saw the blue desert sky and envisioned the trackless void beyond. In the mid-fifties space travel was something most people were careful to avoid talking about in serious tones. But it wasn't far-out to the people at Edwards, and Armstrong soon sensed that the road to space began on the high desert. The air force and NACA were sitting down to design a sleek, winged rocket plane called the X-15, a descendant of the rocket-powered X-1 that Chuck Yeager had used to break the sound barrier here a decade earlier. Carried aloft under the wing of a B-52 bomber and then released, the X-15 would zoom to the fringe of the atmosphere, then glide to an unpowered, or "dead-stick," landing on the lakebed. Armstrong was among the handful of pilots slated to fly it.

In the meantime, there was Sputnik, which set off a chain of events that moved spaceflight toward reality faster than Armstrong could have anticipated. By 1959 NACA had become the National Aeronautics and Space Administration, and Project Mercury was in the headlines. Some—though not all—of the Edwards men snickered at the idea of "astronauts" who would be essentially passengers from start to finish. The NASA pilots knew that *they,* not the Mercury astronauts, were on the cutting edge, and that they would fly

At the controls of the lunar module simulator, Armstrong and Aldrin practice landing on the moon. Outside, teams of instructors fed the pair simulated emergencies ranging from difficult to diabolical.

into space at the controls of the X-15's more advanced successor. Soon that craft too was on the drawing board: Dyna-Soar, a manned glider designed to be lofted by a Titan booster and then land on a runway. Armstrong was named NASA's prime pilot on Dyna-Soar. He had no interest in becoming a Mercury astronaut.

By 1962, in its third year of operations, the X-15 was streaking through the desert sky at unheard-of velocities—more than five times the speed of sound—and record altitudes. That April, on his sixth X-15 flight, Armstrong piloted the sleek, black arrow to 207,000 feet, high enough to see a bright and gently curving horizon beneath a black sky. For a moment he lingered there at the top of his arc, suspended between aviation and spaceflight, before returning to earth.

The view from 207,000 feet only whetted Armstrong's appetite; he felt he had glimpsed his own future. But that future, he realized, was not in the X-15, or in Dyna-Soar, which now seemed doomed, never to leave the drawing board. Already, John Glenn had demonstrated that astronauts were more than passengers by taking over manual control of his Mercury spacecraft when the automatic system malfunctioned. Armstrong realized he'd been hasty in rejecting the astronaut program. It wasn't easy to think about leaving Edwards, but NASA was headed for the moon. The week he flew the X-15 to the heights, NASA put out the call for the second group of astronauts. One day, Armstrong and a fellow NASA pilot named Bill Dana were mulling over their respective plans. Dana said he thought there was a much better future flying the Supersonic Transport that everyone was talking about. Armstrong told him, "You can do whatever you want to about that, but space is the frontier, and that's where I intend to go." ☾

JUNE 1969
FLIGHT CREW
TRAINING BUILDING,
KENNEDY SPACE CENTER

Neil Armstrong and Buzz Aldrin stood side by side in the lunar module simulator. A soft green light filled the enclosure, emanating from dozens of electroluminescent dials and readouts. "Okay, Neil," said a voice in their headsets, "we'll give it to you at pitchover minus 30 seconds." It was the voice of one of the simulation instructors who sat at a console nearby. On the instrument panel, gauges jumped to life and registered engine temperature, chamber pressure, fuel and oxidizer readings, thrust, and the other vital signs of a lunar module in flight. A tape meter scrolled down ever diminishing readings of altitude; another showed their speed of descent. Two "8-ball" indicators showed the LM's orientation in space. Armstrong and Aldrin scanned each gauge, just as they would in the moments before the lander began its final descent to the surface. Suddenly a coded message flashed on the LM's computer display. "P sixty-four," Aldrin announced, and suddenly, through the two small triangular windows, a bright moonscape swung upward into view. Once again, as they had done many times over the past few months, Armstrong and Aldrin were about to confront Apollo's greatest unknown.

"Space is the frontier," Neil Armstrong had said seven years earlier, and with the lunar landing he had found a piloting task at the frontier of spaceflight. Nothing like it had ever been attempted. The liftoff from the moon, on which his and Aldrin's lives would depend, would be a rocket launch that was basically like many others; only the launch site would be extraordinary. But a manned rocket *landing*—that was something for which there was no prior experience. The airless moon offered no alternatives.

The trip from 50,000 feet down to the surface would be, in essence, a long, controlled fall from orbit, a trajectory governed by the same exacting laws of motion as orbit itself. Called the Powered Descent, it would begin with the LM flying horizontally over the moon at a speed of just under 3,800 miles per hour, and would end with a touchdown as gentle as a leap from a bar stool. There were two keys to the Powered Descent. One was an engineer-

In theory, Armstrong knew, he could let the computer fly the LM all the way to touchdown, without touching the controls. In reality he wasn't about to do that.

ing marvel: the world's first rocket engine with a throttle. With it the LM could descend at a whole range of speeds; if necessary it could even hover in the sky like a helicopter. The second was the LM's onboard computer. More than half the lander's weight was fuel for the descent rocket, but the fuel budget was so tight that there would have been little chance of making it to the surface without computer control. The computer wizards at MIT had programmed it to govern the entire descent. It would compute the precise amount of thrust needed from the descent engine at any given moment. It would keep track of the LM's distance to the designated landing site. It would share control of the spacecraft with the pilot, or, if asked, give up control altogether.

In their simulations of the Powered Descent, Armstrong and Aldrin began where Tom Stafford and Gene Cernan left off, 9 miles above the moon. With the lander tipped back horizontal, they lit the descent engine and throttled up to full power, using its thrust to brake their orbital speed. Obeying Kepler's laws, the lander fell moonward as it slowed. In this early phase the craft was oriented with its windows facing the moon. This allowed Armstrong to observe landmarks passing underneath and use them to judge whether he and Aldrin were on the proper trajectory. Then, some 4½ minutes into the descent, the craft rotated until it was windows-up, allowing the LM's landing radar to bounce echoes off the surface, to determine altitude. At about 6 minutes, the computer throttled the engine down to 55 percent thrust, and the long brake continued. Throughout this period, if everything went well, Armstrong and Aldrin did little more than watch the gauges.

Finally, 8½ minutes after engine ignition, the final phase of the Powered Descent began. Some 7,500 feet above the moon the computer pitched the lander forward until it was nearly upright—this was called the pitchover maneuver—so that its engine would not only brake the craft's forward speed but keep it from falling too fast. From then on the lander followed a slanting path to the surface like a car braking on a long, straight mountain road, with the computer carefully riding the throttle like a motorist's foot on the brake. With only 4 minutes from pitchover to touchdown, things happened very fast. In that sense, landing on the moon was similar to Armstrong's X-15 flights. Some X-15 pilots talked of a kind of "fast time" when the clock seemed to race. There wasn't time to read checklists; they had to have the entire flight plan memorized. Pilots sometimes talk about being "ahead of the airplane," a state of mind that enables you to think one step ahead of your craft, the better to cope with emergencies. But that was nearly impossible flying a rocket plane at five times the speed of sound—and it wouldn't be much easier in a lunar module, descending to the moon.

For Armstrong, pitchover was the moment when the piloting part of the landing began. With the cratered plains of the Sea of Tranquillity spread before him, he scanned the advancing moonscape, looking for familiar landmarks he had memorized from photos. And by using a special grid called the Landing Point Designator, he could see for the first time where the computer was aiming to land. The LPD grid was marked off in degrees on the LM's double-pane window, and was used like a gunsight. By sighting along the grid at the proper angle, which was supplied by the computer, Armstrong was able to spot the landing point. If he didn't like what he saw, he could tell the computer to change its aim by nudging the LM's attitude controller with his right hand: A nudge to the left or right shifted the aim point accordingly; tilting the controller backward or forward moved the landing spot up- or downrange.

Meanwhile, Aldrin interrogated the computer, asking it for the LM's height and speed and comparing the results with the data from the radar. Then he relayed this information to Armstrong. Aldrin knew very well that his title of "lunar module pilot" was a misnomer, because all the flying during the landing—and there might not be much of it—belonged to Armstrong.

In theory, Armstrong knew, he could let the computer fly the LM all the way to touchdown, without touching the controls. In reality he wasn't about to do that. It wasn't just that as a self-respecting test pilot he wanted to be at

IN 1969

Surgeons in Houston, led by Denton A. Cooley, perform the first implantation of an artificial heart.

CBS cancels *The Smothers Brothers Comedy Hour* for failure to submit the show for review prior to broadcast.

***Oliver!* wins best picture Oscar; Barbra Streisand in *Funny Girl* and Katharine Hepburn in *The Lion in Winter* tie for best actress.**

1
2

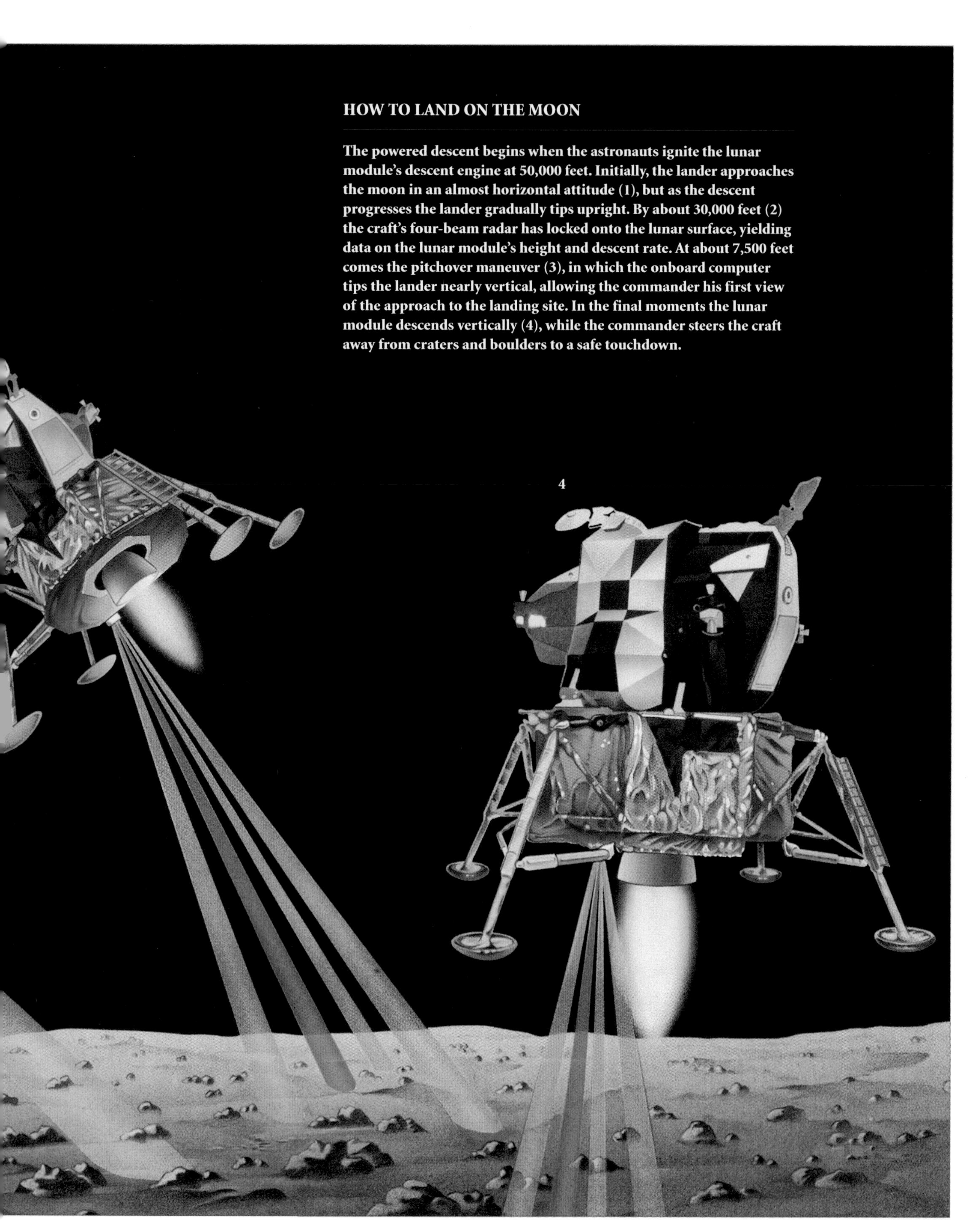

HOW TO LAND ON THE MOON

The powered descent begins when the astronauts ignite the lunar module's descent engine at 50,000 feet. Initially, the lander approaches the moon in an almost horizontal attitude (1), but as the descent progresses the lander gradually tips upright. By about 30,000 feet (2) the craft's four-beam radar has locked onto the lunar surface, yielding data on the lunar module's height and descent rate. At about 7,500 feet comes the pitchover maneuver (3), in which the onboard computer tips the lander nearly vertical, allowing the commander his first view of the approach to the landing site. In the final moments the lunar module descends vertically (4), while the commander steers the craft away from craters and boulders to a safe touchdown.

the controls at the moment of landing; it was because the computer was blind. It might bring him and Aldrin down on an absolutely perfect trajectory, right into a crater or a boulder field. Armstrong planned to take over from the computer when he was about 500 feet above the moon.

Even then, Armstrong would let the computer do most of the work. Flying the LM entirely by hand was so difficult that he wanted to avoid it at all costs, and unless the computer went out, he wouldn't have to. Instead, he would maneuver the LM by tilting it slightly to one side, and let the computer continue to ride the throttle. Armstrong knew there was always the possibility that something would go wrong—from communications failure to a malfunction of the descent rocket—and force him to abort the landing. In that case, he could press the button marked ABORT STAGE, setting off a dramatic chain of events. In an instant, pyrotechnic bolts would sever the connections between the descent and ascent stages. At the same moment the ascent rocket would blast to life, boosting Armstrong and Aldrin back toward orbit. But an abort, especially one at low altitude, carried its own risks. If the ascent engine didn't light, there would be no time to find out why and do something about it; Armstrong and Aldrin would crash in a matter of seconds. Even if it did fire, the ascent stage might not separate cleanly from the descent stage. And even if there were no malfunctions, there was the problem of finding Collins, with the timing for the rendezvous now completely disrupted; Armstrong and Aldrin would face a long and complicated journey back to the command module. And it would have been several hours since Aldrin had aligned the LM's guidance system with the stars. An abort would force the men to fly with a guidance system that could be significantly in error. And so the last thing Armstrong wanted to do was abort—not only because it would mean failure, but because it could be even more risky than the landing itself.

As the LM got very close to the moon, Armstrong and Aldrin would enter the most hazardous portion of the descent. Somewhere in the last 200 feet, they would be too low to abort successfully if the descent engine quit, for the LM would be going too fast for the ascent engine to arrest the lander's plunge and start the ascent stage upward again. The astronauts, borrowing a term from helicopter pilots, called this part of the descent the "dead man's curve."

Even if nothing went wrong, Armstrong would have his hands full. Already, he would have had to find a safe landing spot, free of large craters and boulders. By the time the LM was 100 feet up, Armstrong would have to arrest nearly all of the craft's forward motion and begin a slow, careful vertical descent.

The computer would still control the throttle, but Armstrong would be able to make small adjustments by using a special toggle switch. By clicking the switch up or down he could increase or decrease his rate of descent by one foot per second, repeating this as many times as necessary to get the change he wanted. It seemed to Armstrong to be a strange way to control a craft, and he was skeptical that the switch would prove effective in the actual landing. But he would have to wait and see until he was 100 feet above the moon.

In these final moments Armstrong's gaze would be directed almost entirely at the moon, and he would rely on Aldrin's steady reports on altitude, horizontal speed, and descent rate. The first part of the LM to touch the moon would be three long metal probes attached to the footpads; at that moment, a blue light labeled LUNAR CONTACT would glow on the instrument panel. Aldrin would be watching for the contact light, ready to call it out. At that moment, Armstrong would shut down the engine, and the LM would fall the remaining three feet to the surface.

Perhaps the toughest thing about landing on the moon was that there was no such thing as a second chance. Even in carrier landings—and Armstrong had flown plenty of those in Korea—there was usually the chance to circle and try again if something went wrong the first time. But in the lunar landing he and Aldrin would have to stay on top of any problems that might come up, even as the computer brought them closer to the moon. No surprise that the simulator instructors knew Armstrong and Aldrin as serious students. Aldrin, who was becoming a virtuoso with the LM's computer, was a bundle of self-confidence, ready to debate with instructors and experts on any technical matter.

Armstrong was all business. As the man who Mike Collins would call "far and away the most experienced test pilot among the astronauts," Armstrong had seen his share of close calls, and more than once had faced malfunctions he had never seen in simulations, including one on his Gemini 8 mission that nearly cost him his life. It was March 5, 1966, and Armstrong had just accomplished the first space docking, linking up with an Agena target rocket 185 miles above the earth. He and his copilot Dave Scott were monitoring the instruments on the joined spacecraft when suddenly they began to tumble. The men were sure that the problem was a stuck thruster on the Agena. If they let it continue, Armstrong knew, the tumbling might break the two craft apart. He undocked, and the Gemini began to gyrate faster and faster, spinning a full turn every second, as the earth and sun alternately flashed by the windows. Armstrong's vision

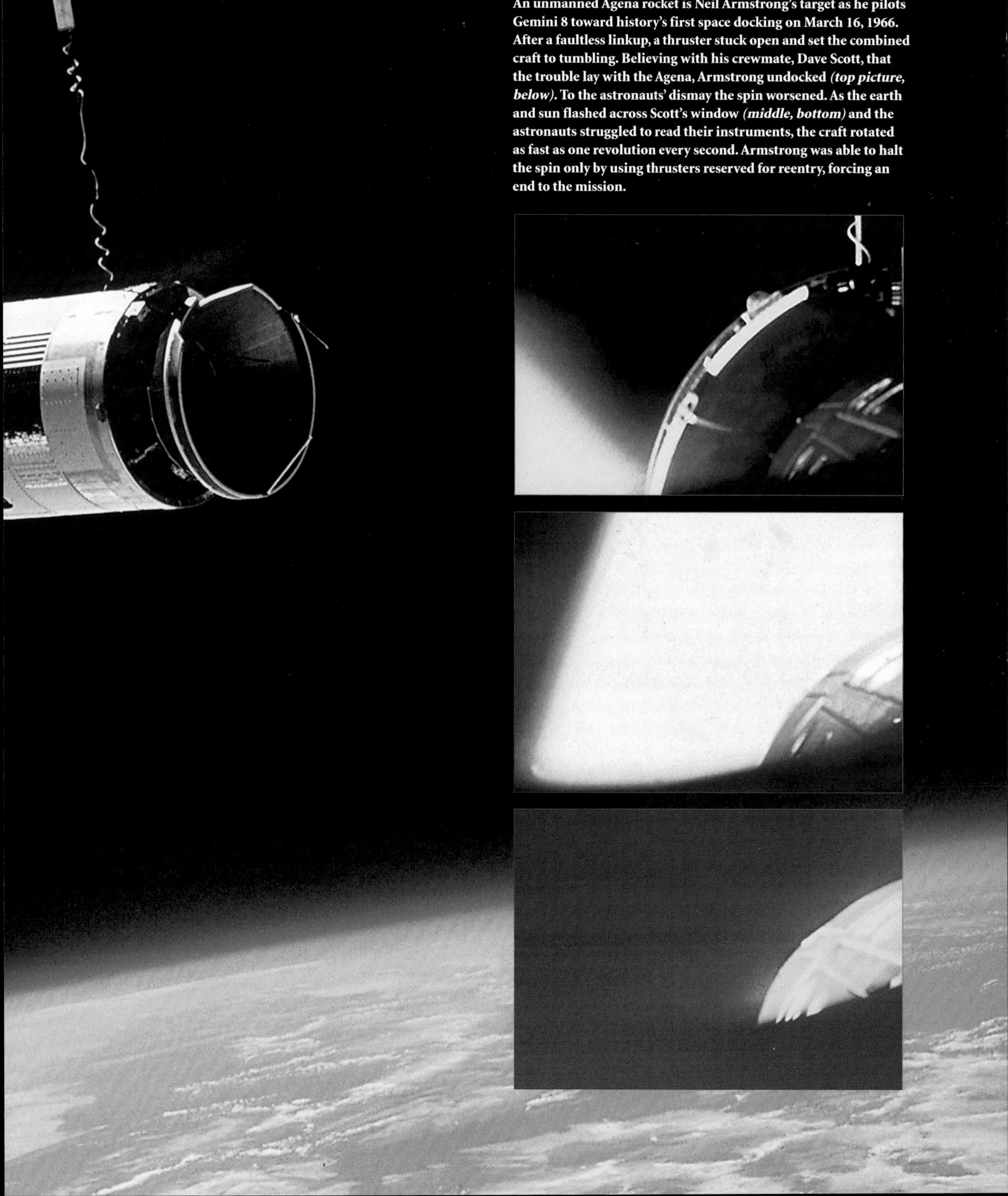

An unmanned Agena rocket is Neil Armstrong's target as he pilots Gemini 8 toward history's first space docking on March 16, 1966. After a faultless linkup, a thruster stuck open and set the combined craft to tumbling. Believing with his crewmate, Dave Scott, that the trouble lay with the Agena, Armstrong undocked *(top picture, below)*. To the astronauts' dismay the spin worsened. As the earth and sun flashed across Scott's window *(middle, bottom)* and the astronauts struggled to read their instruments, the craft rotated as fast as one revolution every second. Armstrong was able to halt the spin only by using thrusters reserved for reentry, forcing an end to the mission.

blurred as he searched the instrument panel and he said to Scott, in a voice tinged with wryness, "I gotta cage my eyeballs." Armstrong brought all his energy to bear on the problem, trying methodically to isolate the stuck thruster, but he could not. If the spin worsened, he and Scott were in danger of losing consciousness; then, it would be only a matter of time before the Gemini broke apart. Armstrong was forced to turn off the main thrusters and switch on a backup system, reserved only for reentry, an action which was tantamount to aborting the flight. Half a minute later he had halted the spin—and lost the mission. ☾

Other than disappointment, Armstrong carried no special psychic burden from that incident. He had no score to settle. He had been in the test flight business too long not to view Gemini 8 as just another page in the history of his profession, another encounter with the unexpected. Now, training to land on the moon, it was the unknowns that Armstrong worried about most.

If the spin worsened, he and Scott were in danger of losing consciousness; then, it would be only a matter of time before the Gemini broke apart.

For the first few weeks of training, the instructors went easy on Armstrong and Aldrin. In one practice landing after another, the two pilots worked out the teamwork they would need to handle the descent. By the time Stafford's crew came back from the moon, Armstrong and Aldrin were landing confidently and smoothly. And then, the instructors decided, the honeymoon was over, and they were nothing short of diabolical in their efforts to push their students to the limit. Now there was a Gemini 8 every day, and Armstrong was unflappable. The instructors would freeze the pointing mechanism for the descent rocket, so that the nozzle could no longer be aimed from side to side, and Armstrong would compensate by tilting the entire lunar module when he wanted to change direction. They would make a maneuvering rocket stick in the on position when he was in his final, hovering descent, and Armstrong would cant the LM against the unwanted thrust as if he were leaning an airplane into a crosswind.

Every so often, in their efforts to reach new levels of ruthlessness, the instructors threw in a set of emergencies so complex, so difficult that no one could have handled them. With other astronauts this usually brought words of protest from the cockpit: "What the hell are you giving me that for? That's *negative training.*" But no matter what they did to Armstrong the instructors never heard an angry word from the simulator. Instead they'd find out only in the discussion afterward, sitting around the console, when Armstrong's clear, pale face would turn red, and his usually restrained voice would crack, and the instructors would know they had pushed the "war" too far this time. Minutes later Armstrong would be back at the controls, taking another try, and he never ended a session without making it down successfully.

By the end of May, the simulator at the Cape was linked via radio to the Mission Control Center in Houston. These so-called integrated simulations were designed to test not only the astronauts but the flight controllers. Integrated simulations were considered essential, and for good reason. Spaceflight is not a solo venture by daring pilots; it is a *partnership* between the astronauts and mission control, and nothing demonstrated this more than landing on the moon. For the dozens of flight controllers who worked for Chris Kraft, the first lunar landing was a coveted goal just as it was for the astronauts.

The flight director who would be in charge during the landing was Eugene F. Kranz, a former air force fighter pilot who had flown F-86s in Korea. With rugged features and a light blond crew cut, Kranz had the look of a drill sergeant. During Gemini, one of his flight controllers nicknamed him General Savage. Inside, Kranz was as sentimental as they come. Like most people at NASA he was an unabashed patriot who could get misty when he heard "The Star Spangled Banner." At the start of the work day he would go into his office and listen to John Philip Sousa marches, to get his blood flowing. He was a devout Catholic and a man of strong beliefs, and at his core he believed in the exploration of space. The space program wasn't just a part of his life; it *was* his life. ☾

At age thirty-five Kranz was already a veteran flight director. He'd begun his tour in mission control in 1960, as Chris Kraft's assistant flight director during Mercury, then as a flight director himself in Gemini; then he had moved on to Apollo. He'd been in the flight director's chair for more emergencies than he cared to think about, including Armstrong and Scott's brush with disaster in Gemini 8. Every nerve was tuned to the process of making decisions under pressure. During the landing Kranz would sit in the third row in the Mission Operations Control Room presiding over a team of flight controllers who would keep track of every system onboard the descending lunar module and monitor its trajectory down to the moon. Each controller

Gene Kranz, seated at the flight director's console in mission control, wears his trademark white vest as he confers with his deputy, Jones Roach *(right)*, while mission planner Bill Tindall *(left)* watches.

would in turn rely on his own team of "back room" experts, down the hall from the MOCR.

Kranz knew his team of flight controllers the way a battle commander knows his troops. These young men, like all the people who worked for Chris Kraft, were exceedingly bright. It was extraordinary what these people could do, together, during a mission. Kranz had seen it happen time and again: in the compressed moments of launch or some other critical phase, they became one giant conglomerate brain, twenty minds wired together in parallel, each focused on some small piece of the whole event. In such situations, they could solve almost any problem that came their way, given twenty seconds to work on it. In twenty seconds a controller could look at the problem, talk to someone in his back room, think, talk to someone else, come back to the first person, and make a decision. And all the while, he would be monitoring the events around him, listening not only to the conversations on the flight director's loop, but the air-to-ground, and perhaps one or two other loops. The amount of information processed by one controller was staggering. And the entire team was trained for that kind of split-computer mentality. With that kind of brain power at work, twenty seconds could be a long time. ☾

Kranz and his team started simulating lunar landings as soon as Apollo 10 ended. At first, they worked on their own, using computers to take the place of simulators, and instructors instead of astronauts. After that they moved on to the integrated simulations. Sometimes Armstrong and Aldrin were in the simulator at the Cape; at other times it might be the backup crew, or the Apollo 12 team. In any case, the simulations were so realistic that it was impossible to tell the difference from a real flight. People who sat in on them experienced gut-wrenching tension.

The simulation instructors in Houston waged war on Kranz and his team relentlessly as their counterparts at the Cape did with Armstrong and Aldrin. Their goal was to find the open seams in their decision making, to trap them right at the instant when two options overlapped, so that if they made the wrong choice, or picked the right one but didn't implement it in time, there would be no recovery. Kranz was the point man. If an emergency came up, he would have to decide whether to continue or order an abort. And that was where things went bad, in late May.

It was difficult for the controllers to master the complexity of the Powered Descent. It was especially tough for them to judge when the astronauts had entered the dead man's curve. Then there was the added wrinkle of the lunar distance. It took 1.3 seconds for a message from the spacecraft to reach earth, and another 1.3 seconds for mission control's reply to reach the astronauts. That delay was also part of the simulations. But the most difficult part was getting the feel of how to handle malfunctions during the Powered Descent. At first the controllers were so conservative they'd abort at the first sign of trouble. Afterward, the simulation instructors would point out to Kranz's men that they could have kept going, if only they had done such and such. . . . And the next time, the team would be so daring that the problems would overtake them, and the LM would crash into the moon. By the first part of June, there had been so many crashes that Kranz wondered if he and his team would ever get it right.

Kranz wasn't alone. During integrated simulations every manager in Houston, Chris Kraft included, was listening to the squawk box in his office as if it were the best radio show on the air. After the first crash, the black telephone behind the flight director's console rang, and it was Kraft. Despite his concern, he almost joked about it—"I see you let it get away from you." The next time the black phone rang, perhaps a week later, Kraft wasn't joking anymore: "What's the matter with you guys?" Kranz wasn't sure he knew. Maybe the answer was that no one had ever tried anything like this before.

Kranz had the distinct impression that the astronauts were getting frustrated too. It was hard enough just getting the complex simulator at the Cape—not to mention the roomfuls of computers in Houston—to work long enough to get in all the integrated simulations required in the training syllabus. What the crew didn't need was a team of controllers who couldn't respond correctly to an emergency. Kranz was beginning to doubt his own abilities.

One day late in June, Armstrong and Aldrin stood side by side in the lunar module simulator, descending to a simulated moon. The pitchover maneuver came right on time. Aldrin scanned the gauges, and called out data on altitude and speed. Armstrong surveyed the moon for a landing spot. Suddenly, it all went bad. The attitude indicator began to tumble; a thruster had stuck on. Aldrin looked out and saw the moonscape tilting crazily, looming ever larger. He was sure that if they continued the descent they would be killed. But Armstrong did not abort. Aldrin hesitated to tell his commander what to do, but he felt like a helpless passenger in an Indy race car. At last he said, through clenched teeth, "*Neil—hit abort.*" Finally Houston gave the order: "Apollo 11, we recommend you abort"—but it was too late. The TV image of the moon froze, and the gauges ceased changing, and Aldrin knew he and Armstrong were scattered across the moon in a thousand pieces. He looked over at Armstrong, who was still absorbed in the problem. ☾

That evening, Aldrin and Collins were having a nightcap in the living room of the crew quarters. While Collins sipped a beer, Aldrin dipped into a bottle of Scotch and began to let off steam about the morning's simulated disaster. Everyone had been listening, he told Collins, and Armstrong knew that. And yet he hadn't aborted; in fact he'd been indecisive. And this was going to be recorded as a *crew failure* . . . As the Scotch bottle emptied, Aldrin's voice grew louder and more strident. Suddenly a door opened and there was Armstrong, in his pajamas, a frosty look on his face. "You guys are making too much damn noise out here," he said. Aldrin realized he had heard every word. Collins excused himself and went to bed, leaving his crewmates to their discussion.

Aldrin talked it out with his commander. Armstrong said he didn't think anyone would see the incident as a black mark on the crew. By the next day, all seemed forgotten, but Aldrin could not deny that a certain coolness had set in between himself and Armstrong since the whole business about who would be first man out of the LM, and this had not improved things any.

Armstrong, for his part, wasn't concerned about having crashed in an integrated simulation. In truth, he'd thought the exercise had been a good test of the mission control, and was anxious to see how they would handle it. And they hadn't. As for being indecisive—he was testing his own limits and abilities in the one place where he could safely afford to. After all, you couldn't kill yourself in the simulator.

●◐◐○○○◑●

Meanwhile, throughout the spring of 1969, Mike Collins was waging his own battle for mastery of the command module simulator. If Armstrong and Aldrin had the landing to worry about, then Collins had just about everything else, for the command module pilot's job now included, in addition to navigation, being in charge of the burns into and out of lunar orbit, as well as the reentry into the earth's atmosphere. Collins too fell prey to a team of ruthless simulator instructors, and when he made a mistake it meant being stranded in lunar orbit, burning up in the earth's atmosphere, or plummeting into the ocean with parachutes still packed away. By June, Collins was steeling himself for the final weeks of battle. He would master his command module and his mission, and that was all there was to it.

Collins had joined the lunar landing crew after the most difficult episode of his career. He recovered from surgery in time to help send his former crewmates to the moon. As Borman, Lovell, and Anders spoke from lunar orbit Collins listened from mission control and felt a special connection to

Mike Collins trims a rosebush outside his house. As command module pilot on Apollo 11, Collins accepted that he would be going only 99.9 percent of the way to the moon's surface.

events, and a special envy. When Apollo 8 splashed down Collins stood amid cheers, waving flags, and cigar smoke, and was overcome by emotion. Leaving mission control, he shed his tears in private. Now, on the lunar landing mission, Collins would have the role of staying behind in lunar orbit while his crewmates made history. Fully aware of how people expected him to feel, he told interviewers with complete honesty that he felt no frustration. He was going 99.9 percent of the way, he said, and that was fine with him.

But in joining the crew of Apollo 11, Collins had taken his place among two men who were in many ways his opposite. He was as easygoing as Aldrin was serious, as accessible as Armstrong was remote. Alone among the three, he professed no love of machines. He barely tolerated the computers that had taken over spaceflight. The instructors would hear his voice from inside the simulator, "All I do is punch buttons!" Collins's tastes ran toward fine wines and good books. He dabbled in oil painting and cultivated roses in his Houston garden. To reporters faced with Armstrong's inscrutability, Aldrin's technical relentlessness, Collins was a breath of fresh air. He fielded their queries with good humor; his face seemed to say that yes, these are interesting questions. When asked, he did his best to explain his two crewmates. He spoke of Armstrong's "almost towering intelligence" and his sly wit, and said with a smile, "He has his own barriers erected." In truth, Collins admired Armstrong and Aldrin for their brilliance and technical skills, and felt fortunate to be flying with them. But he wished he knew them better. He lamented the fact that the three of them seemed to communicate only about technical information. Even now, after months of training together, he felt he barely knew them. He liked Armstrong but didn't know how to close the distance his commander kept. Aldrin, as Collins would later write, "is more approachable; in fact, for reasons I cannot fully explain, it is *me* that seems to be trying to keep *him* at arm's length. I have the feeling that he would probe me for weaknesses, and that makes me uncomfortable." ☾

But Collins understood that close friendship was not required to fly a good mission. What was required, in his case, was knowing how to rescue Armstrong and Aldrin if they got into trouble during the landing or the ascent from the lunar surface. There were eighteen different variations of emergency rendezvous, and Collins had to learn how to fly them all. But even if he never needed his rendezvous "cookbook," Collins would have to take care of docking the command ship with the lander, then removing the docking mechanism from the connecting tunnel. For the latter task, there was a horrendously long and opaque checklist loaded with terms like "capture latch release handle lock." If he couldn't get the probe and drogue out of the tunnel, Collins would later write, "I was supposed to get out the tool kit and dismantle it. Me, who couldn't repair the latch on my screen door. I hated that probe, and was half convinced it hated me and was going to prove it in lunar orbit by wedging itself intractably in the tunnel. . . ." If that happened, Collins knew, Armstrong and Aldrin would have to make a risky space walk to return to the command module. ☾

In Collins's mind, the eighteen rendezvous scenarios and the probe and drogue swirled around twin responsibilities: He must get his crewmates back to earth safely, and he must not do anything to screw up the mission. During the spring of 1969 Collins felt the eyes of the world upon him as he never had before, and his labors were tinged with anxiety. He also knew that Apollo 11 was putting a great strain on his wife, and that was something he wanted to stop as soon as possible. One day when they were flying a T-38 to the Cape together, Deke Slayton offered to put Collins back in the rotation after Apollo 11. Collins knew he'd probably command one of the later lunar landings, but he declined Slayton's offer. He had already decided that, assuming it was successful, this flight would be his last. ☾

●●◐○○○◑●

As the first weeks of June passed, Gene Kranz and his team had turned the corner. They had nearly mastered the challenges of the Powered Descent. They had firmed up the mission rules—critical guidelines for how to handle emergencies—and were becoming increasingly confident. But concern was mounting elsewhere, especially at NASA Headquarters, where managers worried that the astronauts were tiring and would not be ready in time for a July launch. And there were moments when Armstrong had his doubts; it just seemed that there was too much to do before the scheduled launch date of July 16. Should NASA postpone the mission until the next launch window in August? There wasn't much time to decide, because the Saturn V was on the pad, and technicians were getting ready to load propellants onto the lunar module and command ship. The hypergolic fuels used were too reactive to sit in the spacecrafts' plumbing for more than a few weeks without risking corrosion. If NASA wanted to postpone the launch, they would have to decide soon.

Deke Slayton came by the simulators at the Cape one day to find out how the astronauts saw things. Collins told Slayton he was ready, but he knew the burden rested on his crewmates. Aldrin did not want the training to stretch on indefinitely, feeling that they might be only marginally better by August. It came down to Armstrong. He told Slayton it would be tight, but they would be ready. On June 12, after holding a ninety-minute teleconference with Slayton, Wernher von Braun, Chris Kraft, and other Apollo managers, Apollo program director Sam Phillips made it official. Apollo 11 would leave earth on July 16, as planned.

MONDAY, JUNE 16
ELLINGTON
AIR FORCE BASE,
HOUSTON

A machine that looked like a strange, skeletal prehistoric bird ascended into the muggy air at Ellington Air Force Base, whining and belching puffs of vapor like a possessed steam calliope. The machine turned and hovered in midair for an instant, then flew slowly over the concrete apron. This was the moment that made NASA managers chew their fingernails. Neil Armstrong was piloting the Lunar Landing Training Vehicle, one of the most unforgiving flying machines ever built. For Armstrong, it was also one of the most essential: it let him train to land on the moon.

The LLTV looked as if it had been put together at an aerospace garage sale. A crisscross of metal struts with no skin formed the body of the craft, leaving fuel tanks, engines, and plumbing fully exposed. Four legs stuck out like a bedstead. The key to the trainer was that it had two independent means of propulsion. A jet engine supported five-sixths of its airborne weight. Armstrong didn't concern himself with the jet, but maneuvered the LLTV using a pair of rocket engines, powered by hydrogen peroxide gas, that simulated the lunar module's descent engine. Small jets mounted around the LLTV's metal framework mimicked the LM's attitude-control thrusters.

In May 1968, Armstrong had ejected from an earlier version of the trainer when it ran out of attitude-control fuel and became unstable. The accident board traced the incident to a design flaw. No sooner had that problem been solved than NASA test pilot Joe Algranti bailed out of the new LLTV in De-

The Lunar Landing Training Vehicle hovers above the tarmac line at Ellington Air Force Base *(below)*. When Neil Armstrong flew an earlier version of the trainer in May 1968 *(opposite)*, it became unstable *(top)*, causing him to abandon ship in a cloud of ejection-seat exhaust *(middle)*. He parachutes to safety *(bottom)*, narrowly missing the craft's burning wreckage.

cember; this time it was an aerodynamic problem. Bob Gilruth was pressing Slayton not to let any more astronauts fly it, but Slayton resisted, saying he would sure as hell rather risk losing one at Ellington than above the surface of the moon. And Armstrong agreed, knowing that only the LLTV could prepare him for the last few hundred feet of the Powered Descent.

Throughout the past few days Armstrong had piloted the unwieldy craft. At an altitude of a few hundred feet, he activated the lunar simulation mode, and the LLTV flew just as if it were in a vacuum and in the moon's pull. Almost every instinct Armstrong had developed from a lifetime of flying airplanes was wrong now. The LLTV was balanced on the thrust of its rocket engine like a dinner plate on a magician's broomstick. To start moving in one

direction, Armstrong had to tilt the LLTV slightly, letting the rocket engine push him where he wanted to go. Because of the one-sixth g, he had to tilt the craft six times as far as he would have in normal gravity. He had to be careful not to tilt too far, or would risk falling out of the sky. At the same time, with the engine's thrust pointed off to one side, he had to ride the throttle in order to keep from losing altitude. And once he started moving he would keep going, just as if he were flying in a vacuum; to arrest his motion he had to tilt the LLTV in the opposite direction. Simply changing his landing point a few dozen feet became an act of supreme coordination.

The thing that most surprised Armstrong when he first flew the LLTV was how sluggish it was. Each maneuver had to be anticipated well in advance, because it took a surprisingly long time to get going and just as long to stop. To make matters even more difficult, the LLTV carried enough fuel for only about 6½ minutes of flight; each simulated landing lasted 3 or 4 minutes at most. It was precisely because the LLTV was so unforgiving that it was invaluable to Armstrong. It conditioned him to the "fast time" clock once more, for each time Armstrong brought the LLTV down he knew that his fuel supply was dwindling, that each maneuver skirted the edge of the machine's stability, and that he had only one chance to make it. In a setting no more exotic than Ellington Air Force Base, he had to play it for keeps, just as he would in the airless sky of the moon.

The LLTV drifted slowly over the runway, no faster than a car pulling into a garage. To those watching from the flight line it was an eerie sight, as if it were controlled by some force originating beyond the steamy midday air. Finally the machine started straight down, slowly, like a dragonfly deciding whether to alight. At last the spindly craft came gently to rest. Ground technicians informed Armstrong that he had only 90 seconds' worth of fuel left. "Understand," he answered, and lifted off on another run. By day's end he had completed eight flights in the machine. He was confident now that when the time came for the real thing, he would be ready.

●●◐○○○◑●

Ironically, Armstrong and Aldrin had little time to study the world they would visit. Armstrong was clearly interested in the science of his mission, and the geologists had long considered him a promising student. For Apollo 11, he made time in the overcrowded training schedule for a field trip to the mountains of westernmost Texas. The press found out in advance and almost turned it into a circus. They followed the astronauts' cars with a caravan of their own. *Time* magazine hired a helicopter that roared overhead and made it nearly impossible for Armstrong and Aldrin to hear what the geologists were saying. Aside from that one outing, they depended on briefings by geologists from the Manned Spacecraft Center and from the U.S. Geological Survey in Flagstaff, Arizona. Armstrong also asked scientist-astronaut Jack Schmitt to act as their liaison with the scientific community.

As Armstrong made a point of saying at a pre-flight press conference, the surface of the moon was not an unknown place in the strictest sense, thanks to the unmanned Lunar Orbiter and Surveyor probes. But even with the naked eye, it's easy to see two different types of terrain on the lunar near side: the dark splotches called *maria* (the Latin word for "seas") and the bright highlands, which scientists call the *terrae.* A good pair of binoculars reveal that the *maria* are relatively smooth compared to the heavily pockmarked highlands, and by 1969 most geologists agreed that if Armstrong and Aldrin managed to touch down on Mare Tranquillitatis—the Sea of Tranquillity—they would find an ancient lava plain made of basalts like those found in Hawaii and other volcanic locales. And judging by the pictures sent back from the Surveyor landers, they would also find what was apparently a ubiquitous layer of fine-grained debris, probably formed by the continuous rain of meteorites over the eons.

But when it came down to it, the probes could only radio back numbers. They could not completely dispel the aura of mystery around the moon, even in the minds of some scientists, and there were some dire predictions. Cornell astronomer Thomas Gold insisted that the moon was covered by a layer of fluffy powder dozens of feet thick. He warned that the LM would sink out

IN 1969

Georges Pompidou takes office as president of France.

Warren Burger becomes chief justice of the United States.

Queen Elizabeth II names her son, Charles, Prince of Wales, heir apparent to the throne.

Judy Garland dies at age 47.

of sight as soon as it touched down; he told NASA that Armstrong and Aldrin should drop brightly colored weights as they descended and watch to see that they remained in view, or else abort the landing. None of his colleagues—not even pictures from the Surveyors, resting unharmed on the surface—could dissuade him. ☾

Other forecasts were harder to refute. One theory emerged that charged particles emanating from the sun had so altered the rocks over eons that they would burst into flames as soon as they were exposed to oxygen inside the lunar module; no one would know the answer until it happened. And no amount of speculation could answer the question of whether there was life on the moon. No geologist thought there was. It offered an almost unimaginably hostile environment, devoid of air and water, exposed to the vacuum of space, bathed in deadly solar and cosmic radiation, pelted incessantly by micrometeorites. One geologist pointed out that if he wanted to build a sterilization machine, he would construct something like the surface of the moon. And yet, the implications if even one microbe survived the trip back to earth on a space suit or under a fingernail were so unsettling that NASA could not ignore the possibility, especially when voiced by several U.S. government agencies and other countries. NASA constructed an $8 million Lunar Receiving Laboratory at the space center to store the lunar samples in total biological isolation and house the astronauts in the last two weeks of a twenty-one-day quarantine, while officials nervously waited to see if they came down with any symptoms. Quarantine was something Armstrong and his crew would just as soon have done without, but they had no choice. ☾

The weak link in the plan was that no one had figured a way to get the astronauts from the command module to the LRL without exposing them briefly to the environment. A special quarantine trailer would house them for the trip from the Pacific back to Texas, but that left getting from the command module to the trailer. That's where the Biological Isolation Garments came in. After splashdown, the astronauts would open the command module hatch just long enough to let the recovery swimmers toss in three BIGs; after zipping them on the men would climb into the waiting life rafts, scrub each other with disinfectant, and ascend to the helicopter in the recovery net. From then on they would be in quarantine. There would be no speeches on the carrier deck; after a short walk from the helicopter to the quarantine trailer they would close the door behind themselves and view the world through a pane of glass—there and in the LRL—until the quarantine was ended. In the first days of July 1969, that moment seemed all too distant.

NASA
APOLLO 11

SATURDAY, JULY 5, 1969
MANNED SPACECRAFT CENTER, HOUSTON

With just eleven days to go until launch, Armstrong, Aldrin, and Collins spent the July 4 weekend at home, a last visit with their families before heading to the Cape. There had been precious little time off these past seven months; even their time at home was often claimed. A specialist would come by to give an informal briefing; evenings were spent holed up in the study with flight plans and training documents. Most of the time, though, they weren't even home. *Life*'s July 4 issue, with the cover story, "Off to the Moon," included the customary report from the home front—Armstrong fishing with his sons, baking homemade pizza, and playing piano duets with his wife, Jan; Collins trimming roses in the backyard; Aldrin taking his kids to AstroWorld—but the truth was, those things would never have happened if an outing with family hadn't been a PR requirement.

Since January, Slayton had tried to keep the press at a distance, simply because Armstrong's crew had so much training to pack in. Even so, the men had made time for this or that reporter to come by the house and ask questions about their lives and their mission. Finally, Slayton gave in and agreed to a last press conference, and Armstrong's crew spent most of Saturday, July 5, talking to the media. At this point, the men were well into their twenty-one-day pre-mission medical quarantine, and so on this summer afternoon they strolled onto the stage wearing hospital masks and did not remove them until they had taken their places inside a plastic-enclosed booth. A few reporters grinned back at them from behind their own masks. One asked whether any precautions had been taken to prevent the men from catching germs from their own families. Collins answered, "My wife and children have signed a statement that they have no germs. . . . Seriously, there are no special precautions being taken." But the journalists directed few questions to Collins; they were much more interested in his crewmates, and especially, his commander. For seven months now, Armstrong had been telling interviewers that he wished the press would convey that Apollo 11 was a massive group effort, that it was a mistake to focus on him, but he had not been successful. At the press conference one reporter suggested to him that, as the first man to set foot on the moon, he would be so famous that his personal life would cease to exist. He added, "Do you have any thoughts on this prospect?"

On July 5, 1969, eleven days before Apollo 11's scheduled launch, Armstrong, Aldrin, and Collins hold a pre-flight press conference in a special plastic enclosure built to protect them against exposure to illness.

"I suppose," Armstrong said, smiling shyly, speaking in characteristically measured words, "if there is any recognizable disadvantage to being in the position I'm in then that's it. I think that's a fair trade." ☾

On Monday, July 7, Armstrong, Aldrin, and Collins headed back to the Cape for the last time. For the next nine days their world was a high-tech monastery, equally divided between the simulators and the crew quarters. Richard Nixon had planned to visit them here on the night before launch to have dinner, but canceled after Chuck Berry publicly fretted that he might infect the astronauts. Privately, Armstrong and his crew fumed about the gaffe; the president was no more likely to harbor germs than the dozens of people they worked with every day.

For Mike Collins, the incident was a momentary distraction from the "awesome sense of responsibility" for the mission he was about to fly. That pressure was very much on the mind of NASA administrator Tom Paine when he dined in the crew quarters on Thursday, July 10. Over dinner, Paine made an extraordinary promise to Armstrong's crew. Don't take any unnecessary risks to accomplish the mission, he told them. If anything should go wrong, don't hesitate to abort. He would see to it that they would not have to get back in line for another flight; they would be assigned to the very next mission to try again.

For Collins, Paine's promise took some of the pressure off, but not for Neil Armstrong. He was only too aware that the nation's prestige was riding on this mission. It was impossible *not* to be aware, in the fishbowl they'd been living in since January. And although he felt ready, and sensed that Collins and Aldrin did too, he knew also that the landing would test the entire Apollo system—the hardware, the mission control teams, and themselves—to the limit. Even on the morning of July 16, 1969, as Armstrong led his crew out into the TV lights and onto the transfer van, with an old comb and a package of Life Savers in the pocket of his space suit, he had little doubt they would make it safely back to earth. But the landing, in his mind, was still a fifty-fifty proposition.

A seemingly confident Neil Armstrong leads his crew to the transfer van on the morning of July 16, 1969. Mindful of his mission's complexity, Armstrong at this moment is anything but certain that Apollo 11's attempt to land on the moon will succeed.

Packed onto a jetty, spectators raise cameras and binoculars as the Apollo 11 countdown nears zero. An estimated quarter-million people crowded the beaches of Cape Kennedy to witness the event.

Apollo 11 rises from Pad 39A at 9:32 A.M., July 16, 1969. Led by the spacecraft's launch escape rocket, the 36-story vehicle takes 13 long seconds to clear the launch tower, where the camera that took these pictures was mounted.

THE FIRST LUNAR LANDING

APOLLO 11

SATURDAY, JULY 19, 1969 5:30 A.M., HOUSTON TIME 2 DAYS, 21 HOURS MISSION ELAPSED TIME ABOARD APOLLO 11, 177,000 MILES FROM EARTH

I. THE *EAGLE* HAS LANDED

Buzz Aldrin opened his eyes and floated in the darkness, collecting his thoughts, remembering where he was. All was still inside the command module *Columbia;* only the hum of the cabin fans broke the total silence of the void. Every so often strange flashes appeared. Aldrin did not know what they were, but they seemed to be something actually entering the cabin, perhaps a vagabond cosmic particle decaying in the command module's atmosphere. He had seen the flashes last night too; now he made a mental note to mention them to Armstrong and Collins. They were still asleep, down in their sleeping bags. Aldrin had spent the night up in his couch on the right-hand side with a lightweight headset taped to his ear in case Houston tried to call during the night. Nobody kept watches on a spaceflight anymore; ever since Apollo 8 they had done away with that in the interest of getting better sleep. On this mission especially, sleep was important.

They had agreed before they left the earth, he and Armstrong and Collins, to take it easy on the way out to the moon, knowing that it would be a mistake to arrive in lunar orbit tired the way Borman, Lovell, and

An hour into the moonwalk, Buzz Aldrin photographed his own boot and its everlasting imprint on the Sea of Tranquillity.

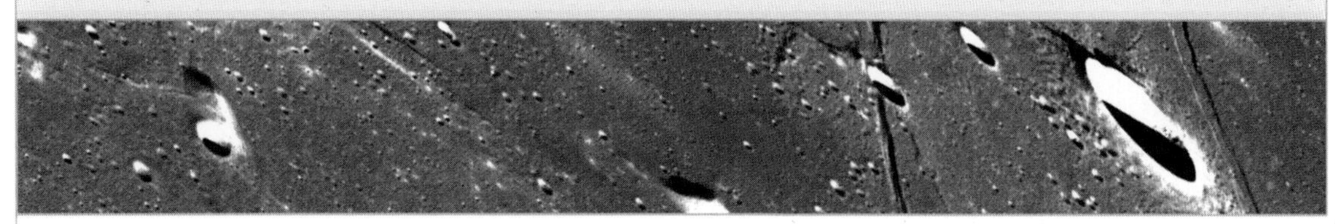

Anders had. It came down to a state of mind: they would convince themselves, as they coasted out to the moon, that the mission had not really begun, and that it would not begin until the moment when Armstrong and Aldrin floated into the lunar module *Eagle,* undocked from *Columbia,* and started the final descent to the moon.

Aldrin was up early—the end of the rest period wasn't for another two hours—but he knew they would be reaching the moon in about seven hours, and he wanted to know whether they would be making a last midcourse correction before they got there. He keyed his mike.

"Houston, Apollo 11."

Aldrin heard only static; he guessed that one of the tracking stations was about to go out of contact, carried by the earth's rotation, and that the Manned Space Flight Network was in the process of switching to another big dish. He lifted the shade from *Columbia*'s right-hand window and looked for earth, but outside there was only dull, starless black.

It would have been a tall order, pretending the mission hadn't started yet, if Armstrong, Aldrin, and Collins hadn't flown before. They knew what it was to work in a strange and hostile environment. None of them had any trouble with motion sickness on the way out. On the contrary—Aldrin thought floating inside the command module was even more fun than space walking, for it offered all the freedom of movement with none of the limitations of a pressurized space suit. With *Eagle* attached to the command module's nose Apollo 11 was like a small, two-room space station. The two joined craft coasted moonward, twirling slowly in the sun's glare.

Aldrin hadn't gone inside *Eagle* until the day before, when he and Armstrong made a scheduled inspection of its cabin. He'd gone in first, making the topsy-turvy passage "up" from the command module and "down" into the lander; he would later call it the strangest sensation of the entire voyage. It was Aldrin's first time inside *Eagle* in two weeks, and it felt good to be back. Armstrong arrived—hanging from the ceiling, by Aldrin's reckoning—with the television camera. Before the flight they had barely enough time to learn how to turn the camera on, never mind learn how to point it accurately or stage anything like a professional-looking show. It was "challenging," to say the least, when mission control came on the air ten minutes before a TV show started and announced matter-of-factly that 200 million people were standing by for the broadcast; they're *all* watching; now—what are you going to show us?

But this one went well. For most of it Aldrin was on camera, giving his audience a look at various items of equipment—a movie camera they would

Outward bound for two days, Buzz Aldrin pauses during an inspection of the lunar module *Eagle.* He and Armstrong also staged a televised tour of the lander that was broadcast around the world.

use to film the Powered Descent, a flight plan, one of the protective visors he and Armstrong would wear during the moonwalk. From what Charlie Duke in mission control said, the TV show was an unqualified success; much of the world picked it up live, and the picture was so clear you could read labels on the instrument panel.

But such diversions aside, Aldrin hadn't felt quite on top of things since they left earth. There was that feeling of being slightly "behind the airplane" that he'd had as a squadron pilot whenever he flew a new jet for the first time. Even though everything was going smoothly, even though he was doing things he had done literally hundreds of times in the simulator, it was the *reality* of it all—seeing the earth beyond the windows, floating free, knowing that the world was listening and watching almost everything they did—that put him just slightly off balance. And there was always another "performance" ahead, whether it was a midcourse correction burn or a television show.

"Houston, Apollo 11." This time there was an answer. It was Ron Evans, the Capcom who was working the "night shift." With the three of them asleep

As Apollo 11 prepared to enter lunar orbit on July 19, the astronauts caught this spectacular glimpse of the moon, eerily lit only by earthshine. The glowing halo around the moon in this painting by Don Davis is the solar corona.

Evans's shift got pretty dull, and he sounded glad to make contact. He told Aldrin the trajectory was so good that there wasn't going to be a midcourse, and he could go back to sleep for another couple of hours. So Aldrin settled in and tried, without success, to sleep again. After a few minutes he was aware of some activity beneath the couches; Armstrong and Collins were awake. They shed their lightweight sleeping bags, emerged from behind the seats, and the three men began the business of their fourth day in space. Fuel cells needed to be purged. There was breakfast to fix. And there was an unspoken but undeniable tension in *Columbia*'s cabin. Each of them could feel it: the ruse was getting harder and harder to pull off. They were about to put their lives and their abilities on the line.

Armstrong, Aldrin, and Collins had not seen the moon on the way out, but according to the flight plan they were supposed to take some pictures of it a few hours before braking into lunar orbit. As they finished breakfast, a sudden darkness came around them, and for the first time in the flight the sky was full of stars, too many to count, each with a steady, gemlike brilliance. They had flown into the lunar shadow. Through the windows of the slowly turning spacecraft they looked out at the place where the sun had once been, and there was the moon: a huge, magnificent sphere bathed in the eerie blue light of earthshine, each crater rendered in ghostly detail, all except for a third of the globe, which was a crescent of blackness. As their eyes adapted to the darkness they saw that the entire moon was set against a gigantic ellipse of pearly white light, the glowing gases of the sun's outer atmosphere, which stretched beyond the moon into the blackness. Somehow in these strange, cosmic illuminations the moon looked decidedly three-dimensional, bulging out at them as if to present itself in welcome, or, perhaps, warning. ☾

7:25 P.M., HOUSTON TIME
3 DAYS, 10 HOURS,
53 MINUTES
MISSION ELAPSED TIME

Even before Apollo 11 left earth, Neil Armstrong knew the approach to his landing site as well as he had known the desert towns along the approach to Edwards. He'd spent some of the quiet hours of the trip out to the moon studying the photographs radioed from the unmanned probes and carried back aboard Apollo 10. Now that he was in lunar orbit, and he could look down on the Sea of Tranquillity with his own eyes, *Eagle*'s landing approach was even easier to recognize than Stafford and Cernan's pictures had suggest-ed. The very moment of Apollo 11's departure from earth had been timed so that tomorrow, during the landing, the sun would be 10 degrees above the

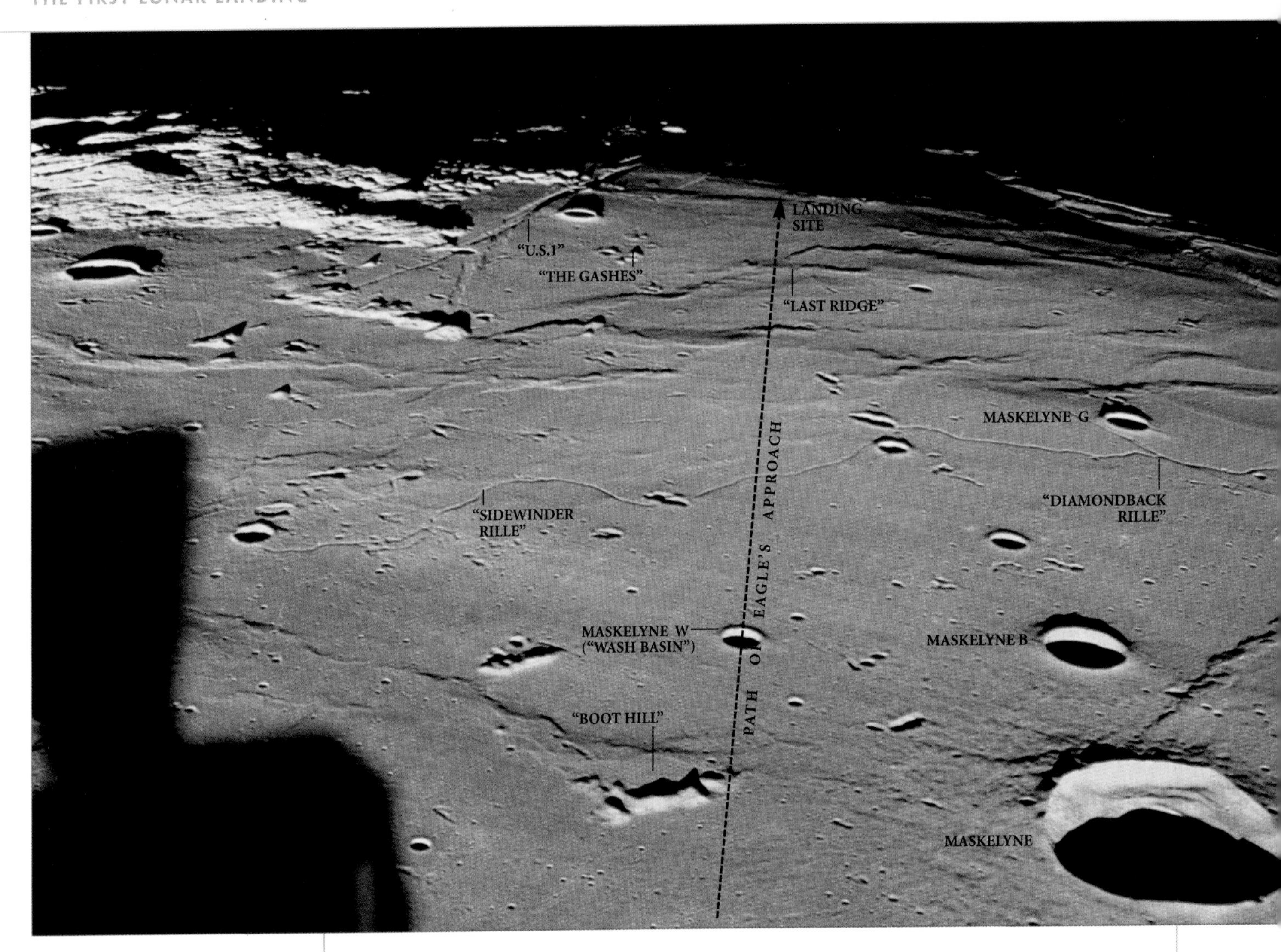

The line drawn on this view of the moon's surface, made from orbit on July 19, shows the path of Armstrong and Aldrin's descent to their landing site near the horizon. Checkpoints along the approach would help Armstrong judge whether *Eagle* was on course.

horizon, throwing the *mare* into relief, so that he would be able to spot rough ground as he flew *Eagle* to a landing.

Landing Site 2 was an ellipse measuring 11½ miles by 3 miles, or about as long as Manhattan Island and half again as wide. There were many places that were more exciting, geologically speaking, but they were also much riskier or harder to get to. On later landings there would be time to explore; for now, the geologists would be happy with any place Armstrong and Aldrin managed to visit. And so Site 2, chosen for its blandness, was a completely unremarkable stretch of *mare,* close to the lunar equator and thus easiest to reach. Only Stafford and Cernan had seen it close up, but that was from 9 miles. What would it look like from 500 feet? Would it offer a safe place to land a lunar module? Armstrong was cautiously optimistic. Soon enough, he would find out for certain.

Late in the day, while Aldrin went into *Eagle* to make systems checks, Armstrong lingered at *Columbia*'s window, hoping to glimpse the landing

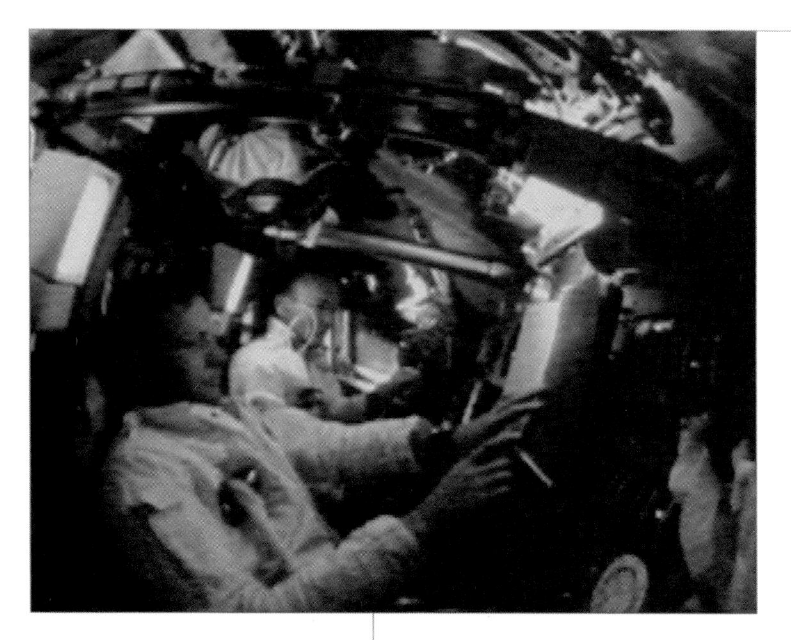

Inside the command module, Armstrong prepares for his upcoming piloting task by studying a photomap of the landing site. In the background is command module pilot Mike Collins.

site itself. Below, a crater called Moltke, only a few miles wide, glowed in the light of lunar dawn. Around it, the *mare* was cut by flat, narrow valleys like desert roads. Just to the north, almost enveloped by night, lay the place where he and Aldrin would try to land. Collins asked, "Can you see the landing site?"

Armstrong peered down among the long shadows. "I'm not sure," he said. Suddenly he and Collins heard Aldrin's voice in their headsets.

"I can see it," Aldrin said from inside *Eagle,* contained excitement in his voice; "I got the whole landing site here." After a moment Armstrong and Collins could see it too, barely emerged from the shadows. From this height, Landing Site 2 was tiny, but it was possible to make out some details—and it didn't look encouraging. In this illumination, with even the tiniest features thrown into jagged relief, it looked so forbidding as to be untenable. It was difficult to accept that while they slept that night, the rising sun would tame this unsettling scene. Right now it looked like the last place anyone would want to land.

8:32 P.M.

"Amazing how quickly you adapt," Collins said brightly at dinner. "Why, it doesn't seem weird at all to me to look out there and see the moon going by, you know?" But behind his calm words, Collins harbored unspoken concern for his crewmates. He could not tell whether they felt anxiety, for they seemed entirely relaxed. Of the three of them, he was least comfortable with the risks they were undertaking, most conscious of the fallibility of complex machines. He had come to see the flight as a long and fragile daisy chain of events, and was only too aware that at any time the chain could break. Now he felt something like an anxious parent with two children about to go away on a long trip. He offered to take the watch for the night, encouraging Armstrong and Aldrin to sleep underneath their couches. "You guys ought to get a good night's sleep, going into that damn LM." As his crewmates readied their supplies for the next morning Collins said flatly, "I thought today went pretty well. If tomorrow and the next day are like today, we'll be safe."

SUNDAY, JULY 20
12:18 P.M., HOUSTON TIME
4 DAYS, 3 HOURS,
46 MINUTES
MISSION ELAPSED TIME

"You cats take it easy on the lunar surface. If I hear you huffing and puffing I'm going to start bitching at you." Collins's gentle admonition came as *Columbia* and *Eagle* coasted through another sun-drenched orbital noon, with undocking only minutes away. When the time came, Collins pushed the button to release the LM and called out, "Okay, there you go! Beautiful!"

Pushed by the spring action of the docking mechanism, the two craft drifted apart. Armstrong carefully steadied *Eagle* with a few quick bursts of thruster fire. For several minutes *Eagle* and *Columbia* drifted in formation, orbiter and lander, robot spider face to face with command ship. Collins fired off pictures as the lander turned a slow pirouette before him.

"I think you've got a fine-looking flying machine, there, *Eagle*," Collins offered, "despite the fact you're upside down."

"Somebody's upside down," Armstrong said.

"You guys take care," Collins radioed.

Armstrong replied simply, "See you later."

Collins gave a short burst of the maneuvering thrusters and *Columbia* pulled away until *Eagle* was merely a point of light. Collins had given some thought to the chances of accomplishing this mission, and he had independently arrived at the same fifty-fifty odds as his commander. For the next few hours, he had work to do, tracking *Eagle*'s flight over the craters and relaying this information to Houston. But when it came time for the Powered Descent, he would simply listen, like everyone else.

2:35 P.M., HOUSTON TIME
MISSION OPERATIONS
CONTROL ROOM, MANNED
SPACECRAFT CENTER

It was quiet in mission control, the way it almost always was during a mission. At the front of the room, a giant screen showed a green-colored map of the moon, with a tiny, moving command module and LM; it showed that *Columbia* and *Eagle* were flying over the far side, out of radio contact. Already, the controllers had passed up the data to *Eagle* for the Descent Orbit Insertion burn and the Powered Descent. Soon the spacecraft would reappear, and if all had gone according to plan, *Eagle* would be on its way down to 50,000 feet.

Gene Kranz and his team of flight controllers—they were called the White Team—sat at their consoles, wearing lightweight headsets, anticipat-

Floating above the Sea of Fertility, *Columbia* pulls away from *Eagle* after undocking. The maneuver marked the beginning of more than 20 hours that Mike Collins would spend alone inside the command module, circling the moon.

ing the Acquisition of Signal. For the moment their television screens, normally full of up-to-the-minute data, were static. Behind the controllers, in a glassed-in gallery that looked out over the control room, NASA managers, current and former astronauts, and other VIPs crowded together to witness the drama to come. Just looking at them drove home the fact that this was not a simulation. ☾

In the third row of the MOCR, Gene Kranz sat at the flight director's console wearing a brand new white vest. When he was on shift he always wore a white vest, made for him by his wife, Marta. This one was white brocade with silver thread; it was a special vest for a special occasion. To Kranz it seemed as though months had passed since the trials of June, but in the last weeks they had finally mastered the Powered Descent. By launch time, Kranz knew he had a winning team. From the time he had awakened this morning, Kranz was sure that Armstrong and Aldrin would land on the moon today. He had gone to church and then to his office, where he had played his Sousa tapes. A few minutes after eight that morning, he took the flight director's chair in mission control to begin the shift for the descent phase. He felt unwavering confidence in the spacecraft, the astronauts, and most of all, in his men. In his mind, it was crucial that he exude that confidence.

Kranz looked down from his console to the first row of the MOCR, the place the flight controllers called "the Trench." There was Bob Carlton, who watched the LM descent engine like a hawk; his call sign on the loop was "Control." Jerry Bostick was "FIDO" (for Flight Dynamics Officer); he would monitor the tracking data as the LM descended. And Steve Bales, the expert on the LM's guidance system, who would be called "Guidance." At twenty-six, Bales was no younger than many of his fellow controllers, but he seemed especially boyish; it was his enthusiasm, his slightly unkempt looks. It was Bales's job to keep tabs on the LM's computer, landing radar, and its trajectory down to the moon.

Off to Kranz's right, in the MOCR's second row, a half-dozen other men rounded out the team, keeping track of the other systems aboard the LM. Each controller, plugged into his own "back-room" specialists, funneled data and advice to Kranz in the third row. And directly in front of him, there was astronaut Charlie Duke, the young South Carolinian whose drawl had answered the excited reports from a barnstorming Tom Stafford and Gene Cernan. Charlie Duke knew more about the lunar module than any astronaut who wasn't already on an Apollo crew; that expertise had prompted Neil Armstrong to ask him to serve as Capcom for the landing.

A short time ago Kranz had ordered Security to lock the doors to the

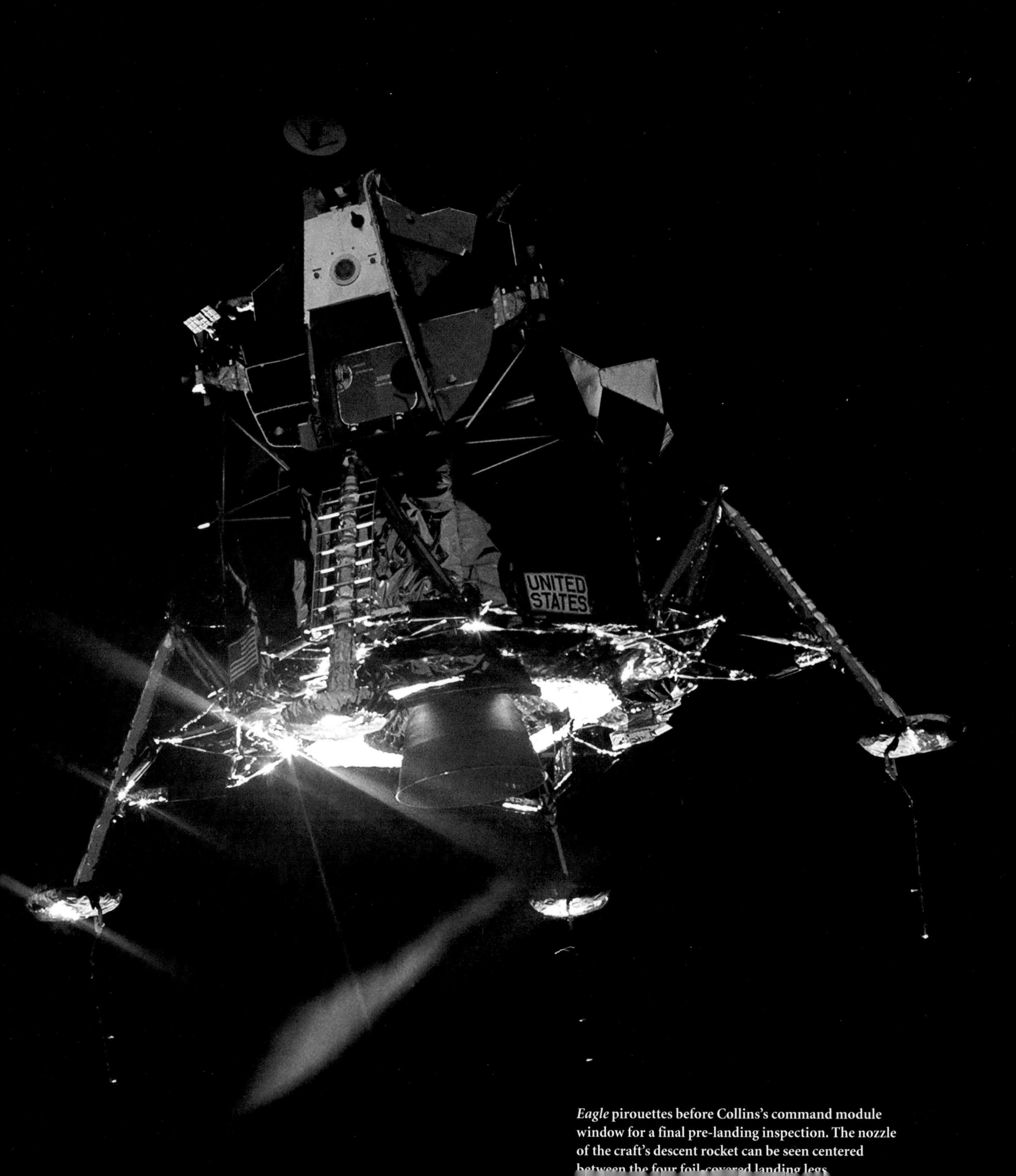

Eagle pirouettes before Collins's command module window for a final pre-landing inspection. The nozzle of the craft's descent rocket can be seen centered between the four foil-covered landing legs.

MOCR. Then he'd switched to a separate communications loop, one that the VIPs and the pool reporter couldn't hear, and talked to his men. "Okay, gang, we've had a good training period," he began. "And today, we're really going to do it, we're going to land on the moon. This is the final exam. . . . " He reviewed the key milestones of the Powered Descent, and made sure his controllers were clear on their own ground rules for feeding him information over the loop. Then Kranz gave a pep talk worthy of a battlefield commander. "This is the best team I've ever worked with," he said. "I have ultimate confidence in you people. . . . What we're about to do now, it's just like we do it in training. And after we finish the sonofagun, we're gonna go out and have a beer and say, 'Dammit, we really did something.' " ☾

2:46 P.M.

Telemetry readouts streamed into the control center as *Columbia* came out from behind the moon. Seconds later, Kranz heard Mike Collins tell Charlie Duke, "Listen, babe, everything's going just swimmingly." Almost two minutes passed before *Eagle* reappeared, and almost immediately communications with the LM became spotty. The signal would drop out and then return. Kranz sweated out the interruptions, knowing that only minutes from now he would have to give Armstrong and Aldrin a go-ahead for the Powered Descent, and when that time came his controllers would have to have good data. How much data was a judgment call that only Kranz could make. If everything was going well, they could get by with only a few seconds' worth. But even if things were bad, Kranz was willing to stretch rules, because he knew the risks entailed in an abort. Several astronauts were seated next to Duke, listening in; one of them was Pete Conrad. He suggested to Duke that *Eagle* yaw slightly to one side to improve the signal strength. That worked. ☾

3:06 P.M.

Inside *Eagle,* 50,000 feet above the moon, Armstrong and Aldrin stood side by side, anchored by harnesses to the LM floor. Four-day beards darkened their faces. Within their bubble helmets, they scanned the instrument panel as the time for Powered Descent approached. Their mouths were dry from the pure oxygen flowing through their space suits. Already they had pressurized *Eagle*'s fuel tanks, called up the proper computer program, and checked their trajectory by sighting on the sun with the LM's navigation telescope. With forty seconds to go, Armstrong made sure that Aldrin had turned on the movie camera to record the descent. Armstrong set the switch to arm the descent engine. Then, seconds after Aldrin pushed the PROCEED button, the two men spoke at once: *"Ignition."*

"Just about on time," Armstrong said. The descent engine came to life so gently that Armstrong and Aldrin heard and felt nothing. Only the gauges told Armstrong it was firing. Less than half a minute later it roared to full thrust and the cabin filled with a soundless, high-frequency vibration.

Once more communications dropped out. "They've lost you," Collins radioed to his crewmates. A moment later, after Aldrin switched to a different antenna, communications returned.

Now Neil Armstrong turned his attention to the moon. *Eagle* was face-down, and through his small, triangular window Armstrong could see landmarks he recognized. Each checkpoint was appearing 2 full seconds ahead of schedule, and since *Eagle* was going nearly a mile a second, that meant they would come down about 2 miles beyond their aim point. Armstrong keyed his mike.

"Our position checks downrange show us to be a little long," he radioed to earth. Now *Eagle*'s engine throttled down, exactly on schedule, and Armstrong realized the computer was not aware of the error.

The long brake continued, and now, 46,000 feet above the moon, it was time for Armstrong to turn *Eagle* over on its back so that its landing radar would point at the moon. When he did, he and Aldrin found themselves looking out at the earth, afloat in blackness.

At 40,000 feet the landing radar came to life, blurting information on speed and altitude to *Eagle*'s computer. From these data the computer continually revised its trajectory calculations, and *Eagle* shuddered with corrective bursts from the maneuvering jets. Armstrong was surprised to hear them fire so often, much more often than in the simulator. No smooth ride at this stage; they were lurching their way down to the moon.

Aldrin, meanwhile, began a running dialogue with the computer, checking its height calculations against the data from the radar. As expected, the two disagreed by several thousand feet. Aldrin knew the radar echoes were more reliable, and he planned to tell the computer to accept those data, but first he wanted mission control to take a look. To do that, he keyed in a command to tell the computer to display the difference, or delta-H, as it was called. Suddenly, the men heard the high-pitched buzz of the Master Alarm in their ears. On the computer display the PROG light glowed amber.

IN 1969

Charles Manson and members of his cult are indicted for the murders of actress Sharon Tate and four others in her Los Angeles home.

Police fire on Catholic demonstrators in Northern Ireland, killing 10. The injured included 514 civilians and 226 police officers.

The Woodstock Music and Art Fair attracts 400,000 rock fans to Bethel, N.Y.

Charlie Duke *(foreground)* mans the capsule communicator post in mission control as *Eagle* nears the moon. Seated to his left are Jim Lovell, backup commander for Apollo 11, and backup lunar module pilot Fred Haise.

"Program alarm," Armstrong radioed. It was the crispness of his words, rather than the tone of his voice, that conveyed urgency.

Quickly, Aldrin queried the computer for the alarm code, and "1202" flashed on the display. Aldrin did not know just what 1202 meant—and this was not the time to dig out the data book to find out—but it had something to do with the computer being overloaded with too many things to do. He had never seen this kind of alarm in a descent simulation; now he wished it would just go away.

In mission control Gene Kranz felt as if he were in one of those bad simulations back in the dark days of early June. He had almost been glad when the problems began as soon as *Eagle* came around the moon, beginning with the spotty communications. Kranz welcomed a little adversity at the starting gun; it would get his men thinking instead of waiting tensely. A few malfunctions made the whole thing seem more like a simulation and less like history. But the problems snowballed. Even as the Powered Descent began, critical data from the spacecraft kept cutting out, then returning for brief moments—just barely long enough for Kranz to let the landing continue. ☾

In the midst of it all Kranz had done what he called "going around the horn," polling his men. He knew their voices so well that he could pick them out in the web of simultaneous conversations filling his headset. As each man spoke Kranz listened for signs of strain; he heard solidity. And none more so than young Steve Bales, the LM computer expert, who came back with such unbridled enthusiasm—*"Go!"*—that Kranz almost burst out laughing. ☾

Now, with the LM's computer threatening to abort the landing, Bales was his most important man. Back in the first week of July the same kind of computer alarms had come up during a simulation with the Apollo 12 crew. Kranz had ordered an abort, and afterward the simulation instructors had really let him have it. If he had only been familiar with the alarms, they told him, he could have kept going. For the rest of that day and that night Kranz sat down with Bales and some of the computer experts from MIT, studying each type of alarm and what to do if it came up. So when Kranz heard Neil Armstrong call "twelve-oh-two," he knew it was serious. Whether they could continue or not was up to Steve Bales.

But the complexities of the LM's computer were too much for one person. Bales wasn't certain what the 1202 alarm meant. He put the question to one of his back-room experts, Jack Garman. Garman knew that for some reason—no one knew why—*Eagle*'s computer was saying, "I have too many things to do in my computation cycle, so I'm going to give up and start at the top of the list." As Kranz waited for an answer from Bales, he heard Armstrong tensely radio, "Give us a reading on the 1202 program alarm." Before Kranz could speak, Bales responded, "We're—we're Go on that, Flight." Garman had told Bales that as long as the alarm was intermittent, not continuous, everything would be okay. But if the alarm returned and didn't go away, the computer could give up working altogether, and that would mean an almost certain abort. ☾

But if the alarm returned and didn't go away, the computer could give up working altogether, and that would mean an almost certain abort.

Inside *Eagle,* Armstrong and Aldrin heard Charlie Duke's urgent but assured words, *"We're Go on that alarm."* Once more, Aldrin queried the computer for the delta-H; once more an alarm rang in their headsets. Again a message from Steve Bales came to them, via Charlie Duke: mission control would keep tabs on the delta-H, alleviating some of the computer's workload. Above the moon, and in Houston, everyone hoped the fix would work.

Halfway through the Powered Descent, right on schedule, *Eagle*'s computer throttled the engine back to half its maximum power. Suddenly Armstrong and Aldrin felt themselves grow lighter. The long brake was over now. It was time for *Eagle* to pitch over from its faceup position and begin the final descent. Armstrong and Aldrin waited intently for the computer to execute the maneuver. Just as planned, 7,500 feet up, *Eagle*'s maneuvering thrusters fired to pitch the craft forward. In *Eagle*'s windows the flat horizon swung upward into view, and Armstrong looked out at the cratered plains of the Sea of Tranquillity, bright in morning sunlight.

Armstrong checked the altitude and speed: 5,000 feet up, 100 feet per second, just as expected. For a moment he took control of his craft, pulsed the maneuvering thrusters, then gave the ship back to the computer, satisfied that *Eagle* would respond when it was time for him to take over. Now his eyes went back to the gauges. Three thousand feet up now, descending at 70 feet per second, about 48 miles an hour. *Eagle* was right on the planned trajectory. He heard Duke say, "You are Go for landing."

This was the time for Armstrong to watch for his landmarks and look for a good place to set down. Before he could do so, Aldrin announced, "Program alarm. Twelve-oh-one." In Houston, Steve Bales was ready. Before Kranz could finish asking him what 1201 meant, Bales shot back, "Same type; we're Go."

Inside *Eagle,* Armstrong's eyes went back to the moonscape. He gazed past the grid on his window and said to Aldrin, "Give me an LPD."

Aldrin queried the computer and told Armstrong, "Forty-seven degrees."

"Forty-seven," Armstrong repeated. He sighted along the window grid, past the 47 degree mark. He could see the target, still more than a mile in the distance, but advancing rapidly. It looked promising. "That's not a bad-looking area," he said blandly. Suddenly the alarm was back. Aldrin had no sooner cleared the alarm than it sounded again. Again, Armstrong's attention was diverted by the threat of an abort, while *Eagle* flew onward. When the alarms quieted down the moon was only 1,000 feet below him, and he did not like what he saw. A crater as big as a football field was just ahead, surrounded by a field of boulders, some as big as Volkswagens. The computer was blindly taking them there, down into the middle of the boulder field. And for an instant, Armstrong weighed the matter. He was all but certain those boulders would prove to be pieces of lunar bedrock. If he could find a safe place to land just short of the boulders, he and Aldrin would no doubt find some prizes for the geologists. But *Eagle* was going too fast; there were just too many rocks. It was time for him to take over. He switched to ATTITUDE HOLD and pitched the lander forward until it was almost level, letting the descent rocket brake its fall without slowing its horizontal flight. Only 350 feet up now, *Eagle* skimmed over the boulders and headed toward safer ground. Armstrong planned to set down on the first clear place he could find.

Buzz Aldrin did not know about the crater or the boulders, and he heard nothing from Armstrong, who was too busy flying to get out more than a few clipped words every now and then. And Aldrin was too busy to look out the window. His eyes went back and forth between the gauges and the computer readout, his hands went to the computer's keyboard to extract the critical information. And his voice, heard on hot-mike by Armstrong and the listening world, was a steady stream of data: "Three hundred and fifty feet; down at four. Three hundred thirty, six and a half down. You're pegged on horizontal velocity." His voice was almost electronic.

In mission control, Kranz and his controllers heard only Aldrin's stream of numbers. Telemetry told them that Armstrong had assumed semimanual control of the lunar module. No one knew about the football-field-sized crater; they knew only that *Eagle* was no longer following the nominal landing profile, that it had slowed its descent and was still moving at a good clip over the moon. Gene Kranz knew then that the partnership had all but dissolved, that the "center of gravity of the decision-making process" was no longer some point midway between himself and the moon. It was Neil Armstrong. Charlie Duke knew it too, and he said over the loop to Kranz, "I think we'd better be quiet." There was nothing to do but listen to Aldrin's voice and hope that the fuel held out.

Armstrong flew onward, sharing control of *Eagle* with the computer. As he cleared the big crater he was careful to pitch *Eagle* back again to avoid building up too much speed. Below, he could see a string of boulders; he banked slightly to the left to get away from them. The response was sluggish and familiar. The lunar module was a much better flying machine than he had expected, easier than any simulation. The little toggle switch to control their rate of descent—the one he had been skeptical of back on earth—was working well. Now Armstrong pushed it several times to slow their fall.

The moon rushed up at him, new terrain advancing quickly over the horizon. He heard Aldrin: "Three hundred feet, down three and a half, forty-seven forward." His heart pounded. Still there was no clear place. The clock seemed to race. He had to buy time to search for a spot. Armstrong slowed *Eagle*'s descent rate.

"How's the fuel?" Armstrong's voice was quiet, even relaxed.

"Eight percent," Aldrin answered. That was less than they'd had in the simulations. Now, at last, Armstrong saw what looked like a patch of smooth ground, just ahead. "Okay," Armstrong said, "looks like a good area here."

Aldrin stole a moment to glance out. On the bright ground 250 feet below was a dark silhouette clearly recognizable as a lunar module, bristling with antennae and landing gear and ringed with a halo of sunlight. Then he went back to the gauges, his voice more insistent as he fed Armstrong data. "Two hundred twenty feet. Thirteen forward, eleven forward. Coming down nicely."

Now Armstrong saw that the place he had selected was no good. "I'm going right over a crater," he said, his words lost in Aldrin's numbers. "I gotta get farther over here." Again he nudged the hand controller forward, leveling the craft, using the last bit of forward speed to clear the crater. And just beyond it he saw where he was going to land. It was a smooth, level place about 200 feet square, bounded on one side by a few large craters and on the other by a line of boulders. He knew they were getting very low on fuel. As if to emphasize this the DESCENT QTY light now glowed on the instrument panel. Ninety seconds of fuel left, and 20 seconds of that had to be saved for an abort. But Armstrong had his landing place. And only 100 feet separated *Eagle* from the moon.

It was crucial to bring *Eagle* straight down, with no horizontal motion; otherwise there was the risk that the touchdown might break off a landing leg. Armstrong trained his vision on a place just beyond the landing point, which he would use as a reference to judge *Eagle*'s height and motion all the way to touchdown. Now he noticed that everything was wrapped in a transparent haze. The blast of the rocket was disturbing the dust of the moon. As *Eagle* descended the haze became a sheet of rushing streaks that flew away from him in all directions, obscuring the surface. They confused his perception of motion like a fast-moving ground fog blowing across a runway. But he could see rocks on the surface, sticking up through the blur like islands, and he fixed his gaze on them.

Armstrong clicked the rate-of-descent toggle until *Eagle* descended no faster than an elevator. He focused on the rocks and on Aldrin's numbers: "Sixty feet, down two and a half, two forward, two forward." He heard Charlie Duke's voice: "Sixty seconds." A minute's worth of fuel left until he and Aldrin would be forced to abort.

In mission control, stomachs tightened. No one knew about the big crater and Armstrong's efforts to avoid it. They knew only that in almost every simulation Armstrong had landed by this point. And everyone, from the controllers riveted to their displays to the VIPs who watched in agonized silence, knew that every second brought Armstrong and Aldrin closer to their abort limit. Even now, it was impossible to know how it would end.

Fifty feet above the moon. Now thirty. *Eagle* was drifting slowly backward and Armstrong did not know why, but he knew he must not land while he could not see where he was going. He pulsed the hand controller, struggling to arrest the unwanted motion. He was displeased with himself, sure that he was not flying *Eagle* smoothly. He wished he could buy more time, but he was too low on fuel to slow the descent any further. Twenty feet to go. He'd stopped the backward drift but still wrestled with a sideways motion that had crept in. They were flying the dead man's curve now, too low to abort if the

IN 1969

North Vietnamese president Ho Chi Minh dies at age 79.

The tanker *Manhattan* becomes the first commercial ship to traverse the Northwest Passage.

In England, Monty Python's Flying Circus makes its first television appearance.

The 100-to-1 shot New York Mets win the World Series behind right-hander Tom Seaver's 90-plus mile-per-hour fastball.

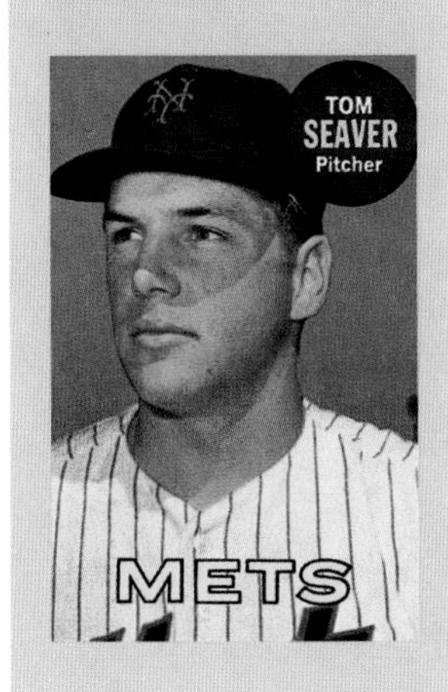

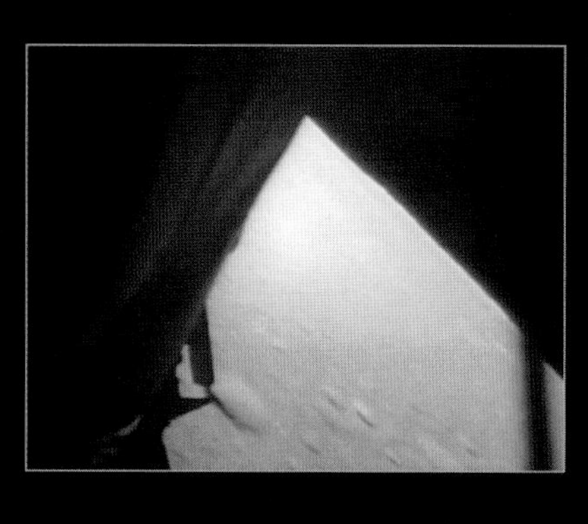

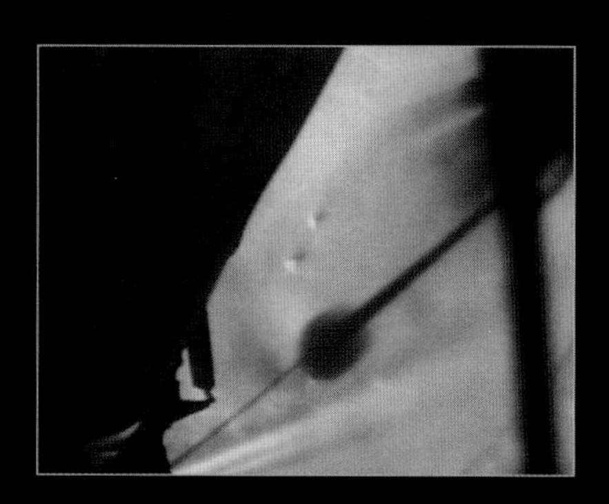

Onboard movies show a view of the moon similar to Armstrong's as he and Aldrin landed. From 20,000 feet *(top)*, the lunar horizon curves across a corner of the picture. At 100 feet *(middle)*, *Eagle* flies over an 80-foot crater. Ten feet above the moon *(bottom)*, the shadow of a lander leg falls across a torrent of dust raised by descent engine exhaust.

engine quit, but in the back of his mind Armstrong knew that if that happened they'd be okay, they would just fall onto the moon. Dust blew furiously. Once more, words of caution came from earth: "Thirty seconds." Then Buzz Aldrin said, "Contact light."

Armstrong had planned to shut the engine down at this moment: the engineers had warned him that if the rocket got too close to the surface, the back-pressure from its own exhaust might blow it up. But he was so absorbed in flying that he forgot about that. With the engine still firing, *Eagle* settled onto the moon so gently that neither man sensed the contact. Quickly Armstrong hit the ENGINE STOP button and said, "Shutdown."

Now there was a blur of activity as Armstrong and Aldrin set switches and keyed numbers into the computer, while Aldrin rattled off the postlanding checklist:

"DESCENT ENGINE COMMAND OVERRIDE, OFF—ENGINE ARM, OFF—413 is in."

Then there was a moment of quiet, and the two men turned to one another in the tiny cabin. Their eyes met, their bearded faces grinned at each other inside bubble helmets, and their gloved hands clasped. Armstrong keyed his mike. "Houston, Tranquillity Base here. The *Eagle* has landed."

The answer from earth was like a sigh of relief. "Roger, Tranquillity, we copy you on the ground," radioed Charlie Duke. "You got a bunch of guys about to turn blue. We're breathing again. Thanks a lot."

They had done it. For an instant, Armstrong and Aldrin savored relief and elation that the greatest challenge was behind them. For Armstrong it was more than a personal high; hundreds of thousands of people had worked for the better part of a decade to share this triumph. And for himself, the landing had been everything a pilot could ask for. It had been a close call, but that just sweetened the victory. There was no way to know exactly how much fuel remained when they touched down—the gauges just weren't that accurate—but it was something like 20 seconds' worth left before the abort limit. Of course, Armstrong knew, 20 seconds is a long time.

II. MAGNIFICENT DESOLATION

TRANQUILLITY BASE
3:17 P.M., HOUSTON TIME
4 DAYS, 6 HOURS,
45 MINUTES
MISSION ELAPSED TIME

Seconds after *Eagle*'s rocket engine shut down the dust particles departed on long, flat trajectories, and the stillness of a billion years returned to the Sea of Tranquillity. Inside *Eagle,* Armstrong and Aldrin put their elation on hold; there was work to do. Now that they were on the moon the most important thing was to get ready to leave it, in case of an emergency. That meant checking *Eagle*'s systems, entering abort data into the computer, and aligning the guidance platform to the weak lunar gravity. That last item was Aldrin's task, and while he attended to it, meticulously sighting on stars with the navigation telescope and conversing with the computer, Armstrong had a chance to survey the place where they had landed.

Eagle had come to rest on a broad, level plain, pockmarked with craters a few dozen feet to a fraction of an inch across, and scattered with rocks and boulders. In the distance Armstrong could see ridges that might have been twenty or thirty feet high, but it was hard to tell: there were no buildings, trees, or any other features normally used to judge size and distance. The lack of atmosphere gave an unreal clarity to the view, better than the clearest day on earth. Hills and boulders at the horizon were as sharp as the rocks next to

In Wapakoneta, Ohio, Stephen and Viola Armstrong rejoice as their son settles his spacecraft safely on the moon. Said Mrs. Armstrong at the moment of landing, "They're on; they're on!"

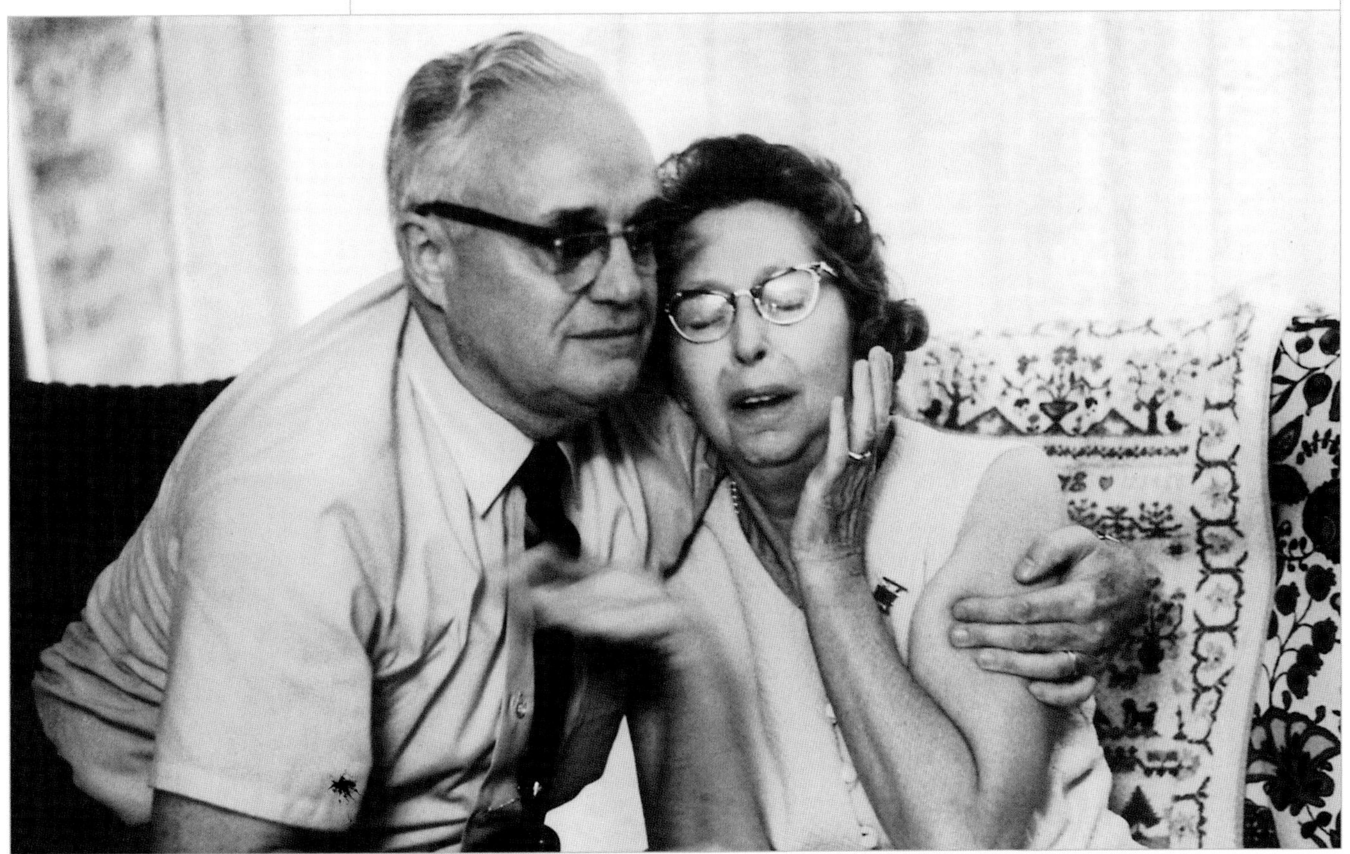

As seen through *Eagle*'s windows soon after the spacecraft came to rest, the Sea of Tranquillity appears as a bright plain scarred by craters and littered with rocks.

Eagle's footpads. Beyond that bright edge, as empty as the margins of a fifteenth-century map, was the blackness of space.

Most amazing to Armstrong was the strange play of light and color. Directly ahead, to the west, the light of the rising sun was brilliantly reflected by a landscape of light tan. This gave way on either side to a dimmer, grayer tan, and when he craned to look off to the side, where the ground was crisscrossed with long morning shadows, he saw an ashen gray.

It was not a hostile scene. Somehow it did not look like a place where an unprotected man would perish in seconds; on the contrary, it seemed inviting, as if he and Aldrin might descend to the craters in beach clothes and get a suntan. As he looked out, Armstrong wondered where he and Aldrin had landed. With the distraction of the computer alarms he'd missed all his checkpoints on the way down. Now he searched the horizon for some feature he might be able to identify, but found none. With a wry smile he radioed Houston, "The guys who said we wouldn't know where we were are the winners today." He knew there must be a small army of controllers and geologists working on it. And he hoped that sometime in the next 21½ hours Mike Collins would be able to spot them with *Columbia*'s navigation sextant.

Wherever they were, Armstrong was ready to explore. To the north, a line of boulders broke the rolling smoothness; they looked to be within easy walking distance. Armstrong suspected they had been ejected from some large crater nearby, perhaps that giant crater he had avoided during the descent. They were probably pieces of lunar bedrock; the geologists would undoubtedly want samples. He might even be able to run back to the crater and take some pictures.

Armstrong turned his gaze to the LM's small overhead window, and he searched the velvet blackness for stars, but saw only the blue earth. "It's big, and bright, and beautiful," he told Charlie Duke. Then, after running through the countdown for a simulated lunar liftoff, he and Aldrin turned off most of *Eagle*'s systems. The lander would be dormant until it was time to leave the moon; that would come early tomorrow afternoon, if all went well.

NASSAU BAY, TEXAS

Outside the Aldrin house, where it had begun to rain into the muggy afternoon air, a group of reporters assembled on the front lawn. Already they had spoken to Jan Armstrong and Pat Collins about this day's incredible events; now they wanted Joan Aldrin's reaction. For some time, they waited for her to emerge.

Savoring her husband's triumph, Joan Aldrin and children Janice and Andy talk to reporters in front of their house shortly after *Eagle* touched down on the moon.

Inside, Joan was savoring relief. Buzz had always said that the moment of truth would be the liftoff from the moon, but in her mind it was the landing. When the Powered Descent began Joan was in front of the television, surrounded by family and friends, her emotions under control. She listened intently to the squawk box, but it was hard to hear what Buzz was saying, and when she did it was so technical that she did not often understand. Fortunately, Jerry Carr and Rusty Schweickart were there to explain what was happening. Then *Eagle* was very low, and she could barely stand the tension. Rusty told her that the fuel was now down to a matter of seconds, and that Neil and Buzz had not yet found a place to land. The words overwhelmed her. She stood in the silence of that crowded room, holding onto a doorway, her eyes brimming with tears, and listened to her husband's voice. Finally, in the midst of the numbers, she heard Buzz say, "Okay, engine stop." Those words she understood: they had made it. She found the embrace of Buzz's uncle, Bob Moon, and then headed for her bedroom. ☾

By the time the reporters and the TV crews gathered, Joan was on a high. After a time, she went out to them, sheltered from the rain by an umbrella held by a NASA public affairs man. All the times that she had faced the press and dutifully concealed her terror were behind her. She radiated life, and she used this moment as if it had been made for her: here was a stage to play on. One of the reporters asked, "What were you doing when they landed?"

"Well," said Joan in a stage whisper, "I was holding onto the wall. I was praying." The reporters did not pick up on her performance. As they asked the same questions they had already asked the other two wives—What are your plans for the moonwalk? Will you let the children stay up to watch?—she wondered, *My God, what is wrong with them?* They seemed drained of energy. She would rouse them.

"Listen!" cried Joan, her eyes wide. "Aren't you all excited?" For a moment she was silent, suddenly in command of her small audience, and then, she let loose her jubilation: *"They did it! They did it!"*

5:11 P.M.

It was hard to believe that two men could land on the moon and go to sleep before setting foot on it, but that was what the conservatively minded flight plan called for: In case Armstrong and Aldrin had to make an emergency liftoff and rendezvous, they would need to be rested. Before the mission, Armstrong had approved the early, four-hour sleep period knowing full well that, barring any problems, he and Aldrin would almost surely reject it on the moon. He wasn't about to say anything to the press—if for some unforeseen reason he and Aldrin ended up sticking to the original plan, they'd write, "Astronauts step on moon, four hours behind schedule." But now, there was no reason to wait. *Eagle* was in perfect order. One-sixth g felt entirely natural; in fact, they liked it better than either normal gravity or weightlessness—it had much of the buoyancy of zero g without the disorienting lack of up and down. And so Armstrong and Aldrin agreed: They would go out early. Armstrong called Houston to suggest that the moonwalk begin at about eight o'clock in the evening, Houston time, some five hours ahead of schedule. Almost immediately Charlie Duke came back with a go-ahead.

"You guys are getting prime TV time," Duke said.

5:57 P.M.

In the history of exploration, one of the few moments that could compare with this one came on May 29, 1953, when Edmund Hillary and Tenzing Norgay became the first humans to stand atop the windswept summit of Mount Everest. In their brief minutes atop that ultimate peak, Hillary's actions were those of the conqueror; he aimed his camera down each ice-crusted ridge at the lands below and snapped the pictures that would prove to the world they had made it. The Nepalese Sherpa Tenzing, meanwhile, hollowed out a place in the snow and filled it with offerings to his God. For him the climb was not a conquest but a pilgrimage.

Now it was Buzz Aldrin who enacted a spiritual observance in a strange and distant place. In the weeks before launch, he had searched for some gesture that would be worthy of the moment, and he had decided to celebrate Communion. Deke Slayton had warned him against broadcasting any religious observance over the air; NASA was still coping with a controversy stirred by the Genesis reading on Apollo 8. Aldrin's Communion would have to be a secret one. ☾

Now that it was clear that he and Armstrong were on the moon to stay for a while, Aldrin took advantage of a quiet moment. He opened the stowage pouch that contained his personal mementos and removed a plastic bag containing a small flask of wine, a chalice and some wafers, and set them on the little fold-down table just beneath the keyboard for the abort guidance computer. He keyed his mike. "This is the LM pilot speaking. I'd like to take this opportunity to ask every person listening in, whoever and wherever they may be, to pause for a moment and contemplate the events of the past few hours, and to give thanks in his or her own way." Armstrong looked on, an expression of faint disdain on his face (as if to say, "What's he up to now?") while Aldrin went on with his ceremony. Released in the gentle gravity the wine poured slowly and curled gracefully against the side of the cup. Aldrin read silently from a small card on which he had printed words from the book of John:

I am the wine and you are the branches
Whoever remains in me and I in him will bear much fruit;
For you can do nothing without me.

In Nassau Bay, Joan Aldrin—who had settled into a warm and wonderful state, playing old Duke Ellington records with one ear tuned to the squawk box—marveled when her husband asked for a moment of silence. Though she did not know he was taking Communion, she saw it as happy evidence of a hidden dimension. All this time she had suspected Buzz was so caught up in the technical side of his mission that he had missed its significance—but now she realized she had been wrong.

IN 1969

Across the country, opponents of the conflict in Southeast Asia stage Vietnam Moratorium Day, with prayer vigils, mass meetings, and candlelight processions.

West Germans elect Willy Brandt chancellor.

The Beatles release *Abbey Road,* the last album the band recorded before breaking up.

If Aldrin's Communion marked a very personal observance, then Neil Armstrong had his own ceremony to think about. Almost from the moment the world learned that he would be the first human being to set foot on the moon he had been asked what he would say as he crossed that threshold. His mail had been full of suggestions, including passages from the Bible, verses of Shakespeare, and countless others. Everyone from the press to the simulator instructors brought it up. Not even by leaving earth could he escape; Collins and Aldrin asked about it on the way to the moon.

If it hadn't been for the fact that everyone made such a big thing of it, Armstrong wouldn't have focused on the matter at all. The landing was the flight's greatest achievement, and in Armstrong's mind, it amounted to the first human contact with the moon. ☾

But to a public estranged from the technology of this journey, a landing was less meaningful than a footstep. If it was natural for them to want historic words for historic occasions, they were nevertheless asking them of a man who does not deal liberally in words. But now, on the moon, Armstrong knew he could delay no longer. As he thought about the first step he would take from *Eagle*'s footpad he pondered the inherent paradox—a small step, yet a significant one—and he knew what he would say.

7:21 P.M.

The moon cast its light into *Eagle*'s cabin as Armstrong and Aldrin began the most critical and the most tiring part of the entire moonwalk: suiting up. They were already behind schedule simply because there had been things to do that had never been part of the practice runs—such as stowing the trash from dinner—and they took longer than the men expected. But this was no time to rush, and Armstrong and Aldrin worked with the care of skydivers packing their chutes, following the checklist to the letter. First they pulled on the lunar overshoes, whose rubber soles had coarse treads designed to give sure footing on alien soil. Next they strapped on the Portable Life Support System—the massive backpack that had been the bane of their existence during training. On the moon each one weighed just over twenty pounds; Armstrong and Aldrin had no trouble hefting them with one hand. Still, they felt the *mass* of the packs—that was undiminished—and despite their efforts to avoid bumping into control panels and each other in the cramped cabin, that happened more than once.

Oxygen hoses were next; their metal fittings locked into receptacles on the front of their suits. Then came hoses for their water-cooled underwear. Water from the backpacks would circulate through a network of tiny tubes woven into the undergarment. This method of cooling a man inside a space suit was so effective that there was almost no way for him to become overheated, the way Gemini spacewalkers had; tests had showed that he would tire himself out before that happened. Armstrong and Aldrin carefully locked each hose in place, and then locked the *locks* in yet another level of security. Both men were on hot-mike now, and the radio transmissions that came down from the moon sounded like strange, high-tech poetry: "Locks are checked, blue locks are checked. Lock-locks, red locks, purge locks . . ."

Onto each man's clear bubble helmet went a special outer helmet equipped with a gold-plated visor to reflect the sun's unfiltered glare. On their chests, they wore small control units for their radios and to display readings on the backpack. Each methodical step brought them closer to setting foot on the bright ground beyond *Eagle*'s windows.

"All set for the gloves," Aldrin said, and each man pulled on a space age version of a knight's gauntlet, with coverings of woven steel-fiber, and rubber fingertips that afforded some measure of dexterity. With the flick of a switch, each man started the pumps and fans in his backpack and heard the familiar, reassuring hum of machinery that would keep him alive, and felt the whoosh of oxygen past his face. Their ears registered increasing pressure as the suits inflated to 3.5 pounds per square inch. Now Armstrong and Aldrin were self-contained, mobile spacecraft.

9:28 P.M.

All that remained was to vent *Eagle*'s oxygen into space, but even that took longer than expected. Aldrin opened the valve and the men watched the pressure reading creep downward. After three minutes it was four-tenths of a pound; a minute later, two-tenths.

"Let me see if it will open now," Aldrin said, reaching for the hatch handle. It stayed firmly shut. The pressure read one-tenth of a pound and holding. Neither man wanted to tug on the thin metal door for fear of damaging it. Finally, Aldrin peeled back one corner to break the seal; that did it.

"The hatch is coming open," Armstrong radioed, excitement creeping into his voice. As it did so, the last wisps of *Eagle*'s atmosphere rushed outward in a flurry of ice particles, and the two men stood in the vacuum of space.

While Aldrin held the hatch open, Armstrong sank to his knees and carefully moved his suited bulk through the opening. He moved onto a large platform called the porch, with large handrails on either side, that bridged the hatchway with the ladder. When his boots met the top rung, he grasped the handrails and raised himself upright. After five days of floating within the confines of a spacecraft, the change in visual scope was profound. The sensation of height, absent in deep space or in orbit, returned to him. Before him, the shadowed, foil-clad bulk of his lunar module; beyond, a pristine wilderness.

He could not descend yet; for one thing, the world was waiting to see the event. Armstrong pulled a D-ring on *Eagle*'s side and an equipment stowage tray lowered like a drawbridge. On it a small TV camera began transmitting

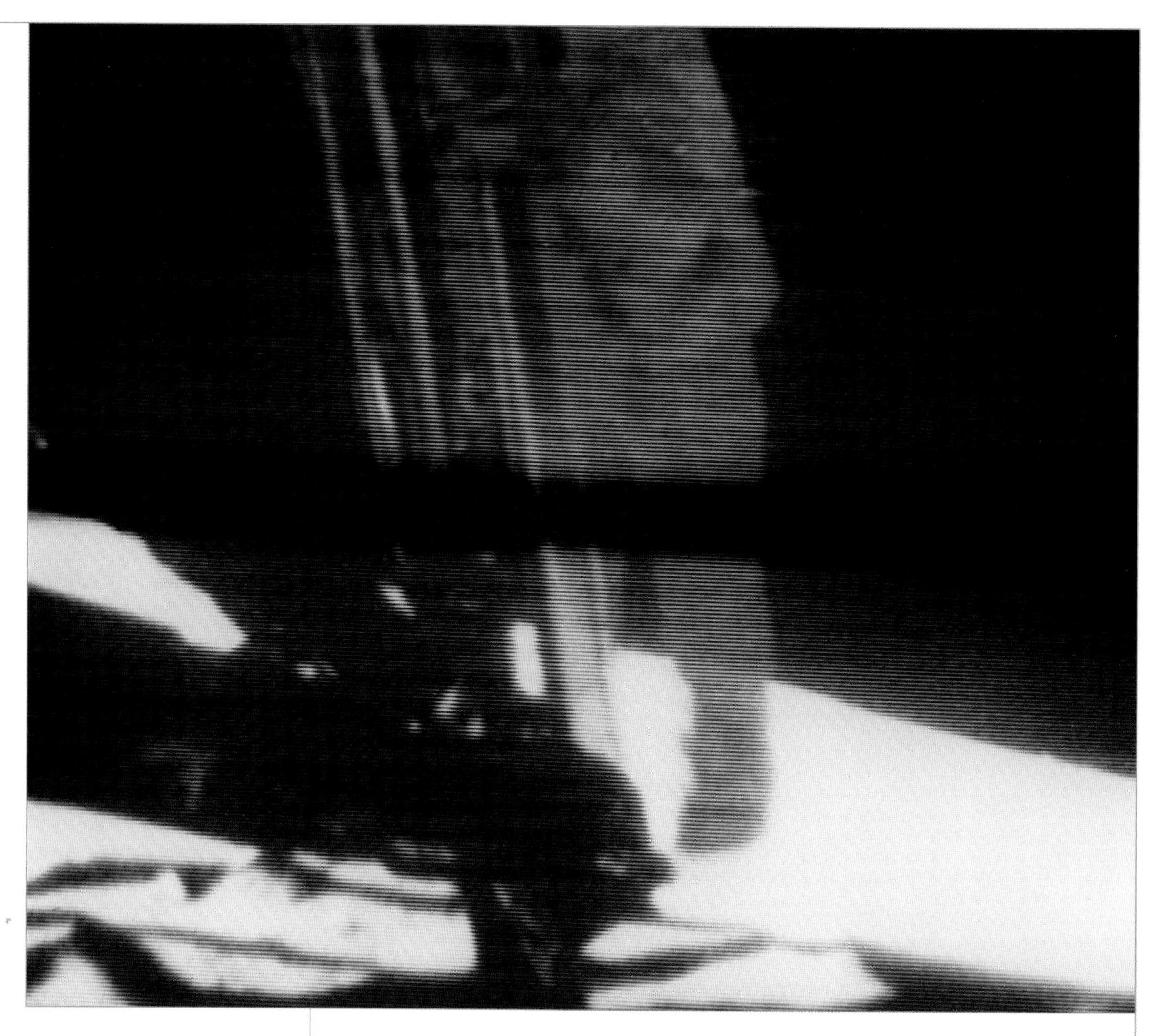

Televised by a camera housed in the descent stage, the ghostly figure of Neil Armstrong descends the ladder on *Eagle*'s front leg. Many millions of people around the world saw this historic image on television, transmitted live from the moon.

to earth, where Cliff Charlesworth and his team of controllers listened and waited. Moments passed, and then Armstrong heard Capcom Bruce McCandless radio, "We're getting a picture on the TV!" On the big screen in mission control a strange, almost abstract black-and-white image flickered into existence. The front leg of the lunar module slanted across a tableau of black sky and bright ground, and at the top, the shadowy form of Neil Armstrong descended, one rung at a time, toward the moon.

When Armstrong reached the bottom rung he paused. The legs were designed to compress with the force of landing, bringing the ladder closer to the surface in the process. But *Eagle* had touched down too gently for that to happen; Armstrong was still more than three feet up. For a moment he dangled his foot in space, then launched himself into a slow-motion fall, landing on both feet inside the foil-covered footpad. Before he went any further, he wanted to be sure he could get back up. He sprang upward and almost missed the bottom rung, but at last managed to steady himself. Satisfied, he descended once more.

Standing in deep shadow, Armstrong looked down at the soil just beyond the footpad, and as he had done many times in training, he described what he saw for the benefit of the scientists on earth. "The surface appears to be very, very fine grained as you get close to it; it's almost like a powder . . ." In simulations his voice had been decidedly matter-of-fact; now it was laced with excited curiosity.

"That's one small step for man—
one giant leap for mankind."

–Neil Armstrong

Grasping the ladder with an upraised glove, Armstrong turned to his left and leaned outward. "Okay," he said, "I'm going to step off the LM now." Silently, carefully, he raised his left boot over the lip of the footpad and lowered it to the dust. Immediately he tested his weight, bouncing in the gentle gravity, and when he felt firm ground, he was still, one foot on the last vestige of earthly things, the other on the moon. He spoke: "That's one small step for man"—now a pause—"one giant leap for mankind." Again he tested his weight and was reassured to find that his boot penetrated only a fraction of an inch. Still holding on, he stretched out his toe and dragged it backward several times, furrowing the soft ground. Dust clung like soot to the light-blue sole of his boot. Having made this first, tentative exploration, Armstrong lowered his right foot and stepped sideways, both hands resting on the big horizontal strut of *Eagle*'s landing gear. And at last, after bouncing up and down a few more times, he let go of *Eagle* and stood on the moon.

Armstrong moved away from the lander with the halting steps of a man learning to walk again. He moved with a shuffling, stiff-legged gait; it was difficult to bend at the knee and movement came mostly from the ankles and the toes. But he felt buoyant, something between walking and floating. Heavy and light were redefined: his space-suited body, 348 pounds on earth, now weighed only 58 pounds. It was almost familiar—the simulations were that

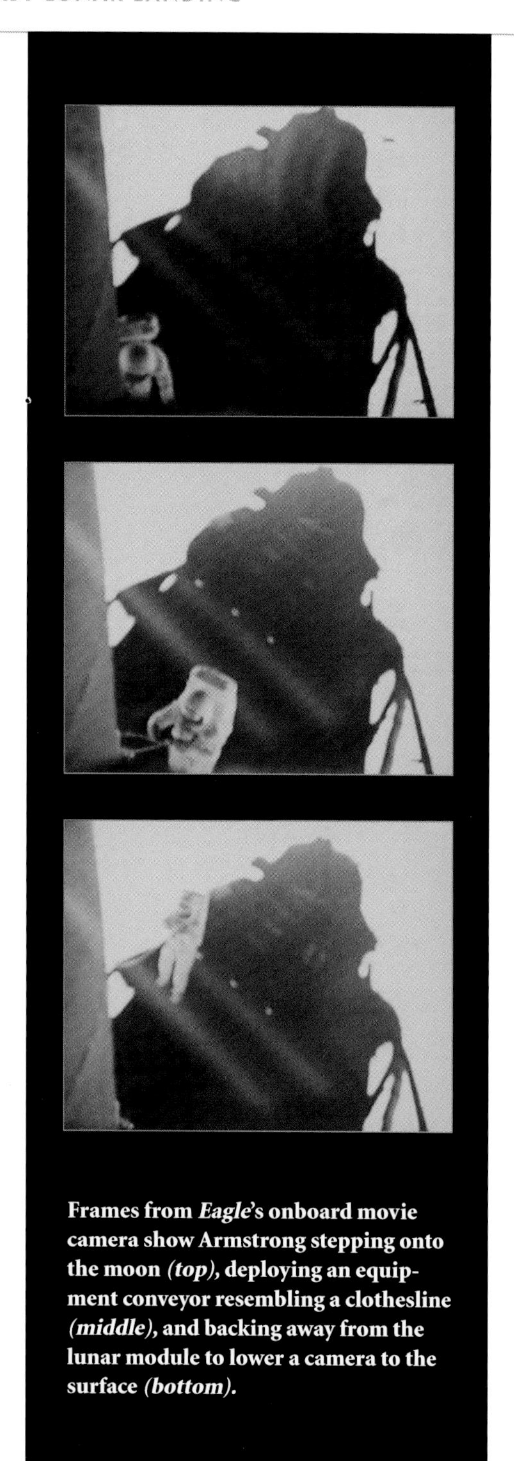

Frames from *Eagle*'s onboard movie camera show Armstrong stepping onto the moon *(top)*, deploying an equipment conveyor resembling a clothesline *(middle)*, and backing away from the lunar module to lower a camera to the surface *(bottom)*.

good—and it was even easier to move around than he had expected.

He knew that the first order of business was to collect a small bag of soil, called the contingency sample, that would serve as the scientists' hedge against an aborted moonwalk. But he would do that in sunlight, and he wanted to take care of getting a Hasselblad down to the surface while his eyes were still adapted to the darkness. This he accomplished with some effort as he and Aldrin operated a special conveyor line. Then, with the camera mounted on the control unit on his chest, still standing in the lander's shadow, Armstrong snapped the first pictures taken on the surface of another world.

But pictures were not supposed to be his first priority, and after a minute McCandless reminded him about the sample. Aldrin added his own reminder. "Right," Armstrong said quickly, finishing his panorama, then he reached into a pocket on his thigh and pulled out a collapsible handle with a detachable bag at one end. He moved into sunlight for the first time, the glare penetrating his mirrored visor like a thousand-watt spotlight. Turning away, Armstrong began to dig into the surface, and what he found surprised him. Everywhere there was the same soft powder, and yet here and there he met resistance. He managed to scoop up enough dust to fill the bag, and even managed to snare a couple of small rocks; the geologists, he told himself, would get their money's worth.

"That looks beautiful from here, Neil," Aldrin said. He was talking about the sample, but Armstrong responded as if he had meant the moon. "It has a stark beauty all its own," he said, excitement finally invading his voice. It *was* beautiful. It had the serenity of the high desert of Edwards, only here was the ultimate desert, complete in its stillness, and in its starkness. When he turned, he saw the same peculiar transformation from bright tan to ashen gray he'd seen from the LM windows. And when he held the contingency sample in his hand, the mystery of the moon's color deepened: The soil in the bag was almost black, like powdered graphite. ☾

Holding the now unneeded collector handle, Armstrong considered throwing it like a javelin, but thought better of it, and instead gave it an underhand toss. It sailed away on a long, lazy trajectory, spinning in slow motion in the sunlight and traveling an impossibly long distance before landing in the dust.

"I didn't know you could throw so far, Neil," joked Aldrin. The man on the surface of the moon answered with a delighted laugh in his voice, "You can really throw things a long way up here!" Armstrong was elated, and for good reason. The moon offered him firm ground and good footing. Working in one-sixth gravity, in contrast to the grueling training sessions, was easy. Barring a major problem with equipment or with *Eagle,* the first moonwalk was bound to be as successful as anyone had hoped.

In his first assignment on the moon, Armstrong uses a special scoop to collect a contingency sample of lunar dust and rocks, which he deposited in a pocket strapped to the leg of his space suit.

Photographed by Armstrong, Buzz Aldrin emerges from *Eagle*'s hatch onto a platform called the front porch. The tread clearly visible on Aldrin's boots is designed to give good footing on lunar dust.

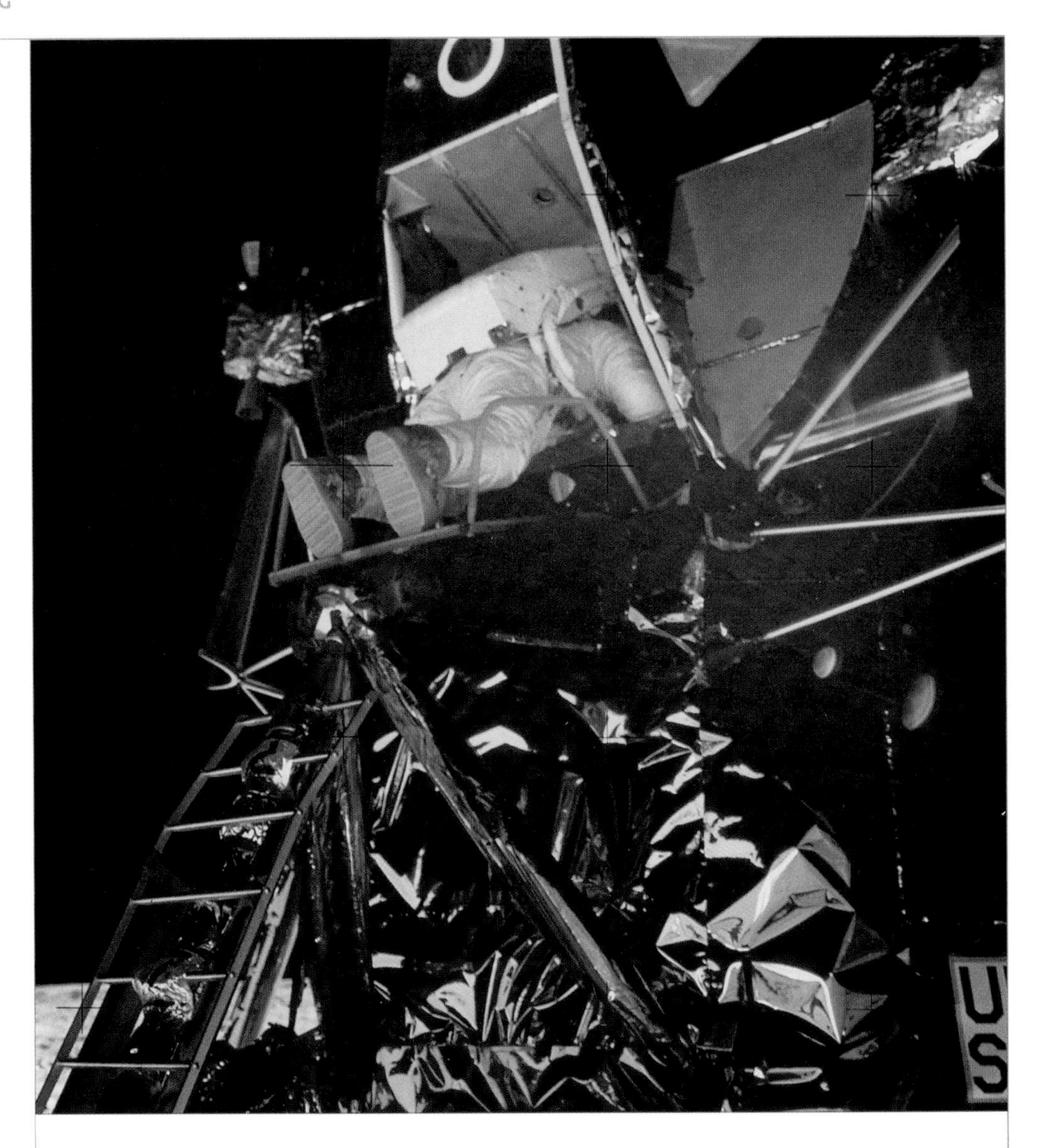

10:10 P.M.

"Are you ready for me to come out?" Not a trace of eagerness sounded in Buzz Aldrin's voice, even though he had been watching Armstrong walk on the moon for fourteen minutes. Now it was his turn. While Armstrong radioed guidance, he emerged onto the porch. He offered a mild joke about making sure not to lock the hatch on his way out, which got a laugh from his commander. Then, ever methodical, Aldrin made his own descent, describing his progress to earth. But when he was standing in the footpad, looking out at the moon, his powers of description momentarily left him. He saw disorder, and yet there was a precision, he would say later, the precision of rock and dust. There must be some combination of words that would describe it, but Aldrin could only utter, "Beautiful view!"

Armstrong agreed, "Isn't that something? Magnificent sight out here." Hearing this, Aldrin suddenly had the words he was looking for. With quiet wonder in his voice, he said, "Magnificent desolation." Holding the ladder with both hands, Aldrin swung both feet out of the footpad and onto the moon.

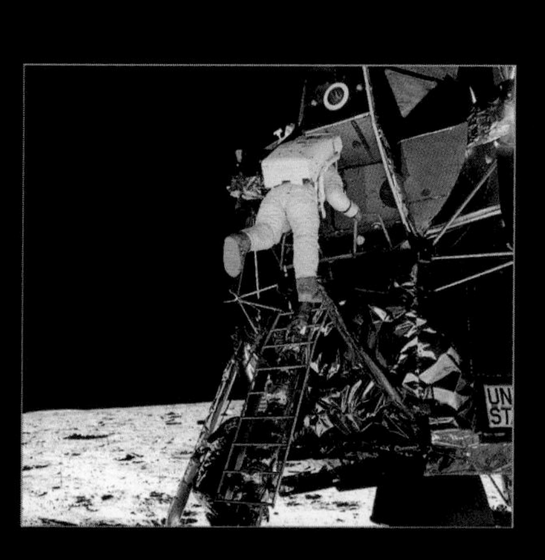

Aldrin descends *Eagle*'s ladder one rung at a time *(top)*, pausing on the last rung *(middle)* before dropping to the footpad *(bottom)*.

The checklist called for Aldrin to check his balance and stability, and that he did, twirling and leaping like a dancer in slow motion, feeling the strange inertia of his backpack. To compensate for the mass of the pack, he had to lean forward at a seemingly impossible angle; on earth, he would have fallen over. But the pull of this small world was so mild that he could not easily tell when he was standing exactly upright. Looking into the distance, Aldrin scanned the plains of the Sea of Tranquillity. The land curved gently but noticeably away from him, all the way out to the horizon, which was only a mile and a half away. He could actually *see* that he and Armstrong were standing on a sphere.

Aldrin's eyes went to his feet, where a fascinating display of motion took place every time he took a step. Each footfall launched a spray of particles that sailed outward in perfect arcs, unhindered by an atmosphere, all coming to rest roughly the same distance away. Intrigued, he kicked his foot like a child on a playground, sending streams of dust flying gracefully into space. He looked at his own footprints and marveled at their sharpness, as if he had placed his foot in talcum powder. And always, he radioed his observations to earth.

And so, two men at the edge of human experience went about their work, their faces hidden by mirrors, their voices so unrevealing that most of the time only people who knew them well could hear the excitement in them. They talked about the mechanical behavior of the soil and the appearance of the rocks, and it was all very technical, all under control. The first men on the moon were not about to indulge in excited exclamations or elaborate statements of wonder—not just because the first lunar landing was so laden with history, but because it was not in their nature.

10:40 P.M.

By international agreement no nation could claim the moon, even one that managed to go there. That was reflected in the plaque on *Eagle*'s front leg, bearing the inscription, "We came in peace for all mankind." But it was the United States that had accomplished the feat, and NASA had decided that the Stars and Stripes would be raised during the moonwalk. Already, Armstrong had mounted the TV camera on a stand about sixty feet from *Eagle*, where it would broadcast the rest of the activities. Now he and Aldrin unfurled an American flag, stiffened with wire so that it would fly on an airless world, and struggled to plant it in the dust. As hard as they tried they could push the flagpole only six or eight inches into the ground. For a moment it seemed the flag would fall over in front of the worldwide audience, but at last the men managed to steady it; then they backed away. ☾

Posing for Armstrong's camera, Aldrin looked at the banner and felt a swell of patriotic pride and humility come over him. He thought of the thousands of people who had helped get it to the moon, and the millions who must be watching him and Armstrong at this moment. He had an almost mystical sense of the unity of humankind, so strong that he felt as if he and Armstrong were not alone. Aldrin marveled at the paradox: No one had ever been farther from earth, and yet no one had ever been the object of more attention. ☾

On the other end of that paradox, an estimated 600 million people, a fifth of the world's population, were indeed watching and listening, the largest audience for any single event in history. Across the United States it was a hot July evening, and in department stores keeping summer hours, and at "moonwalk parties," and in bars suddenly visited with an unaccustomed silence, a fantastic, high-tech stage play was unfolding on every working television set. It was a scene of utter stillness, except for two figures who bounded and leaped like snowmen brought to life, with *Eagle*'s spidery form as a backdrop. The picture seemed ghostly, as if it had lost some of its substance crossing the quarter-million-mile distance to earth. In all, the images from the moon were like a window on a dream. ☾

In Nassau Bay, Joan Aldrin watched in quiet amazement. Earlier, when Buzz first appeared, she kicked her feet and blew kisses at the screen. She laughed and felt ready to cry at the same time. As she watched Neil and Buzz move jerkily about, she thought of a silent movie, or an old cartoon. It couldn't be real, she thought. And yet, here it was: men were walking on the moon, and one of them was her husband.

But walking, strictly speaking, was not the right word for what Buzz Aldrin was doing now as he took center stage to "evaluate the various paces that a person can use traveling on the lunar surface." Aldrin took off on a slow-motion jog, heading for the TV camera. Each step launched him into space, his body suddenly a projectile on a ballistic arc, suspended in mid-stride, until he landed in a spray of powder. Time slowed; he was at the top of the arc waiting to come down. The mass of the backpack required him to anticipate changes in direction well in advance—just as Armstrong had done, flying the LM—and as he ran he kept his eyes out four or five steps ahead, watching for rocks or craters. Now he bounded across the moonscape on two feet. "So-called Kangaroo Hop does work, but it seems that your forward mobility is not quite as good as it is—as it is in the conventional—more conventional one foot after another." As he ran he looked like a science-fiction version of Eadweard Muybridge's turn-of-the-century movies of the human figure in motion. Aldrin fully expected that when he was back on earth the engineers would use the videotape to make careful measurements of his motions (much as Muybridge had done so long ago) to aid future moonwalkers. Instead they would be content simply to hear him tell about it.

Aldrin was in the middle of his experiments in locomotion when he heard McCandless say, "Neil and Buzz, the president of the United States

Eagle's onboard movie camera captures Armstrong _(left)_ and Aldrin setting up the flag of the United States. The compacted lunar soil made it difficult for the men to drive the flagpole into the ground.

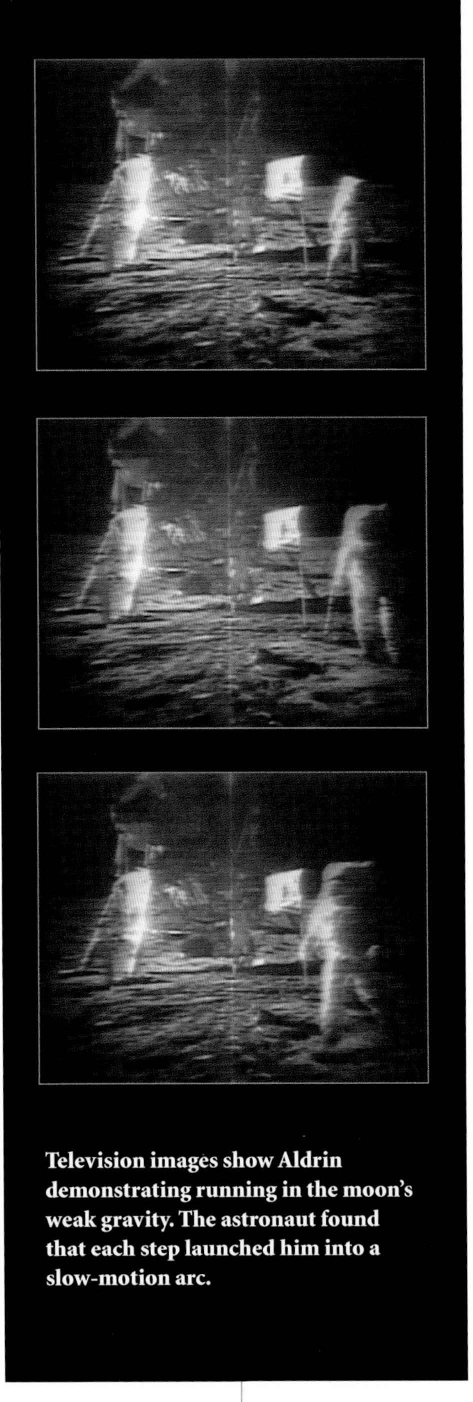

Television images show Aldrin demonstrating running in the moon's weak gravity. The astronaut found that each step launched him into a slow-motion arc.

is in his office now and would like to say a few words to you." Armstrong responded formally, "That would be an honor."

Aldrin suddenly felt his heart pound with anticipation. He was taken by surprise; later he would learn that Armstrong had known the president might call, but had not mentioned it. The two men faced the TV camera and stood still; moments later they heard Richard Nixon's voice:

"Hello, Neil and Buzz, I'm talking to you by telephone from the Oval Room at the White House. And this certainly has to be the most historic telephone call ever made from the White House. . . . " Throughout the moonwalk, Aldrin had the slightly discomforting sense of being a part of something bigger than himself. He noticed a kind of detachment from the event, as if he were watching it unfold before him, somehow beyond his control. And it seemed especially so in these moments, standing before the flag, listening to the president. As he listened he wondered what he might say in response; he decided he would not say anything.

"For one priceless moment, in the whole history of man, all the people on this earth are truly one. One in their pride in what you have done. And one in our prayers that you will return safely to earth."

There was a silence, and then Armstrong responded, "Thank you, Mr. President. It's a great honor and privilege for us to be here, representing not only the United States but men of peace of all nations . . . men with a vision for the future. . . ." To some listeners, Armstrong's voice seemed thick with emotion, as if he were on the verge of tears. Years later, Armstrong would say wryly that in answering the president with a few hundred million people listening he was probably concentrating on trying to say something that made sense.

"Thank you very much," Nixon said, "and all of us look forward to seeing you on the *Hornet* on Thursday."

"I look forward to that very much, sir," Aldrin said. The two men raised their gloved hands in salute, then turned away from the camera and went back to work.

If Armstrong and Aldrin had climbed back into *Eagle* at that moment and blasted off to rejoin Collins, their mission would have been accomplished. The flag was up, and there was in Armstrong's pocket a small bag of the moon. He and Aldrin had already demonstrated that future explorers would be able to work in this alien environment. Anything from now on was frosting on the cake. On earth, this was the point at which many moonwalk parties started to break up. But in Houston a team of now quite frustrated geologists watched like children looking at a toy store on closed-circuit television. Armstrong would not let them down. He'd been preparing to collect samples when the president called; now, while Aldrin set about inspecting and photographing *Eagle,* Armstrong grabbed a long-handled aluminum scoop and began prospecting.

There wasn't much time. Armstrong was allowed only about ten minutes to gather enough rocks and soil to fill one of two aluminum sample containers,

Television pictures from the moon had a strange, ghostly transparency. For better images, the busy astronauts would have had to interrupt their work to set up a special auxiliary antenna.

Aldrin deploys a sheet of foil designed to trap the sun's emission of charged particles, known as the solar wind. Grooves in soil at the bottom of the picture were made by the cable linking the TV camera to the lander.

or "rock boxes." The geologists called this the "bulk sample" and it was intended to be a fairly quick grab. Later, he and Aldrin were to spend time carefully collecting and photographing the so-called documented sample. But Armstrong wasn't at all sure what the rest of the moonwalk would bring, and in case he didn't get a chance for the documented sample he wanted to select as varied a collection now as he could.

Aside from the stiffness of his suit, which fought almost every movement, there was one-sixth g to contend with: no matter how careful he was, when he lifted the scoop from the ground half the contents went sailing away like pieces of Styrofoam. Simply getting the sample over to the LM was a real challenge, but Armstrong persevered, and after several minutes the box was full. Now the task was to seal the box, which would preserve the samples in a lunar vacuum for passage to earth. But that proved to be a struggle, and when he was finished, the entire bulk sample operation had taken longer than planned.

Roughly an hour was left in the moonwalk, and there was still the work of setting up the two scientific experiments. Armstrong realized there would not be nearly enough time for all the exploring he wanted to do. Already he realized the moon was far more interesting than he'd expected. As he accompanied Aldrin on an inspection of *Eagle* his attention was constantly drawn to another interesting feature. Some of the small craters had at their centers bits of something shiny, with a beautiful metallic luster. He had no idea what they were, but they looked just like blebs of molten solder on a workshop table. He wished he still had the scoop in his hand. Here and there he saw what looked like transparent crystals lying in the dust; the biggest was the size of a walnut. He would have to come back for these things later, during the documented sample, if there was time. ☾

11:39 P.M.

The Sea of Tranquillity was more rugged than Armstrong had expected—all bumps and hollows—and not an ideal place to set out a pair of scientific instruments. But about 50 feet from *Eagle* he and Aldrin managed to find a fairly level spot to deploy a solar-powered seismometer to detect moonquakes and an array of prisms that would serve as a reflector for a laser beam from earth, to help scientists measure the precise distance from the earth to the moon. But now, as he and Aldrin finished with the experiments, Bruce McCandless had good news: Mission control was offering a fifteen-minute

Aldrin carries two packages of scientific instruments, called the Early Apollo Scientific Experiments Package (EASEP). In his left hand is a solar-powered seismometer; his right hand holds a laser-ranging experiment to measure more precisely the distance between earth and the moon.

extension. It wasn't much, Armstrong knew, but it would surely help. He'd already abandoned thoughts of inspecting the boulders to the north; like everything else on the moon, they were farther away than he'd thought. So was the giant crater he'd avoided during the descent; he'd fully expected to see its boulder-strewn rim to the east, but it was over the horizon. But there was the smaller crater he'd flown over just before touching down; it was definitely reachable. According to the timeline, he and Aldrin were to start the documented sample now, but Armstrong figured a quick reconnaissance of the crater would be more valuable than the one or two rocks they could pick up in the same amount of time. In any case, he'd already covered his bets with the bulk sample. Without a word to Houston, while Aldrin made his way back to *Eagle,* Armstrong took off running.

Long, loping strides carried Armstrong into the sun's glare to the edge of a pit that looked to be 80 feet across and 15 or 20 feet deep. *Eagle* was nearly 200 feet away, looking like a scale model. Armstrong wished he could climb down to the crater floor and pick up a piece of lunar bedrock, but he knew he mustn't try; if he got into trouble Aldrin's helping hands were a long way off.

Late in the moonwalk, Armstrong ran to the 80-foot crater he had flown over before landing and snapped this series of photographs, arranged here as a panoramic mosaic. The crater appears at far right; the object in the foreground is a stereo close-up camera for photographing small patches of the moon's surface.

He clicked off a series of pictures, hoping to document on film what he had no time to investigate or even describe; then he headed back to the LM. Armstrong had been gone for only about three minutes, but it was the only real exploring he would have a chance to do. ☾

The rest of the moonwalk passed in a rush of activity. There was no time for both men to collaborate on a documented sample. Instead, mission control put Aldrin to work hammering a metal tube into the ground to obtain a core sample, a task that proved even more difficult than planting the flag. Meanwhile, Armstrong scurried about with a pair of long-handled tongs, in search of rocks that would best represent this locale of the Sea of Tranquillity. He wished he had time to collect some of those mysterious, shiny blebs he had seen earlier, or one of the clear crystals he had spotted in the soil; he couldn't find any of those things now. Even as he worked, Bruce McCandless was telling him to press on; time was short. Time, Armstrong would later note, was a strange commodity on the moon. While their mission proceeded with an accuracy of minutes or seconds, he and Aldrin were on a world where a

day lasts a month, where time seems to crawl. Looking at this landscape of craters, rocks, and dust he had the feeling that he was seeing a snapshot of a world in steady-state, that if he had been here a hundred thousand years ago or if he returned a million years from now he would see basically the same scene. But after two hours and thirty-one minutes, he had barely come to know the place. And yet, the knowledge that would emerge made these two and a half hours precious beyond measure. ☾

MONDAY, JULY 21
12:12 A.M., HOUSTON TIME

With a loud and welcome noise oxygen rushed into *Eagle*'s cabin. If the dire predictions about lunar dust catching fire when exposed to oxygen were true, Armstrong wryly mused, then his whole suit was going to burst into flames, because he was covered with grime. Neither man was surprised when nothing happened. But when they took off their helmets, they immediately noticed a pungent odor that reminded Armstrong of wet ashes in a fireplace and to Aldrin smelled just like spent gunpowder; it was the smell of moon dust.

Armstrong and Aldrin took pictures of each other's smiling, bearded faces, and of Tranquillity Base, now looking very much like an expedition site. Beyond the flag, standing in its frozen wave, was the television camera on its stand, and still farther away were the two scientific experiments, everything as still as a ghost town. The ground near the LM was covered with their footprints, each with its sharply chiseled pattern of treads. There was in them something akin to immortality: those prints would remain fresh for perhaps a million years, subject only to the constant rain of micrometeorites from space.

After eating a late dinner they added to their expedition's legacy, opening the hatch once more to toss out the backpacks and a bag of other unneeded gear. When Armstrong learned from mission control that the seismometer had picked up the jolt of the backpacks hitting the surface he teased, "You can't get away with anything anymore, can you?" After fielding questions on the geology of the area, Armstrong and Aldrin prepared for a rest. Even though they had been up since 5:30 A.M., Houston time, and it was closing in on 3:30 in the morning, they were still keyed up, and Armstrong doubted they'd actually sleep. He also knew that a very full, very critical day lay ahead: the second half of John Kennedy's challenge had yet to be fulfilled.

IN LUNAR ORBIT

Every two hours, *Columbia* circled the moon in silent passage with its lone occupant, Mike Collins. The command module was close quarters for three men, but with Armstrong and Aldrin gone it was almost roomy. He'd folded up the center couch and stowed it underneath the left-hand seat, so that now there was a clear aisle from the side hatch to the lower equipment bay. The extra room would be needed in case there was a problem with the docking mechanism, forcing Armstrong and Aldrin to make an emergency space walk from the LM to the command module. For now, it gave Collins unaccustomed freedom. He scurried from his couch to the lower equipment bay and back again, checking systems, making navigation sightings, and attending to a host of housekeeping chores. Command module number 107, the spacecraft he had nursed through its checkout at Downey, California, and piloted across nearly a quarter of a million miles, was purring along without a single malfunction. Collins felt so confident that when one of the command module's cooling circuits grew too cold he chose not to follow mission control's advice to go through a lengthy malfunction procedure. Instead, in his solitude over the far side, some instinct prompted him to see if the machine might cure itself. He checked his switch settings and waited—and the coolant temperature promptly rose to normal. *Columbia* would give Collins no worry in his 22 hours alone. His anxieties were focused elsewhere, with the two men on the surface of the moon. ☾

Buoyed by their safe return to *Eagle* after their two-and-a-half-hour moonwalk, Armstrong and Aldrin took pictures of each other *(Armstrong is at top)* and of the flag they would leave behind.

When his crewmates began the Powered Descent, Collins was listening in, his "cookbook" of rendezvous scenarios at the ready in case they had to abort. In the final minutes he heard them grapple with computer alarms and wondered how serious they might be. As *Eagle* flew onward Collins was, like everyone else, a spellbound listener. When it was over he heard Charlie Duke tell Armstrong and Aldrin there were smiling faces in mission control and all over the world, and Collins radioed, "Don't forget one in the command module."

But no one, not Armstrong and Aldrin nor anyone in mission control knew just where *Eagle* was. The location would be a helpful, though not essential, piece of information for his computer to have during tomorrow's rendezvous. It fell to Collins to try to find the LM on the surface, using the command module's 28-power sextant. This task was a little like looking down on Manhattan from a height of 69 miles, trying to spot a single Greyhound bus with a pair of binoculars—and all the while, moving at 3,700 miles per hour. It would have been pointless to try if not for the command module's computer, which could aim the sextant precisely at any feature under *Columbia*'s path and keep it fixed on the target, compensating for the command module's swift motion. Two hours after *Eagle* touched down, Houston had radioed up a set of coordinates and Collins was at the eyepiece. When he arrived over the landing site the sextant whirred into position, and Collins searched frantically for a glint of light. *Columbia*'s speed allowed him only about two minutes to search any given area within the long ellipse of Landing Site 2, and the sextant's field of view was so narrow that he could scan only one square mile at a time. Two frantic minutes later, Collins had come up empty. Each time he went around from the far side, mission control had a new set of coordinates for him to try, but on his map one guess was as much as 10 grid-squares away from the last. It didn't take Collins long to realize that no one had a handle on the problem. His search continued fruitlessly for the rest of his 22 solo hours.

Alone in lunar orbit, Collins experienced an ultimate state of isolation. He relished the experience, a good part of which he spent at the eyepiece of *Columbia's* 28-power sextant trying to glimpse *Eagle* on the moon below.

If Collins could not see his crewmates on the surface, he could still hear them, via a special moon-earth-moon relay link set up by mission control. For some reason it went off sometime after the landing, leaving him feeling distinctly left out. As Armstrong and Aldrin prepared to go outside Collins asked once again to listen in, and mission control restored the relay. He'd hoped to be listening when Armstrong set foot on the moon, to finally hear his long-awaited words. But the timing didn't work out; Armstrong was just wriggling through *Eagle*'s hatch when *Columbia* slipped behind the far side; by the time he reappeared Armstrong and Aldrin were putting up the flag with 600 million people as witnesses. If, as he suspected, the TV commentators were describing him as a lonely man, they were wrong. He was savoring something as unique as a moonwalk: the experience of the solo moon voyager.

Collins moved through a continual succession of sun-drenched lunar day, soft earthlight, and unyielding blackness. For 48 minutes out of each orbit, from Loss of Signal to Acquisition of Signal, he knew a solitude unprecedented in human history. Before the flight he had been asked more times than he could count whether the thought of being alone in lunar orbit worried him. No, he had answered, I *like* being by myself. To a fighter pilot it was the essence of flying: alone in your craft, in control of your craft. It was nothing less than the purest form of freedom. There were lonelier places than the far side of the moon, and Collins had seen them; he'd taken an F-86 over the Greenland icecap in the dead of winter, hundreds of miles from rescue in the event of an emergency, and felt more anxiety than he did right now. His minutes over the far side were his quiet time, a respite from the constant chatter of mission control on the radio. He was anything but lonely.

Collins would later write of his far-side passages, "I am alone now, truly alone, and absolutely isolated from any known life. I am it. If a count were taken, the score would be three billion plus two over on the other side of the moon, and one plus God knows what on this side. I feel this powerfully—not as fear or loneliness—but as awareness, anticipation, satisfaction, confidence, almost exultation. I like the feeling." As if to capture it, during a quiet period, Collins took the movie camera, held it out at the end of his reach, and turned it on his own bearded face for a few moments, like a man sailing around the world alone.

Into the small hours of Monday morning, Mike Collins circled and worked and waited, waited for tomorrow's moment of truth, when the real success of this mission—getting Armstrong and Aldrin back—would hang in the balance. As his crewmates settled in for the night in *Eagle,* Collins was finishing up his own very long day, covering the windows, turning out the cabin lights, and thinking of his days as an altar boy in the National Cathedral in Washington, D.C., when he used to snuff out the altar candles after a service. Before the flight he'd thought he might have some misgivings about going to sleep if there were a problem onboard. But *Columbia* was working like a marvel, and he drifted easily into weightless slumber.

MONDAY, JULY 21
12:52 P.M., HOUSTON TIME
TRANQUILLITY BASE

Armstrong and Aldrin stood side by side at *Eagle*'s controls, helmets and gloves locked in place. The first launch from another world was two minutes away. Armstrong had spent the night perched on the ascent engine cover, but had not slept at all. Aldrin, who had curled up on the floor, had managed only a few hours of fitful dozing. The problem was that the LM was no bedroom. Moonlight flooded the cabin through the translucent window shades; the instrument panel was aglow with luminescent switches and dials. And it was cold. The men hadn't anticipated that with the shades in place and all the systems turned off there would be no source of heat in the cabin. By the time they realized what was happening it was too late; there was no way to fix it. The oxygen flowing into their space suits only made them colder; they lay in their suits shivering. Hoping the cabin oxygen might be warmer, they took off their helmets; that only let in the high-pitched whine of the LM's coolant pumps. By the time Ron Evans gave the wakeup call they had given up trying to sleep. They were still too keyed up to be tired, and their thoughts centered on one thing: getting off the moon. ☾

The ascent engine, hidden under the can-shaped cover behind them, had only 3,500 pounds of thrust, but that was enough to propel the ascent stage from the lunar surface into orbit. It was another of Apollo's engineering marvels, for it was even simpler in design than the Service Propulsion System engine. Like the SPS it burned hypergolics that ignite on contact, eliminating the need for an ignition system. Once the valves opened, fuel would flow into the combustion chamber, and the engine would fire. That would have to happen two minutes from now.

Before the flight Neil Armstrong had worried about those valves, and he'd suggested to the engineers that they consider replacing the electrical actuating system with a mechanical one that he or Aldrin could trigger by hand if the normal method failed. The engineers considered and rejected the idea; they had high confidence in the electrical system. And Armstrong knew there were several redundant ways to fire the engine; if necessary they could bypass the computer. One of those would work; there was no other way to think about it. ☾

To Aldrin the thought of being stranded on the moon forever simply didn't exist. To conjure that dark thought would have been to go against the whole philosophy behind the mission: Everything had been stacked to ensure their survival. And now, as he and Armstrong followed the checklist through

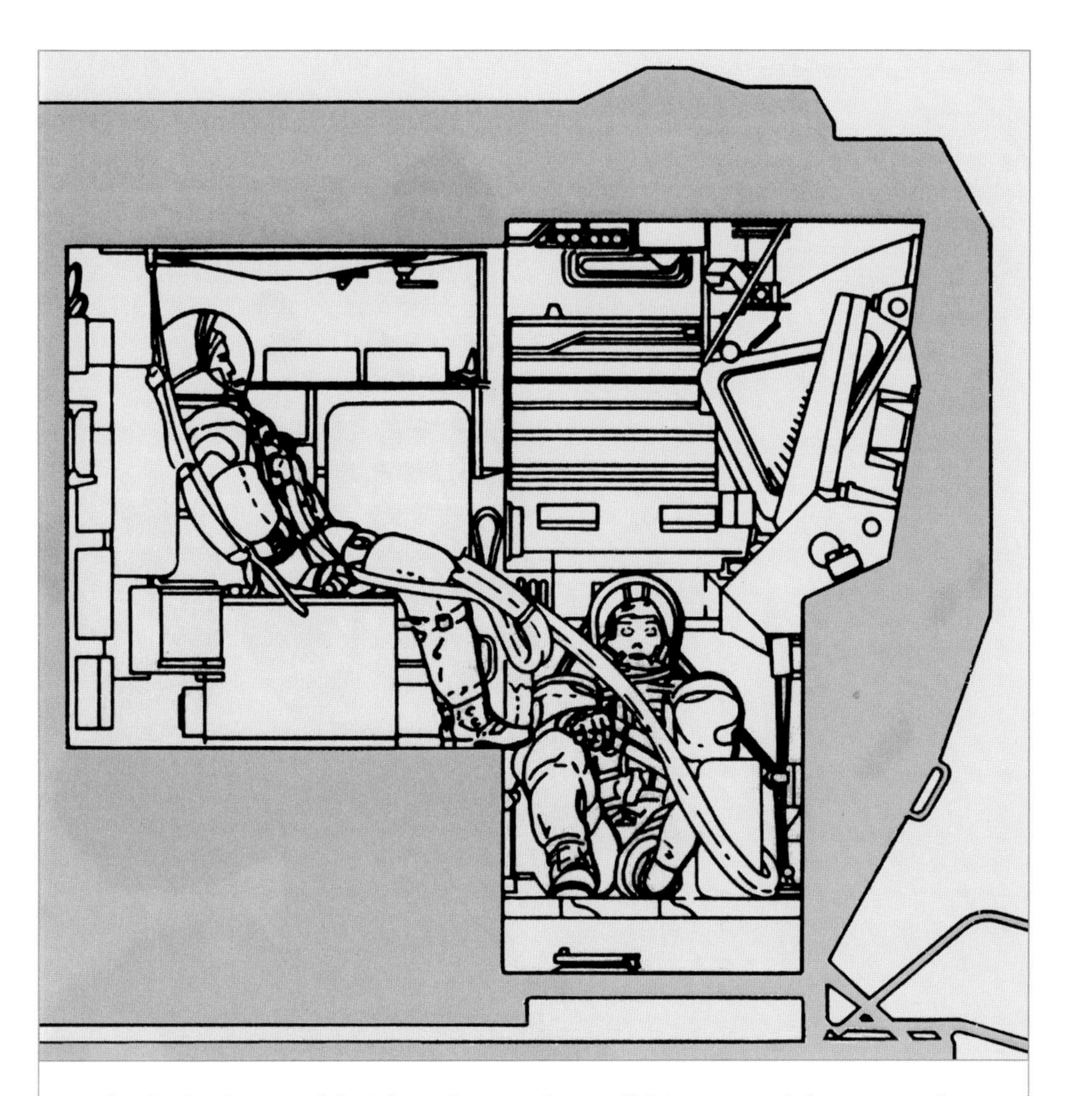

Sleeping accommodations aboard *Eagle* were Spartan at best. As shown in this illustration, Armstrong tried to sleep sitting on the cover for the ascent engine, while Aldrin lay on the cabin floor.

the final minutes of their launch countdown, Aldrin assumed that at zero the lunar stillness would yield to the power of a rocket come to life.

But inside *Columbia,* Mike Collins could not be so confident. Lunar orbit seemed remarkably safe compared to the spot Armstrong and Aldrin were in. Perched motionless on the surface with a single rocket engine to get them off, they belonged to the moon. The engine must work, and it must work long enough for *Eagle* to reach some kind of orbit. Collins was prepared to rescue them if they couldn't make it all the way up to 69 miles. He could drop down to 50,000 feet, but not too much lower than that; some of the lunar mountains were 20,000 or 30,000 feet high. And now, as he waited for liftoff, Collins could no longer push aside his darkest fears. "My secret terror for the last six months," he would later write, "has been leaving them on the moon and returning to earth alone; now I am within minutes of finding out the truth of the matter. If they fail to rise from the surface, or crash back into it, I am not going to commit suicide; I am coming home, forthwith, but I will be a marked man for life and I know it. Almost better not to have the option I enjoy." As

Armstrong and Aldrin's liftoff from the moon is captured in Alan Bean's painting, "The *Eagle* Is Headed Home." The blast from *Eagle*'s ascent rocket sent pieces of foil insulation arcing gracefully above the moonscape.

Armstrong and Aldrin made their final preparations for leaving the moon, Collins listened and sweated out the most anxious moments of his career. ☾

12:54 P.M.

With 45 seconds to go, Armstrong reminded Aldrin of the last actions they would take on the surface of the moon: "At five seconds I'm going to get ABORT STAGE and ENGINE ARM. And you're going to hit PROCEED."

"Right," Aldrin said.

"And, *that's all,*" Armstrong added wryly. If everything worked, he and Aldrin would just be along for the ride from the moment of liftoff until they reached orbit. Now Aldrin began the final countdown: "Nine, eight, seven, six, five, ABORT STAGE; ENGINE ARM, ASCENT; PROCEED—"

He pushed the button. For a fraction of a second there was stillness, and then, suddenly, there was a muffled bang of pyrotechnic bolts, and then a smooth, steady push, like a high-speed elevator, as *Eagle* ascended from the moon.

"We're off," Aldrin exulted. "Look at that stuff go all over the place." Outside, a spray of gold foil and debris from the descent stage flew away in all directions. The flag toppled to the dust. And the Sea of Tranquillity fell away as *Eagle,* ascending in an unreal quiet, headed for lunar orbit.

Joan Aldrin collapses in relief as she hears that *Eagle* has lifted safely from the moon. Said her husband, "We're off. Very smooth ride, very quiet ride."

3:54 P.M.

For the first time in the flight Mike Collins let himself believe they were really going to pull it off. He had spent the past three and a half hours laboriously punching data into his computer, ready to take over if necessary, but his "cookbook" of emergency rendezvous procedures had gone unneeded. And there, in the eyepiece of his sextant, Collins could see a small black dot: *Eagle,* climbing up from the craters so steadily that it seemed to be riding up to him on rails. It was the happiest sight of the whole mission.

Collins floated back to his couch. Through the rendezvous window he could see *Eagle* slowly closing in, its thrusters spitting flame as Armstrong braked for the final approach. Even as he steered *Columbia* into position for the docking Collins raced from one window to the other, taking Hasselblad pictures and movies. And as the last steps of the dance were played out, Collins suddenly called out, "I got the earth coming up behind you—it's fantastic!" Collins captured the sight on film—*Eagle,* the moon, and the tiny blue and white world. He would always remember the moment: all of humanity captured in a single photograph, minus only himself, the photographer.

Eagle rises steadily from the moon's far side _(below),_ then closes in for docking, while the earth rises behind it _(opposite)._ For Mike Collins, the sight of the returning lander was his happiest experience on the mission.

●◐◐○○○◑◑●

"Get ready for those million-dollar boxes," Armstrong yelled up the tunnel to Collins. As he handled the two weightless containers, snugly zipped into white cloth bags, he could feel the mass of the rocks inside them, and he was careful not to move too quickly as he passed them through to Collins. When they were safely stowed in *Columbia* he passed up a small white pouch and told Collins, "If you want to have a look at what the moon looks like, you can open that up and look. Don't open the bag, though." Collins unzipped the pouch and saw a small Teflon bag filled with black soot. Armstrong laughed, "You'd never have guessed, huh?"

"What was that bag?" Collins asked.

"The contingency sample," Armstrong said.

"Any rocks?"

"Yes, there's some rocks in it too. You can feel 'em, but you can't see 'em. They're covered with that—graphite."

And there was plenty of that "graphite" on their space suits; Armstrong thought they looked like chimney sweeps. Before he and Aldrin could rejoin Collins they tried to vacuum it off, not just to be tidy, but as part of the procedures to prevent "moon germs" from reaching earth. With no real vacuum cleaner they had to use a brush attached to one of the LM's air hoses; as they had suspected it turned out to be a vain attempt. Lunar grime had worked its way into the fabric; the suits would never be clean again.

When the cleanup was done and all the unneeded gear was piled in the tiny cabin, Armstrong and Aldrin exited to *Columbia* and closed the hatch. *Eagle* was now dead weight. Collins flipped a switch and the ascent stage drifted away. For his part, Collins was glad to get rid of the craft that had been nothing but a worry to him for six days, but in Armstrong and Aldrin he noticed a quiet sadness. Without a heat shield, there was no way to bring *Eagle* home; no museum would ever put it on display. It would linger in lunar orbit while mission control monitored each component's final hours of life. Long after it had become a dead ship *Eagle* would spiral downward until it crashed, blasting a modest new crater in the dust.

Looking like aliens, Armstrong, Aldrin, and Collins, clad in biological isolation garments, cross from the recovery helicopter to the quarantine trailer waiting for them aboard the aircraft carrier *Hornet*.

USS HORNET
HORNET + 3

Accomplished .....
July , 1969
APOLLO 11

Jubilation erupts in mission control as elated flight controllers watch the astronauts' recovery. The success of the mission was a moment not only of supreme technological achievement, but also of unabashed patriotism.

TUESDAY, JULY 22

Just after midnight, Houston time, the SPS engine roared to life, and three minutes later Armstrong, Aldrin, and Collins were on their way home. The burn was as flawless as the one that had put them in lunar orbit two days before, and in the middle of shutting down systems and reading trajectory data off the computer, Mike Collins spoke for all of them: "Beautiful burn, SPS, I love you, you are a jewel! *Whoosh!*"

THURSDAY, JULY 24
1:16 P.M., HOUSTON TIME
ABOARD THE CARRIER *HORNET*

Armstrong, Aldrin, and Collins stepped out of the helicopter onto the lower deck of the carrier *Hornet* looking like men from another world. Outfitted from head to toe in gray-colored Biological Isolation Garments, they peered through face masks clouded with perspiration and waved to a crowd of sailors and visiting dignitaries whom they saw only dimly. Despite rubbery legs unaccustomed to earth's gravity they made their way quickly to the open door of the silvery quarantine trailer. A NASA doctor followed them and closed the thick, windowed door behind them.

Three and a half days later they arrived in Houston sealed within the trailer as if they themselves were lunar samples. For the next two weeks they lived within the Lunar Receiving Laboratory, recounting all aspects of the flight in minute detail. Aldrin described the strange flashes he had seen on the way to and from the moon; no one had an explanation, but after a time Aldrin noticed that Armstrong seemed annoyed whenever the subject came up. ☾

As the days passed the men joked about being jailed, and at the close of one debriefing session they called out to the engineers on the other side of the glass, "You know where to find us! We're not going anywhere!" In the off hours there were movies, like *Goodbye, Columbus,* and card games; Collins beat Armstrong repeatedly at gin rummy.

Kate Collins, hugging her pet bunny, Snowball, takes in the recovery on television, as her mother, Pat *(left)*, and other astronaut family members and friends raise a glass of champagne to celebrate Apollo 11's success.

President Richard Nixon talks to the elated astronauts through the window of their sealed trailer aboard the *Hornet.* Still ahead for the astronauts was the long journey back to Houston for the remainder of their 21-day quarantine.

They had company, including doctors, a NASA public affairs officer, and some unexpected arrivals—a few scientists who were accidentally exposed to lunar samples. And though the LRL wasn't a bad place—it had a bar, and an exercise room—time dragged. At the end of one debriefing, when asked, "Any other comments?" Collins said quietly, "I want out."

That would come on a hot August night, when the men would be released into a world changed, for at least a time, by what they had done. Armstrong hoped that the first lunar landing would inspire people to believe that

seemingly impossible problems could be solved. As for its impact on their own lives, neither Armstrong nor his crewmates could guess what lay ahead. Until now, they hadn't had time to think about it. But there would be months on the banquet circuit, including a world tour; then each man would find his way into a new life. ☾

For now, sitting in the LRL, Buzz Aldrin had time to ponder the significance of what he and his crewmates had been a part of. Back on the *Hornet,* they had watched videotapes of the news coverage of Apollo 11. There was Walter Cronkite, exulting at the lunar touchdown. Then awed crowds gathered around TV sets, witnessing the first footsteps on another world. For the first time, Aldrin sensed the emotional impact of the first lunar landing. For a man attuned to irony, here was something worth pondering: While the three of them were a quarter of a million miles away, much of humanity had been spellbound by a midsummer miracle. What a moment that must have been. Aldrin turned to Armstrong and said, "Neil, we missed the whole thing."

In one of the best-known of pictures brought back from the moon, Buzz Aldrin's gold-plated sun visor reflects the Sea of Tranquillity, the leg of the lunar module, and the photographer, Neil Armstrong. The foil-covered tube in the foreground is one of the lunar-contact probes attached to three of *Eagle*'s footpads.

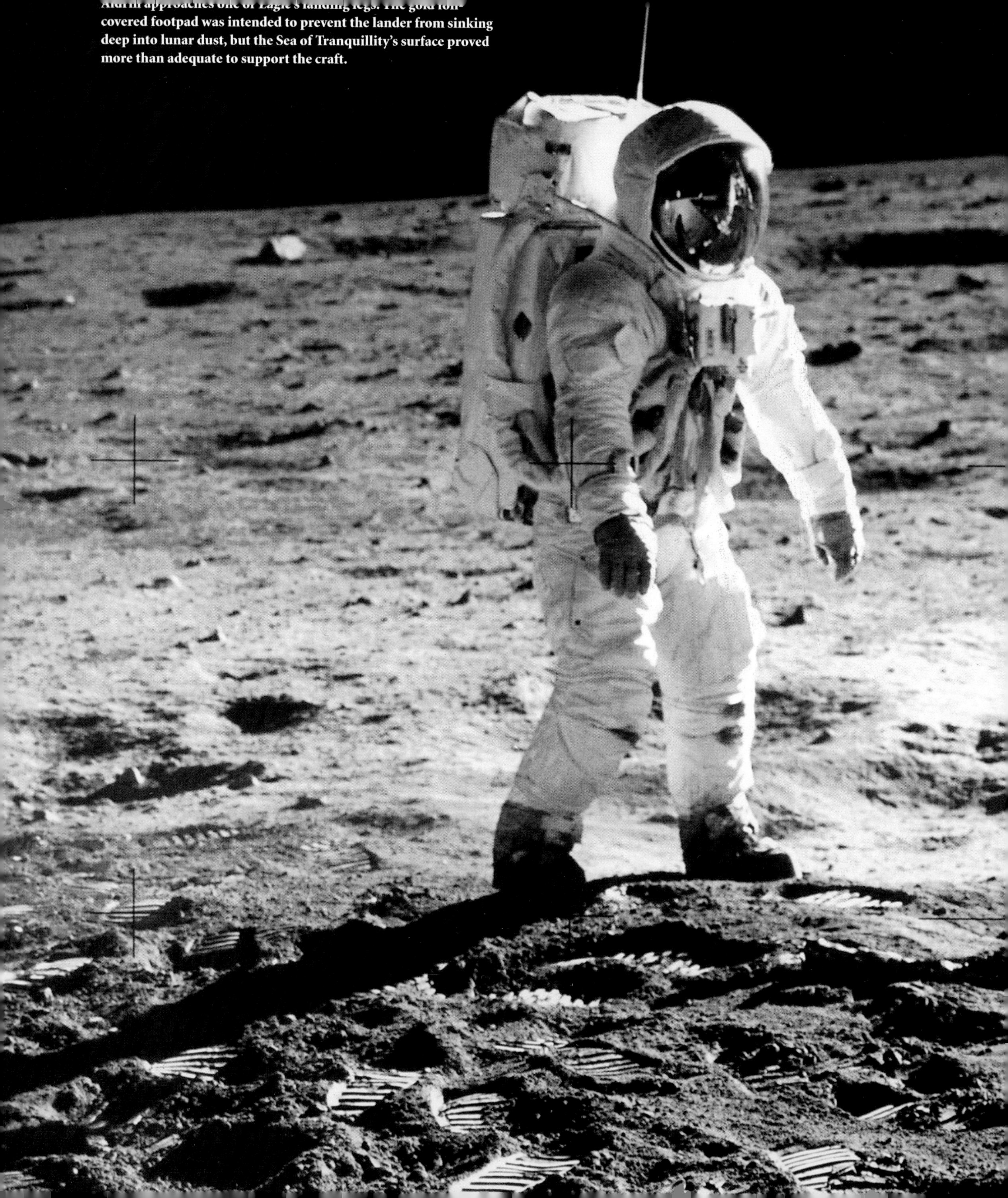

Aldrin approaches one of *Eagle*'s landing legs. The gold foil-covered footpad was intended to prevent the lander from sinking deep into lunar dust, but the Sea of Tranquillity's surface proved more than adequate to support the craft.

Aldrin stands next to the seismometer set up to monitor moon tremors, its six solar panels angled upward to catch the sun. Between the seismometer and the flag in the distance sits the apparatus for the laser-ranging experiment to measure the distance between earth and the moon.

Aldrin stands before the U.S. flag that he and Armstrong had laboriously planted in the lunar soil near the beginning of their walk on the moon. Because the moon lacks an appreciable atmosphere, the flag is stiffened with wire so that it appears to be fluttering in a breeze.

The imprint of Buzz Aldrin's boot compresses the fine lunar dust of the Sea of Tranquillity. Eroded only by a constant rain of micrometeorites from space, the moonwalkers' footprints are expected to endure a million years or more.

AUTHOR'S NOTES

PROLOGUE

16 *Within days after Shepard's flight, he had made his decision:* Kennedy's decision is covered in detail in John Logsdon's *The Decision to Go to the Moon: Project Apollo and the National Interest* (Chicago: University of Chicago Press, 1976) and Murray and Cox, *Apollo: The Race to the Moon.*

17 *the technological hurdles that would have to be cleared to build the Apollo spacecraft:* In 1960, among NASA's planned projects was a three-man spacecraft that would circumnavigate the moon. According to the NASA book *Origins of NASA Names,* by Ulrich et al., Abe Silverstein, NASA's Director of Spaceflight Development, proposed that it be called Apollo. The agency had already set a precedent by naming Project Mercury after a mythological figure, and Apollo had favorable connotations as the Greek god who pulled the sun across the sky in his golden, horse-drawn chariot each day. Apollo was also the god of archery, prophecy, poetry, and music. In 1961, following John Kennedy's challenge, Project Apollo became the name for the lunar landing program (*Origins of NASA Names,* p. 99).

19 I*n 1959, Conrad had been one of sixty-nine young fliers:* Carpenter et al., *We Seven,* p. 6.

CHAPTER 1: "FIRE IN THE COCKPIT!"

28, 30 Skepticism about Grissom and the Mercury hatch: In his book, *Schirra's Space* (p. 75), Wally Schirra points out that on his own Mercury mission he purposely blew the hatch after his spacecraft was lifted onto the deck of the recovery carrier. The recoil of the actuating handle was so violent that Schirra's hand was cut, even through his space suit glove. Grissom had not sustained even a minor injury anywhere on his body—strongly supporting his claim.

31 *the command module simulator here at the Cape was a constant source of difficulty:* So many changes were being made to Grissom's spacecraft that the simulator crews could not keep up; the result was that the simulator never accurately reflected the precise configuration of the real command module. At the same time, the simulator's computer software underwent frequent changes, and often there were bugs which caused the simulator to break down (Murray and Cox, *Apollo: The Race to the Moon,* p. 186).

31 *Finally the problem was solved, and at 2:45 P.M.:* Sequence of events is from Report of the Apollo 204 Review Board, section 4.

31 *seated next to Roosa at the Stony console, Deke Slayton listened:* Helpful sources for Slayton's actions before, during, and after the fire were his witness statement before the Apollo 204 Review Board, made on February 8, 1967, and "The Ten Desperate Minutes," *Life,* April 21, 1967.

33 *Years later, he would still wonder whether or not he made the right choice:* Slayton had thought he might have been able to notice the spark thought to have triggered the fire, and to extinguish it before anything could happen. In hindsight, it seems a long shot, given the speed with which the fire spread.

33 *Block I . . . was never built to go to the moon:* Because Block I had no docking system, it could not link up with a lunar module, the spacecraft that would be used to land on the moon. It also lacked the proper guidance and communications systems for the lunar voyage.

33 *there would be only one Block I mission, Apollo 1:* Until December 1966, there had been two manned Block I flights scheduled: Grissom's, and a second earth-orbit test commanded by Wally Schirra. (At that time, Tom Stafford was on the backup crew for the second mission, along with Frank Borman and Mike Collins.) In December 1966, Schirra's flight was canceled on the grounds that it would be an unnecessary duplication of Grissom's. In its place, NASA planned the first Block II mission, with Jim McDivitt,

Dave Scott, and Rusty Schweickart as the crew, and Stafford, Young, and Cernan as their backups. But on January 27, 1967, the Block II spacecraft had not yet been built.

36 *Wires were constantly being rerouted, black boxes replaced:* This information on the second Block I spacecraft is from Collins, *Carrying the Fire,* p. 256.

36 *The McDonnell Aircraft Corporation . . . had forged a harmonious relationship with NASA:* This was not the case at North American, whose relationship with NASA had been strained by personality conflicts from the beginning of Apollo. Charlie Frick, the man who preceded Joe Shea as head of NASA's Apollo Spacecraft Program Office, was overbearing and even belligerent in his approach to North American. He and North American manager John Paup were at odds from the beginning. This conflict, and the events at North American leading up to the Apollo 1 fire, are described in Gray, *Angle of Attack.*

37 *But now Slayton heard another voice, clearly frantic:* One astronaut who listened repeatedly to the voice tapes from the fire says he never heard panic in Chaffee's voice; instead there was a sense that they were being rapidly engulfed and things were happening very fast. He adds, "You hear it time and again in [the test flight] business. . . . They may have known there wasn't any hope."

43 *Much of the once spotless cabin was covered with soot:* The appearance of the interior of the spacecraft is described in the Report of the Apollo 204 Review Board, section 5.

43 *"If there's ever a serious accident in the program, it's probably going to be me":* Grissom and Still, *Starfall,* p. 172.

43 *"If we die, we want people to accept it":* Armstrong et al., *First on the Moon,* p. 49.

45 Weight and complexity of an oxygen/nitrogen mixture: Charlie Feltz, who was North American's chief engineer in the early days of Apollo, says that initially he pushed for a two-gas system precisely because pure oxygen was so hazardous, but that he was overruled by NASA. And according to Chris Kraft, the test-flight veteran who became one of NASA's central figures in Apollo, there were daunting technical obstacles to using an oxygen/nitrogen mixture. Instrumentation precise enough to monitor the composition of the mixture had not yet been developed.

45 *there was a spark inside Apollo 1:* For an in-depth discussion of the causes of the fire, see *Apollo: The Race to the Moon,* pp. 190-91 and 214.

45 *Velcro fasteners . . . exploded in a shower of fireballs:* At the time of the fire, there was ten times as much Velcro in the Apollo 1 cabin as originally specified, because the astronauts, who had always customized their spacecraft, wanted more of it (*Apollo: The Race to the Moon,* p. 214).

45 B*ut the Apollo managers . . . always had sound reasons for vetoing the change:* Slayton was one of several astronauts who told the author that even before the fire there were plans to change to a one-piece hatch. Furthermore, Charlie Feltz says that his original design called for an outward-opening hatch, but that NASA vetoed it in favor of the inward-opening design.

45 *Each pound of payload cost many times its own weight:* Adding weight to the command module was complicated by the fact that its designers had given it an offset center of gravity, as a means of generating lift—and therefore, trajectory control—during reentry into the earth's atmosphere. Much of the spacecraft's weight was concentrated on the side away from the hatch. Making the hatch heavier required additional weight on the opposite side as ballast. Otherwise, the center of gravity would be shifted toward the middle of the command module, lessening its lift-generating ability.

46 *White and his backup, Dave Scott, used to practice opening the hatch for exercise:* The hatch weighed about 90 pounds, but lifting it was even harder because the astronaut had to reach back over his head.

47 *The greatest irony was that Gus Grissom:* The irony is even greater considering the fact that

NASA managers vetoed the outward-opening hatch largely *because* they wanted to avoid a repeat of Grissom's Mercury incident.

CHAPTER 2: THE OFFICE

60 *Conrad ended up carrying Glenn's bags as well as his own:* This situation is described in Wolfe, *The Right Stuff*, p. 387.

60 *an extra $16,000 per year:* The amount of money for each man declined to around $10,000 when nineteen new astronauts were selected in 1966. Field Enterprises withdrew from the contract before the first Apollo flight. *Life* published its last astronaut-written stories in 1970, after the Apollo 13 mission.

The *Life* contract was a source of much controversy and brought NASA criticism from other news organizations. NASA administrator Jim Webb was against it. The astronauts, for their part, were grateful to have it: for their families, it offered a shield from the media; for the pilots themselves, it was an excuse to turn down innumerable requests for interviews. Says Walt Cunningham, "You can't imagine the pressure from the media chasing us for stories." The ethical question of whether the astronauts should be paid for their stories was settled, at least officially, when John Glenn spent a weekend with John Kennedy and his family at the presidential retreat at Hyannis Port, Massachusetts. Glenn argued that the astronauts deserved to be compensated for the invasion of their privacy that was required to satisfy the enormous public interest in them—which was over and above the news coverage of their missions. Kennedy agreed, as described in "Heroes, not of their own accord," a 1977 master's thesis by Perry Michael Whye, Iowa State University.

75 *announced that Gus Grissom was going to command Gemini 3 and that his copilot would be John Young:* According to other astronauts, the process of picking Grissom's copilot had a few twists and turns. In late February or early March 1963 Slayton called the astronauts into the briefing room at Ellington and told them that Al Shepard and Tom Stafford would fly Gemini 3. Three weeks or a month later, Shepard was grounded. Everyone assumed Stafford would be named to fly with Grissom. But Frank Borman tells the story that he, not Stafford, was told he would be Grissom's copilot, and went over to Grissom's house to meet with him. The next thing he knew, Borman says, John Young had been named in his place. Evidently Grissom had stepped in and made his own choice.

78 *Al Shepard, who had been appointed chief of the Astronaut Office:* Just as Slayton's role was analogous to that of a wing commander, Shepard functioned much like a squadron commander, and reported directly to Slayton. While their duties overlapped, Slayton generally handled the astronauts' dealings with NASA management and with contractors. Shepard ran the internal affairs of the Astronaut Office, including scheduling travel for business and public relations, working out training schedules, and so on.

79 *In the Astronaut Office, Shepard usually kept a chilly distance from his troops:* Shepard's secretary, Gay Alford, kept a sign near her desk that displayed her boss's "Mood of the Day." The sign featured an appropriate picture—a scowling orangutan, a blissful, flower-wielding hippie, and so on—that was visible to any astronaut walking past her desk. Thus warned, the pilots could, if necessary, perform an evasive maneuver.

81 *flying wasn't the basis of this competition:* The competition definitely spilled over into flying, however. On cross-country flights, if two teams of astronauts were flying two T-38's, they would race. If the conditions were just right, and if the pilots didn't waste any fuel taking off, they could make it all the way from L.A. to Houston without refueling. Even then, some astronauts landed with the last wisps of fuel burning in the engines.

83 *There were five others like him among the Fourteen:* Besides Cunningham and Schweickart, the other non-test pilots were Buzz Aldrin, Gene Cernan, and Roger Chaffee.

85 *paid little attention to these events:* However, civil rights had briefly become an issue within the astronaut corps in 1963. A black pilot named Ed Dwight, who had barely graduated from Chuck Yeager's space school, had applied for the astronaut program and been rejected. The astronauts were on a desert-survival course in Nevada when Slayton was summoned for a phone call from Washington. When he returned, he told the other astronauts that he had just spoken to Attorney General Robert Kennedy, who wanted

NASA to accept Dwight. Slayton told the pilots, "I just spoke for all you guys. . . . I said if we had to take him and he wasn't qualified, then they'd have to find sixteen other people, because all of us would leave."

In his autobiography, Yeager wrote that during that time there were still relatively few black air force pilots, but that Dwight "sure as hell didn't represent the top of the talent pool. I had flown with outstanding [black] pilots like Emmett Hatch and Eddie Lavelle; but unfortunately, guys of their quality didn't apply for the [space school]. Dwight did." Yeager says that despite the fact that Dwight lacked the flying experience and engineering background to be admitted to his school, the White House pressured the air force to enroll him. "The only prejudice against Dwight," Yeager wrote, "was a conviction shared by all the instructors that he was not qualified to be in the school." With tutoring, Dwight graduated. But he was not selected as an astronaut. In 1967, the air force selected a black graduate of the space school, Robert Lawrence, for its Manned Orbiting Laboratory program. Shortly afterward, Lawrence was killed in the crash of his F-104 jet.

89 *They wryly called themselves the "Original 19":* The name was coined by John Young, who had served on the astronaut selection board in March 1966 (Collins, *Carrying the Fire,* p. 127).

CHAPTER 3: FIRST AROUND THE MOON

I: The Decision

97–98 Technical problems with the lunar module: For a detailed account of the lunar module's development, see Brooks et al., *Chariots for Apollo: A History of Manned Lunar Spacecraft.*

98 *It was the brainchild of George Low:* After the Fire, Low was named to replace Joe Shea as head of the Apollo Spacecraft Program Office.

100 *he and his deputy George Mueller:* Mueller pronounced his name "Miller."

100 *Webb . . . already knew his tenure would end when Johnson left office:* Webb went to see Johnson in September 1968 to discuss resigning; to his surprise, Johnson was ready to accept his resignation that day. Webb left NASA on October 7, only days before the first manned Apollo flight (Murray and Cox, *Apollo: The Race to the Moon,* p. 323).

103 *Bob Gilruth, the head of the Manned Spacecraft Center:* Gilruth, a pioneer from the early days of flight testing, was one of the giants of the manned space program. His close association with the Original 7—whom he sometimes called his "boys"—was said to be the source of much of the Mercury astronauts' power within NASA.

105 *Jim proudly proclaimed the "flight" a qualified success: Life,* September 27, 1963, p. 86a.

109 *a single meeting one August afternoon in the office of Chris Kraft:* In truth, this meeting could only lay out the basic objectives for the mission. The detailed planning of every aspect—from control of the spacecraft's trajectory to the astronauts' sleep schedules—was conducted in a series of weekly Flight Operations Plans, chaired by engineer Rod Rose. For a definitive account of this and the other Apollo missions from the perspective of mission planners and controllers, including the personalities involved, see Murray and Cox, *Apollo: The Race to the Moon.*

113 *The translunar crossing of some 234,000 miles:* The moon's distance from earth varies from 221,500 miles to 252,700 miles. At the moment Apollo 8 went into lunar orbit it was 233,900 miles.

113 *mascons had to be understood:* Planners knew that the closer a spacecraft got to the moon, the greater the effects of any gravitational irregularities. The Lunar Orbiter probes had orbited at much greater altitudes, generally several hundred miles, than those planned for Apollo. The trajectory people could only guess how severely mascons might affect a command module 69 miles above the moon, or a lunar module at 50,000 feet, heading for a landing.

118–119 *Even when he wasn't in the simulator, Anders was learning his machine:* Anders was the systems expert, but to the press, he was also the rookie—something that irritated him no end. Borman and Lovell picked up on it, of course, which bothered him even more. Anders would only respond, wryly, "When it comes to going to the moon, everyone's a rookie."

119 *The odds of an astronaut's survival: Apollo: The Race to the Moon,* p. 102.

121 *After one test, Eisele quipped:* Baker, *The History of Manned Spaceflight,* p. 310.

124 *Whether another Soyuz was being readied for a circumlunar mission, no one knew:* In 1988 veteran cosmonaut Aleksey Leonov, the first man to walk in space, told the author that in the fall of 1968 he had been training to command the first circumlunar mission, with Oleg Makarov as his flight engineer. The men were to fly a spacecraft called Zond, which was a variant on the Soyuz design. For information on the Soviet plans see note for page 135 (Chapter 4, Part I).

124 *Borman cut him off:* According to Borman, there was talk of having one of the astronauts climb outside and do a spacewalk, with all the risks that entailed. Borman vetoed the idea.

124 *behind that macho, take-charge exterior was a very apprehensive astronaut:* As evidence that his behavior had nothing to do with apprehension about the risk of going to the moon, Borman says he was just as rigid and impatient after the Fire, during the recovery efforts at North American. For example, on those occasions when an astronaut appeared uninvited at Downey to champion some pet improvement for the command module, Borman sent him home. Finally he called Slayton in exasperation and said, "No more astronauts!" Borman's behavior raised more than a few hackles in the Astronaut Office. But they couldn't afford to make a "gold-plated" command module, he said; they had no time to waste.

125 Lindbergh's visit: Some details are provided in "The Heron and the Astronaut," by Anne Morrow Lindbergh, *Life,* February 28, 1969, p. 19. Other information in Jim Lovell's letter to the author, 1990.

127 *most powerful thrust machine ever flown:* At this time, the Soviets were trying to develop two heavy-lift boosters: the G-l, which is said to have been slightly less powerful than the Saturn V, and the N-l, which would have been more powerful. Neither of these was successfully launched, however. See also note for page 135 (Chapter 4, Part I).

II: A Hole in the Stars

137 *the launch pad's automatic sequencer took over:* The Terminal Countdown Automatic Sequencer, located within the mobile launch platform, was an electromechanical device that had no computer software.

143 *some 24,226 miles per hour, the speed necessary to reach the moon:* This was slightly less than the so-called *escape velocity* of 25,020 miles per hour, necessary for an astronaut to blast free of the earth's gravitational pull altogether. The Apollo astronauts did not need to go that fast; they needed only to go fast enough to reach the moon's sphere of gravitational influence.

147 *simply by floating around, an astronaut would push his vestibular system over the edge:* According to Jim Lovell, even on Gemini many astronauts "were nauseated at the beginning of their flights, but no one wanted to admit it." One of them was Gene Cernan, who told the author that he felt ill for the first day of his Gemini 9 mission. Lovell himself says he did not experience any discomfort on his two Gemini missions, but did feel some nausea when he first began to move around inside Apollo 8 soon after launch.

154 *chase down stray bits of vomit and feces with paper towels:* Today, Anders says he can't stand the smell of airline towelettes because it reminds him of this whole episode.

155 Borman's conversation about his illness: Most astronauts were reluctant to make public any details of their in-flight medical problems. This conversation is not recorded in the air-to-ground transcripts and was not available on tape from NASA. It was, however, described by Sam Phillips in his article "Apollo 8: A Most Fantastic Voyage," *National*

Geographic, May 1969, p. 613. The exchange between Borman and Mike Collins is taken from the recording *To the Moon,* produced by Time-Life Records, 1969.

156 *he conversed with Capcom Ken Mattingly:* Mission controllers, including Capcoms, worked in eight-hour shifts. Mike Collins, Ken Mattingly, and Jerry Carr were the three main Capcoms for Apollo 8.

157 *When they turned the camera on the brilliant blue and white planet:* To the astronauts, the earth was some forty times brighter than the full moon appears to us.

158 Not much chance of that, *Anders thought:* Another source of Anders's worry was the newly designed, outward-opening side hatch. Every now and then he would glance warily at it. True, that hatch would have saved their lives in the event of a fire on the launch pad or some other emergency on the ground. But with the vacuum of space on the other side of it, a hatch that was easy to open just didn't give him a very warm feeling. Anders hoped the designers had chosen wisely; he would worry about that outward-opening hatch for the rest of the flight.

158 *Could Jules Verne have imagined the view from Apollo 8?:* As Lovell himself was well aware, there were eerie similarities between Apollo 8 and the flight described in Jules Verne's *From the Earth to the Moon* a century before. Verne's moon voyagers were three in number. Their names even bore a slight resemblance to those of the Apollo 8 crew: Barbicane, Nicholl, and Ardan. They were launched from Florida, in December, and recovered in the ocean. One significant difference was that Verne left his moon voyagers stranded in lunar orbit. After his readers protested, Verne brought his astronauts back to earth in the sequel, *Around the Moon.*

III: "In the Beginning . . ."

164 *its rendezvous with the moon at more than 5,000 miles per hour:* Apollo 8's speed relative to the moon reached 5,700 miles per hour at the time of engine ignition. But some 40 percent of that speed was due to the fact that the moon was racing along in the opposite direction at 2,300 miles per hour.

165 *"Alright, alright, come on," Borman said:* On the onboard voice tapes, there was about a minute's worth of conversation between Anders's "*Look* at that" and Borman's "Alright, alright, come on. . . ." During that time Borman tried to steer Anders back to the reading of the checklist, which he did for a short time, until his curiosity got the better of him and he had to take a second look. *"Fantastic,"* he said. "But you know, I still have trouble telling the holes from the bumps—" At which point, Borman cut him off.

167 *beach sand darkened by the cold embers of bonfires:* When Anders repeated that description in a TV transmission during the second orbit, he won himself a storm of hate mail from poets.

168 *"What does the ol' moon look like from sixty miles?":* NASA figured distance in nautical miles (one nautical mile is equal to 1.15 statute miles). The figure heard in Apollo mission dialogue is not 69 miles but 60 nautical miles.

169 *it was still Arthur C. Clarke's moon:* The moon as it appears in *2001* arguably belongs as much to the film's producer, Stanley Kubrick, but it is Clarke that Anders mentions.

171 *Lovell, following the explorer's prerogative, had named them:* Anders went even further, giving names to dozens of craters on the far side. His choices were made to honor national leaders, NASA managers, and flight controllers. And he named three modest-sized craters for Borman, Lovell, and himself. He chose their locations near the boundary between the far side and the near side, so that when future astronauts flew over them they would be in radio contact with earth: "Roger, Houston. We're now over Anders . . ." Unfortunately, Anders's names were not approved by the International Astronomical Union, the recognized authority for all named features on the moon and planets. The IAU christened three craters for Borman, Lovell, and Anders, but to Anders's irritation they are located within a region of the far side that he and his crewmates never saw: it was in shadow during the flight.

174 *"How fast are they going now?"*: "Christmas cheers on the Apollo 8 home front," by Dora Jane Hamblin, *Life*, January 10, 1969, p. 79.

183 *with the help of a friend in Washington, Borman found something:* Borman sought the advice of a friend in the U.S. Information Agency, a man named Simon Bourgin who had been a traveling companion on the world tour that followed Gemini 7. Sometime later he got a letter from Bourgin with an idea for the broadcast, courtesy of a Washington newspaperman Bourgin knew. The newspaperman, Joe Laitin, had in turn gotten the idea from his wife.

184 *the moon's . . . appearance near lunar sunrise and sunset:* Apollo 8 orbited from east to west; this meant that, for example, the astronauts experienced sunrise over the part of the moon where sunset was occurring.

IV: "It's All Over but the Shouting"

189 *"Please be informed there is a Santa Claus":* When he composed that line, Lovell was thinking of the turn-of-the-century newspaper columnist who reassured a little girl of the existence of St. Nick: "Yes, Virginia . . ."

191 *if they made a single mistake on the rest of the flight the brandy would get the blame:* Even without brandy, Lovell's fatigue got the better of him and he called up the wrong program on the computer, wiping out all its stored information about Apollo 8's orientation in space. An anxious half-hour followed while Lovell worked to realign the command module's navigation platform. For a while afterwards, Borman worried that somehow the episode might have affected the part of the computer's memory that would handle the reentry maneuvers; Houston assured him there was nothing to worry about.

195 *Every so often what looked to be a fist-size chunk shot by:* Later, Anders would learn that those "chunks" had been mere pea-sized bits, each surrounded by a halo of glowing gas.

CHAPTER 4: "BEFORE THIS DECADE IS OUT"

I: The Parlay

211 *the Soviets were now talking about missions in earth orbit:* In their congratulatory telegram to the Americans following Apollo 8's success, the Soviets went so far as to say that the moon race had never existed. Not only was this not true, but it is now clear that the Soviets continued to plan a moon program for some time after Apollo 8, even after it was clear they would lose the space race with the Americans. Their lunar landing attempt hinged on the giant N-1 booster, designed to deliver 10 million pounds of thrust. Four unmanned launches ended in disaster when the rocket exploded. The third of these occurred on July 3, 1969, only thirteen days before the scheduled launch of Apollo 11. The fourth took place late in 1972. For more information, see *Red Star in Orbit* by James Oberg; *Men from Earth* by Buzz Aldrin and Malcolm McConnell (second edition); "Russians Reveal Secrets of Mir, Buran, Lunar Landing Craft" by Craig Covault, in *Aviation Week & Space Technology,* February 10, 1992, pp. 38-39; and the PBS television *Nova* documentary "The Russian Right Stuff."

229 Preliminary timeline for moonwalk, showing lunar module pilot out first: According to the NASA history *Chariots for Apollo,* such a checklist was written in 1964. Bill Anders also recalls that sometime before 1968 he and Al Bean were assigned to help write a preliminary checklist for the first moonwalk. "Almost as a joke," Anders says, on the slim chance that he might find himself on the first landing, he steered the writing of the checklist so that the words "LMP EGRESS" came before "CDR EGRESS"; in other words, that Anders would be out some number of minutes ahead of his commander. Later, Anders had a good laugh over the caper. Bean, who was just happy to be assigned to Apollo lunar activities, was oblivious to such maneuvering.

230 *Armstrong was anything but an exercise fanatic:* On that subject, there was a bit of mythology, a quote, often incorrectly attributed to Armstrong: "I believe the Lord gives us a finite number of heartbeats, and I'll be damned if I'm going to waste mine on exercise." In fact, Armstrong had once spoken those words, but he was quoting someone else,

and he wasn't endorsing that view. Nevertheless, as one astronaut said, "Neil was no jock."

230 *More than one astronaut remembers that Aldrin paid a visit, checklist in hand:* Today, Aldrin is understandably sensitive about this issue, which was emphasized by the media before, during, and after Apollo 11. In 1987 he told the author that before Apollo 11 his feelings about being first to walk on the moon had been "totally mixed." On the one hand, his dislike for publicity made him lean strongly the other way. And he knew it would be awkward if he received more attention after the flight than his commander, who would rightfully deserve accolades for flying the lunar landing. On the other hand, he said, for an aggressive person like himself, in the competitive arena of the Astronaut Office, it was only natural that he explore the possibility of being first—especially when there were operational grounds for doing so. Today, Aldrin maintains that the other astronauts have blown his efforts out of proportion. He had no more desire to be first on the moon, he says, than any of them.

230 *Mission planners had quietly come to the same conclusion in February:* Based on the author's interview with John Covington, who worked in the Crew Systems Division on timelines for the moonwalk, and on *Chariots for Apollo,* p. 322.

230-231 *to keep the first steps on another world free of any militarism:* Julian Scheer, who headed the public affairs office at NASA Headquarters in 1969, says that he felt strongly that a civilian, not a military officer, should be the one to represent NASA, a civilian agency, in making the first lunar footsteps. He voiced this opinion to Sam Phillips a few months before the launch. He adds, "It was nothing against Aldrin" personally.

In 1986, Phillips told the author that Armstrong's civilian status was a factor in the decision—which he said he left to Slayton and the other managers in Houston—but that he did not consider it an important factor. Asked whether NASA as an agency was consciously ruling against a military man as first on the moon, Phillips said, "No, absolutely not."

II: "We Is Down Among 'Em!"

231 One source used in the writing of this chapter was "Our Happy Moon Trip," *Life,* June 20, 1969.

232 *Apollo 10's new TV camera:* Color TV from space was Stafford's idea. He felt the public deserved a more vivid picture of spaceflight than the black-and-white pictures transmitted from Apollo 7 and 8. When he heard that engineers at Westinghouse were developing a spaceworthy color camera, he fought to have it ready in time for his mission.

232 *George Mueller . . . pushed for a landing on Apollo 10:* The idea for a dress rehearsal before the lunar landing attempt had been in the works since June 1967. By January 28, 1969, when Sam Phillips formally approved the mission for Apollo 10, planning was already well along. No one was surprised that Mueller was making such suggestions; according to one Apollo planner "Mueller was always floating new ideas. . . . He shot from both hips."

232 *Stafford's lunar module . . . was too heavy to land:* Originally Stafford had been assigned the first LM capable of making a landing, but in the wake of the delays that prompted the Apollo 8 decision, his lightweight LM went to Neil Armstrong, and Stafford took over Frank Borman's lander, which had been built before Grumman's super-weight-saving program took effect.

235 Vibrations during Apollo 10 Translunar Injection: According to Stafford, the vibrations were the result of a misadjusted pair of pressurization valves on the hydrogen tank within the Saturn's third stage. The valves were adjusted in such a way that as they vented excess pressure, they set up a kind of synchronized pulsing that increased in intensity. By readjusting the valves on later boosters, the problem was avoided.

237 *the astronauts called it "the lem":* This pronunciation is a holdover from the early 1960s, when the lander's official designation was lunar excursion module (LEM). When that was later shortened to lunar module (LM), the old pronunciation stuck. Although there were abbreviations for the command module (CM) and service module (SM), and for the joined pair (CSM), astronauts usually called the entire command ship "the command module."

237 *Once, a workman accidentally dropped a screwdriver inside the cabin:* Collins, *Carrying the Fire,* p. 324.

III: Down to the Wire

246 *NACA's High Speed Flight Station:* Pronounced "N-A-C-A," not "naka."

247 *a postflight party in full swing usually saw Armstrong at the piano:* The late Milt Thompson, a fellow NASA X-15 pilot, described an incident from one flight party that illustrates Armstrong's understated brand of humor. The bar was full of NACA and air force fliers, and Armstrong was at the piano. A group of men from the air force's missile-testing lab across the dry lake came in. In the minds of the fliers at Edwards, these non-pilots—who called themselves "Missileers"—were from the wrong side of the dry lake, so to speak. And yet, here they were, with their official badges pinned to their chests, looking pleased with themselves. One went over to Armstrong and requested a song. Armstrong glanced at the man, saw that he was wearing a Missileer badge instead of a set of wings. Without missing a note, Armstrong deadpanned, "Gee, I don't know any Missileer songs"—at which point the pilots in the bar laughed themselves onto the floor, while the ragtime rolled on. (For more information on the X-15 and its pilots, see *At the Edge of Space: The X-15 Flight Program,* by Milton O. Thompson, Smithsonian Institution Press, Washington, D.C., 1992.)

249 *"space is the frontier, and that's where I intend to go":* When Armstrong made this statement he had already participated in a number of projects that anticipated his work in Apollo. In 1959 Armstrong and another NASA engineer were studying whether a manned booster could be flown into orbit. In the early 1960s, he participated in the initial development of the Lunar Landing Research Vehicle (LLRV). As an astronaut, Armstrong would use the LLRV and its descendant, the Lunar Landing Training Vehicle (LLTV), to practice flying the lunar module to a landing.

255 *Armstrong was all business:* Even in the simulator, however, Armstrong could display his sly wit. While training as the backup crew for Gemini 11, one day Armstrong and his copilot Bill Anders were practicing rendezvous maneuvers. Just minutes away from a critical rocket firing, Anders, busy with the onboard radar and the navigation charts, looked up from his work to find Armstrong asleep. When Anders tried to wake his commander, Armstrong looked at his watch, shot his copilot a narrow-eyed glance and went back to sleep. "Okay," Anders thought, "I'll let him screw up." But then, at just at the right moment, Armstrong sprang awake, reached for the hand controller, and executed a flawless maneuver. Anders realized he'd been had. Later, when he queried Armstrong about the joke, Armstrong just laughed.

259 *one of his flight controllers nicknamed him General Savage:* The nickname was after the hero in the popular TV show "Twelve O'Clock High."

262 *These young men:* There were no women on the flight control teams for any Apollo mission; this was because there were few women engineers. For Apollo 11, a simulation instructor, Ann Accola, was one of a group of people working in a support "back room" to help establish the location of the lunar module from the astronauts' descriptions of the terrain. Today, women comprise almost a third of the flight control rosters (Gene Kranz letter to the author, 1993).

263 *Suddenly, it all went bad:* This simulation probably took place on June 27, which was a day full of simulated emergencies. According to the simulation log for Apollo 11, Armstrong and Aldrin spent a total of six hours in five separate simulated landings. Four of the five runs were aborts.

265 Collins's comments about Aldrin: From *Carrying the Fire,* p. 434.

266 *"I hated that probe, and was half convinced it hated me": Carrying the Fire,* p. 339.

266 *During the spring of 1969 Collins felt the eyes of the world upon him:* Somehow, in the rush of activity, Collins found time to design a mission emblem, using an idea from Jim Lovell: an eagle coming in for a landing above a field of craters, with the earth suspended in the black sky beyond. There would be no names on the patch, only the words, "Apollo

11." The name *Eagle* was a natural choice for the lunar module, while Collins's command module would be called *Columbia,* a name that evoked not only national identity, but Jules Verne's mighty cannon, the Columbiad.

270 *Thomas Gold insisted that the moon was covered by a layer of fluffy powder dozens of feet thick:* One geologist recalled showing Gold a picture from one of the Surveyors and saying, "Tommy, look: They didn't sink!" Another geologist insisted that if Gold was an irritant, he was a beneficial one: by voicing his controversial theories, he forced the other scientists to examine their own ideas more thoroughly.

271 *if he wanted to build a sterilization machine, he would construct something like the surface of the moon:* The geologist was NASA's Elbert King. If anything, King and other geologists argued, a storage facility was necessary, not to keep the earth safe from "moon germs," but to protect the lunar samples from terrestrial contamination (King, *Moon Trip,* p. 61).

271 *Quarantine was something Armstrong and his crew would just as soon have done without, but they had no choice:* Mike Collins saw the contamination issue as a question of probability: a very small number (the chance of life on the moon) multiplied by a very large number (the implications if it did) still produces a finite number. On that basis, he and his crewmates accepted the need for quarantine as more than just a political necessity (*Carrying the Fire,* p. 317).

273-274 *"any recognizable disadvantage . . . the position I'm in":* Frank and Susan Borman recalled that they tried to warn Armstrong and his crew about what lay ahead. Shortly after Armstrong, Aldrin, and Collins were named to the landing crew, the Bormans invited them and their wives for dinner. The Bormans had just returned from their Apollo 8 world tour, and they described with amazement how, in France, one official tried to give Borman a car and another offered Susan a fur coat, both of which they had politely but adamantly refused. "If you think that's bad," the Bormans told the three men and their wives, "you're going to have all of Europe thrown at you, on sterling."

CHAPTER 5: THE FIRST LUNAR LANDING

I: The *Eagle* Has Landed

285 View of the moon in earthlight: Seen from earth at this time, the moon was a slender crescent. The rest of the near side, which was illuminated by earthlight, was also dimly visible. Inside Apollo 11, Armstrong and his crew saw the moon from a somewhat different angle. To them, the moon's sunlit portion was out of view. They could, however, see part of the lunar far side, which was in complete shadow; hence the "crescent of blackness."

288 *Gene Kranz and his team of flight controllers called the White Team:* The practice by flight directors of adopting colors began with the final Mercury flight and continued through Apollo. Each new flight director chose a new color for his team of controllers. When Kranz became a flight director in 1965, he chose white. The controllers that made up each flight director's team would change from one mission to the next, but thanks to their intensive training before each flight, they always functioned as a coherent whole (Murray and Cox, *Apollo: The Race to the Moon,* pp. 286, 288).

292 *"Dammit, we really did something":* This speech is derived from a combination of the author's interview with Gene Kranz and the quote on p. 348 of *Apollo: The Race to the Moon.*

292 *suggested to Duke that Eagle yaw slightly to one side:* Because of *Eagle*'s face-down orientation, its antenna was not able to point in such a way as to enable clear transmissions to earth. By yawing the LM—that is, turning it slightly to one side—Armstrong and Aldrin improved the antenna's aim without affecting the pointing of their descent rocket.

295 *more like a simulation and less like history:* Kranz remembered that one controller did indeed comment, "It's just like a simulation." Kranz says, "I think several controllers relaxed with this comment; I know I certainly did. . . ."

295 *Kranz almost burst out laughing:* Kranz says Steve Bales "was so loud that he startled the entire room. He did not need a communications loop."

296 *He put the question to one of his back-room experts, Jack Garman:* The details of Bales's exchanges with Garman are described in *Apollo: The Race to the Moon,* p. 353.

296 *"I have too many things to do in my computation cycle":* The reason for the computer overload, it was later determined, was the fact that *Eagle*'s rendezvous radar had been left on "automatic," as a precaution for an emergency rendezvous with *Columbia.* But this meant that as *Eagle* descended, its computer was spending part of every computation cycle analyzing the signals from the rendezvous radar. By setting the rendezvous radar to "manual"—an instruction radioed to Armstrong and Aldrin only 30 minutes before their liftoff from the moon—the alarms were avoided during the ascent and subsequent rendezvous, a time when a healthy computer was even more critical than during the descent. See *Apollo: The Race to the Moon,* pp. 365-67.

II: Magnificent Desolation

305 *Inside, Joan was savoring relief:* The accounts of Joan Aldrin's activities on the day of the landing are based on the author's interview with her, as well as information in Armstrong et al., *First on the Moon,* and Mailer, *Of a Fire on the Moon.*

305 *She found the embrace of Buzz's uncle, Bob Moon:* Remarkable but true: Moon was the maiden name of Aldrin's maternal grandmother.

306 *NASA was still coping with a controversy stirred by the Genesis reading on Apollo 8:* According to a book by a former NASA public affairs officer, the Genesis reading brought "a shrill protest from agnostics who tried to convince the federal court that astronauts had no right to express religious sentiments in outer space. That backfired . . . when thousands of God-fearing people petitioned NASA to allow the astronauts freedom to do as they wished." Quoted in "Heroes, not of their own accord," a 1977 master's thesis by Perry Michael Whye, Iowa State University.

307 *in Armstrong's mind, it amounted to the first human contact with the moon:* For the landing, Armstrong gave some thought to quotes; before the flight he and Aldrin decided that if they reached the lunar surface they would use the call sign "Tranquillity Base"—"base" to connote exploration. They told only Charlie Duke, lest the first words from the moon take him by surprise—"Say again, Apollo 11?" And when it finally happened, Armstrong found himself adding quite spontaneously, "The *Eagle* has landed."

312 *The soil in the bag was almost black, like powdered graphite:* In retrospect, this should have come as no surprise; astronomers had long pointed out that the moon, which appears bright in the night sky, is actually a very dark object, reflecting on average only 7 percent of the sunlight striking it—a reflectivity comparable to that of asphalt.

316 *By international agreement no nation could claim the moon:* Equal access to the moon by all of humanity was a provision of the Space Treaty of 1967.

316 *Posing for Armstrong's camera:* There are no photographs of Edmund Hillary on the summit of Everest, simply because Tenzing did not know how to use a camera, and as Hillary said, "Everest was no place to teach him." Coincidentally, there are no good pictures of Neil Armstrong on the moon. The only clear Hasselblad photo shows Armstrong with his back to the camera, working at *Eagle*'s equipment storage tray. He also shows up in somewhat fuzzier motion picture footage taken from the lander's cabin. Buzz Aldrin, who has been asked about this more times than he probably likes, has explained that it was not intentional; the way the timelines were worked out, Armstrong had the camera most of the time. During the brief periods when Aldrin did have the camera, he was focused on such operational tasks as photographing the lunar module and the surrounding terrain. In addition, Aldrin snapped the famous picture of his own footprint—not for the aesthetic merit of the image, but for the scientists who were interested in the mechanical properties of lunar dust.

316 *the images from the moon were like a window on a dream:* In the pages of science fiction there had been almost as many versions of the first moonwalk as there were science fiction writers, but few included live television pictures of the event. To the author's knowledge, the first was Arthur Clarke's *Prelude to Space,* penned in 1947 (Arthur Clarke letter to the author, 1993).

321 *Armstrong realized there would not be nearly enough time:* In truth, Armstrong and Aldrin still had enough oxygen in their backpacks to stay out much longer than the time allotted, but the conservatism of this first landing dictated that they not use it.

324 Eagle *was nearly 200 feet away, looking like a scale model:* One geologist, watching the moonwalk at home, saw Armstrong run out of the field of view and thought, "Oh, my God, where is he going?" A number of observers were startled by Armstrong's apparent deviation from the flight plan. In truth, Armstrong did not violate any mission rules by running back to the crater. Mission planners had specified a maximum distance from the LM of 300 feet. However, they said, it was advisable, for conservatism's sake, that they remain within 100 feet.

325 *Armstrong had been gone for only about three minutes:* Aldrin had no idea of Armstrong's brief exploration. He was focused on his own work back at the LM, preparing the equipment for the documented sampling. The thought of running off to some interesting crater, he said years later, was not something he could allow himself to consider. In an extraordinary situation, he noted, it was useful to have a kind of tunnel vision: focus on the job at hand; don't deviate from the checklist; the better to avoid getting behind.

326 *But after two hours and thirty-one minutes:* The time of 2:31 refers to the interval between opening the hatch and closing the hatch. Armstrong was actually on the surface for a total of 2:13; Aldrin 1:42.

III: "Before This Decade Is Out"

327 *its lone occupant, Mike Collins:* Much of this section is drawn from *Carrying the Fire.*

330 *Moonlight flooded the cabin:* When Armstrong had settled into his makeshift bed inside *Eagle,* he realized another light was shining on his face. When he opened his eyes he was staring right at the earth, shining like a blue light bulb in the eyepiece of the navigation telescope on the other side of the cabin.

330 *there was no other way to think about it:* When Armstrong and Aldrin entered *Eagle* after the moonwalk they discovered a piece of plastic lying on the floor; it was the top of the circuit breaker used in arming the ascent engine. Without realizing it, Aldrin had broken it off with his backpack hours earlier, during the preparations for the moonwalk. No one was particularly concerned, simply because there were ways to work around the problem if the switch could not be used. But Aldrin managed to push the breaker in with a pencil, and in mission control, where controllers studied the telemetry from *Eagle,* it was clear that the fix had worked.

331 *"My secret terror for the last six months": Carrying the Fire,* pp. 411-12.

340 *Aldrin described the strange flashes:* The flashes were determined to be caused by high-energy cosmic rays entering the spacecraft and passing through the astronauts' eyes (*Biomedical Results of Apollo,* p. 355).

341 *That would come on a hot August night:* Armstrong didn't want any reporters, photographers, or other hoopla when the astronauts got out of the LRL. In a discussion with Chuck Berry on the day before the release, he jokingly threatened that he and his crew might feign illness: "I can't guarantee that the people won't limp or have contractions of some sort. If you want to take that kind of a chance, just have those cameras out there." He also said he wished he had a bottle of gentian violet so that he and his crew could paint little spots on their faces.

ACKNOWLEDGMENTS

The editors of this book wish to thank the following persons and institutions for their assistance.

Joan Aldrin, Glendale, Calif.
Tom Crouch, Smithsonian Institution, National Air and Space Museum. Washington, D.C.
Maureen M. Dilg, National Geographic Society, Washington, D.C.
Mike Gentry and Kathy Strawn, NASA/Media Services, Houston, Tex.
Sherie Jefferson and David Sharron, Information Dynamics Incorporated, Houston, Tex.
Eric Jones, *Lunar Surface Journal,* Los Alamos, N.Mex.
The Lovell Family, Lake Forest, Ill.
Lynn McDonald and Larry Felieu, Northrop Grumman History Center, Bethpage, N.Y.
Brian Nicklas, Smithsonian Institution, National Air and Space Museum, Washington, D.C.
Margaret Persinger, Kennedy Space Center, Fla.
Gwen Pittman, NASA Headquarters, Washington, D.C .
Gary Pressel, Adtech Photo Imaging, Houston, Tex.
Kipp Teague, Project Apollo Archive, Lynchburg, Va.

PICTURE CREDITS

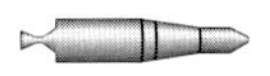

The sources for the illustrations in this book appear below.
Credits from left to right are separated by semicolons, from top to bottom by dashes.

Spine: NASA—NASA #S6750531. **Cover and Dust Jacket:** NASA #AS08142383; plaque created by John Drummond © Time-Life Books, Inc.; NASA #AS11405874 (inset); author photograph courtesy Andrew Chaikin.

1: NASA #AS08142838; plaque created by John Drummond © Time-Life Books, Inc.; NASA #AS11405874 (inset). **2:** Plaque created by John Drummond © Time-Life Books, Inc. (inset). **2, 3:** NASA #AS08142383. **4:** Marvin Chaikin. **12:** CORBIS/Galen Rowell. **14:** NASA #S6719620. **15, 27, 53, 97, 211, 281:** Detail from NASA #S6818733. **16:** NASA #S7834519. **17:** TASS from AP/Wide World/*Life* Magazine © Time Inc. **18:** NASA #612727. **20, 21:** U.S. Navy, courtesy Pete Conrad. **24-30:** Ralph Morse/*Life* Magazine © Time Inc. **32:** NASA #S6102819. **34, 35:** Ralph Morse/*Life* Magazine © Time Inc. **37:** NASA #S6640760. **38, 39:** NASA #S6721294. **40:** NASA. **44:** Ralph Morse/*Life* Magazine © Time Inc. **46:** CORBIS. **48, 49:** Rowland Scherman. **50, 51:** Francis Miller/*Life* Magazine © Time Inc. **52-55:** Ralph Morse/*Life* Magazine © Time Inc. **58:** NASA #S6208774. **60:** CORBIS/Bettmann. **62, 63:** Ralph Morse/*Life* Magazine © Time Inc. **64, 65:** Otis Imboden/National Geographic Society Image Collection. **66-68:** Ralph Morse/*Life* Magazine © Time Inc. **70:** NASA #G6002461—Ralph Morse/*Life* Magazine © Time Inc. **72:** Courtesy Air Force Flight Test Center History Office. **75:** NASA #S7423458. **76, 77:** Printed by permission of the Norman Rockwell Family Trust Copyright © 1965 the Norman Rockwell Family Trust. **78:** Barry Kaplan. **80:** NASA #S6561755. **82:** NASA #6528713. **84:** NASA, courtesy Andrew Chaikin. **87:** Opera Mundi, all rights reserved. **88:** Photofest, New York. **92, 93:** NASA #S6530428. **94, 95:** NASA #S6563197. **96:** NASA #S6855420. **98:** Drawing by George Bell © 1983 Time-Life Books, Inc. **99:** Novosti, London. **102:** Ralph Morse/*Life* Magazine © Time Inc. **106:** NASA #S6561830. **108:** NASA #S6851093. **110:** NASA #S6818733. **112:** Art by John Drummond, Time Life Inc., based on NASA diagram, courtesy Andrew Chaikin. **115:** Art Rickerby/*Life* Magazine © Time Inc. **116, 118:** Ralph Morse/*Life* Magazine © Time Inc. **121:** NASA. **122, 123:** NASA #AS07041582 (inset); #KSC68PC187. **124:** CORBIS/Bettmann. **126-129:** Ralph Morse/*Life* Magazine © Time Inc. **132-135:** NASA/National Geographic Society Image Collection. **136:** NASA #KSC68P620. **138:** Yale Joel/*Life* Magazine © Time Inc. **139:** NASA #KSC68PC341. **140:** NASA #S6610996. **142:** NASA. **145:** NASA #AS08162583. **146:** David Meltzer/National Geographic Society Image Collection. **151:** NASA #AS08162593. **152, 153:** NASA #AS8162605. **154:** CORBIS/Bettmann. **157:** NASA/National Geographic Society Image Collection. **160-163:** NASA, courtesy Andrew Chaikin. **165:** Bill Eppridge/*Life* Magazine © Time Inc. **166:** NASA. **168:** CORBIS. **172, 173:** NASA #AS080142392. **175:** NASA #S6818733. **178:** Joseph Scherschel/National Geographic Society Image Collection. **180, 181:** NASA #AS8132225. **184:** Bruce Dale/National Geographic Society Image Collection. **186:** Laufer/Globe Photos, New York. **188:** *Apollo 8 Coming Home* by Robert McCall, courtesy National Air and Space Museum, Smithsonian Institution. **190:** Lynn Pelham/*Life* Magazine © Time Inc. **194, 195:** NASA #S6915592. **197:** NASA/National Geographic Society Image Collection. **198, 199:** Francis Miller/*Life* Magazine © Time Inc. **200:** © Otis Imboden. **202, 203:** NASA #AS8142383. **204, 205:** © John Olson. **206, 207:** © Ernest Withers, courtesy Panopticon Gallery, Boston, Mass. **208, 209:** Joseph Louw/*Life* Magazine © Time Inc. **210:** Ralph Morse/*Life* Magazine © Time Inc. **212:** NASA. **215:** Courtesy Aldrin Family. **217:** CORBIS/Bettmann. **218:** NASA #AS9203064. **220, 221:** NASA. **222:** Ralph Morse/*Life* Magazine © Time Inc. **224:** NASA #AS9213212. **226:** NASA #S6931048. **228, 229:** NASA #S6939269. **230:** TRH Pictures, London. **232:** NASA, courtesy Andrew Chaikin #S6934076—#S6934075—#S6934077. **234:** NASA #S6934075, courtesy Andrew Chaikin. **238-239:** Yale Joel/*Life* Magazine © Time Inc. **240-243:** NASA. **244-248:** Ralph Morse/*Life* Magazine © Time Inc. **251:** Terence Spencer, Odiham, Hampshire, England. **252, 253:** Art by Jeffrey Oh © 1989 Time-Life Books, Inc. **256, 257:** NASA. **260, 261:** NASA #S6944155. **265:** Ralph Morse/*Life* Magazine © Time Inc. **268:** J. R. Eyerman/*Time* Magazine. **269:** NASA. **271:** Hershenson-Allen Archive, West Plains, Mo. **272:** NASA #S6939269. **275:** Leonard McCombe/*Life* Magazine © Time Inc. **276, 277:** Bill Eppridge/*Life* Magazine © Time Inc. **278-279:** Ralph Morse/*Life* Magazine © Time Inc. **280:** NASA #AS11405880. **283:** NASA #AS11365390. **284:** Don Davis. **286:** NASA #AS11375437. **287:** NASA, courtesy Andrew Chaikin. **289:** NASA #AS11375445. **291:** NASA #AS11446576. **293:** © Tom Miner/The Image Works, Woodstock, N.Y. **294, 295:** NASA #S6939601. **299:** Courtesy Jeff Durand, Jeff's Baseball Cards and Nostalgia Emporium, Springfield, Va. **300:** NASA, courtesy Andrew Chaikin. **301:** Michael Mauney/*Life* Magazine © Time Inc. **302, 303:** NASA #AS11375458. **305:** Courtesy Joan Aldrin. **307:** Michael Ochs Archives, Venice, Calif. **310:** NASA #S6942583. **312:** NASA. **313:** NASA, courtesy Andrew Chaikin. **314:** NASA #AS11405863. **315:** NASA #AS11405866—#AS11405868—#AS11405869. **317:** NASA #S6940308. **318-319:** NASA, courtesy Andrew Chaikin. **320:** NASA #AS11405872. **322, 323:** NASA #AS11405944. **324:** NASA #AS11405961; #AS11405960. **324, 325:** NASA #AS11405959. **325:** NASA #AS11405957; #AS11405956; #AS11405955. **327:** NASA #AS11375528—#AS11375530—#AS11375523. **328-331:** NASA, courtesy Andrew Chaikin. **332:** Alan Bean, courtesy Overton Park National Bank. **333:** Lee Balterman/*Life* Magazine © Time Inc. **334:** NASA #AS11446622; #AS11446625; #AS11446629. **335:** NASA #AS11446642, courtesy Andrew Chaikin. **337:** NASA. **338, 339:** Otis Imboden/National Geograpic Society Image Collection. **340:** B. Peterson/*Life* Magazine © Time Inc. **341:** NASA #S6921365. **343:** NASA #AS11405903. **344, 345:** NASA #AS11405902. **346,347:** NASA #AS11405948. **348, 349:** NASA. **350-351:** NASA #AS11405878.

INDEX

Numerals in italics indicate an illustration of the subject mentioned.

I

J

K

L

M

N

O

P

Q

R

S

ANDREW CHAIKIN, A MAN ON THE MOON
VOLUME I: ONE GIANT LEAP

Grateful acknowledgment is made for permission to reprint an excerpt from "Fly Me to the Moon (In Other Words)." Words and music by Bart Howard. TRO-© Copyright 1954 (renewed) Hampshire House Publishing Corp., New York, N.Y. Used by permission.

First published in 1994 by Viking Penguin, a division of Penguin Books U.S.A. Inc., as Book 1 of *A Man on the Moon: The Voyages of the Apollo Astronauts* by Andrew Chaikin.

This Time-Life edition is published by arrangement with Viking Penguin.

First printing. Printed in U.S.A.

School and library distribution by Time-Life Education, P.O. Box 85026, Richmond, Virginia 23285-5026

Library of Congress Cataloging-in-Publication Data
Chaikin, Andrew, 1956-
A man on the moon / by Andrew Chaikin and the editors of Time-Life Books.
p. cm.
Includes index.
Contents: 1. One giant leap — 2. The odyssey continues — 3. Lunar explorers.
ISBN 0-7835-5675-6
1. Project Apollo (U.S.)—History. 2. Space flight to the moon—History. I. Title.
TL789.8.U6A5244 1999
629.45'4'0973—dc21 99-15449
CIP

TIME® LIFE BOOKS

Time-Life Books is a division of Time Life Inc.

TIME LIFE INC.
PRESIDENT and CEO: George Artandi

TIME-LIFE BOOKS
PUBLISHER/MANAGING EDITOR: Neil Kagan
SENIOR VICE PRESIDENT, MARKETING: Joseph A. Kuna
VICE PRESIDENT, NEW PRODUCT DEVELOPMENT: Amy Golden

EDITOR: Lee Hassig
DIRECTORS, NEW PRODUCT DEVELOPMENT: Elizabeth D. Ward, Paula York-Soderlund

Design Director: Cynthia T. Richardson
Assistant Art Director: Janet Dell Russell Johnson
Senior Marketing Manager: Paul Fontaine
Project Manager: Karen Ingebretson
Associate Editor/Research and Writing: Ruth Goldberg,
Page Makeup Coordinator: Kimberly A. Grandcolas
Editorial Associate: Patricia A. Whiteford

Special Contributors: Andrew Chaikin (captions); Samantha Fields, Marilyn Murphy Terrell (research); Christine Stephenson (text); John Drummond, Janet Johnson, Mary Gasperetti (design and production); Marianne Dyson, Amanda Stowe (picture research); Antheus L. Bowden (picture coordination); Sunday Oliver (index)

Correspondents: Maria Vincenza Aloisi (Paris), Christine Hinze (London), Christina Lieberman (New York)

Director of Finance: Christopher Hearing
Director of Book Production: Patricia Pascale
Director of Imaging: Marjann Caldwell
Director of Publishing Technology: Betsi McGrath
Director of Photography and Research: John Conrad Weiser
Director of Editorial Administration: Barbara Levitt
Manager, Technical Services: Anne Topp
Page Makeup Manager: Debby Tait
Senior Production Manager: Ken Sabol
Production Manager: Virginia Reardon
Quality Assurance Manager: James King
Chief Librarian: Louise Forstall

Separations by the Time-Life Imaging Department

One Giant Leap is the first volume of the three-book set
A MAN ON THE MOON
The other titles are:
The Odyssey Continues
Lunar Explorers